TURNING AND MECHANICAL MANIPULATION . HOLTZAPFFEL

TURNING AND MECHANICAL MANIPULATION. BY CHAELES HOLTZAPFFEL, ASSOCIATE OF THE INSTITUTION OF CIVIL ENGINEERS, LONDON; HONORARY MEMBER OF THE ROYAL SCOTTISH SOCIETY OF ARTS, EDINBURGH; CORRESPONDING MEMBER OF THE AMERICAN INSTITUTE OF NEW YORK; ALSO OF THE FRANKLIN INSTITUTE, PHILADELPHIA, ETC. ETC. VOL. L MATERIALS; THEIR DIFFERENCES, CHOICE AND PREPARATION; VARIOUS MODES OF WORKING THEM, GENERALLY WITHOUT CUTTING TOOLS. VOL. II. THE PRINCIPLES OF CONSTRUCTION. ACTION, AND APPLICATION, OF CUTTING

TOOLS USED BY HAND; AND ALSO OF MACHINES DERIVED

FROM THE HAND TOOLS.

,--, VOL. III. ABRASIVE AND MISCELLANEOUS PROCE8SE8, WHICH CANNOT BE ACCOMPLISHED WITH CUTTING TOOLS.

Vol. rv.

THE PRINCIPLES AND PRACTICE OF HAND OR BIMPLE TURNING. VOL. V. THE PRINCIPLES AND PRACTICE OF ORNAMENTAL OR COMPLEX TURNfNO. VOL. VI. THE PRINCIPLES AND PRACTICE OF AMATEUR MECHANICAL ENGINEERING. ETery Volume la complete in Ucclf.

TURNING

AND

MECHANICAL MANIPULATION.

INTENDED AS

A WORK OP GENERAL EEPERENCE AND PRACTICAL INSTRUCTION,

ON THE LATHE,

AND THE VARIOUS MECHANICAL PURSUITS

FOLLOWED BY AMATEURS.

Br

CHARLES HOLTZAPFFEL,

ASSOCIATE OP THE INSTITUTION OF CIVIL ENGINEERS, LONDON;

HONORARY MEMBER OF THE ROTAL SCOTTISH SOCIETY OF ARTS, EDINBURGH;

CORRESPONDING MEMBER OF THE AMERICAN INSTITUTE OF NEW YORK;

ALSO OF THE FRANKLIN INSTITUTE, PHILADELPHIA,

ETC. ETC.

TO BE COMPRISKD IN SIX VOLUMES.

VOL. I.

MATERIALS; THEIR DIFFERENCES, CHOICE, AND PREPARATION;

VARIOUS MODES OF WORKING THEM GENERALLY

WITHOUT CUTTING TOOLS.

Tlluttraled by upwards of Three Hundred wnnnrın, J

LONDON:

PUBLISHED FOR THE AUTHOR,

BY HOLTZAPFFEL & CO., 64, CHARING CROSS, AND 127, LONG ACRE.

And to he, had of all Bookmllert.

a 1866.

FOURTH THOUSAND.

LONDON: BRADBURY, EVANS, AND CO., PRINTERS, WHITKV'RIARS

The author was led to undertake the present work ori Turning and Mechanical Manipulation, from the circumstance of there being no general treatise in the English language for the guidance of the Amateur in these pursuits. The original works by Plumier and Bergeron, although they were suited to the periods at which they were produced, are neither of them sufficient to convey instruction adequate to the present state of the Art; and the more recent French works leave unnoticed a large portion of the machines and instruments now used by Amateurs.

The greatest difficulty the author has encountered in his task, has been that of selection and arrangement; so as to produce, from materials so numerous and dissimilar, a work of general reference and practical instruction, at once sufficiently copious and accessible. But he hopes this difficulty has been satisfactorily met, by the division of the work into six volumes, upon parts of the subject which are broadly distinguished, and which thereby renders the volumes in a great measure independent of each other. This plan is also carried out in the subdivision of the volumes into chapters, which may be considered severally to include all that was deemed necessary to be stated upon the respective subjects; or to be, so far as they extend, distinct treatises; and which, in cases of doubt, he has not hesitated to submit to various practical friends for confirmation or extension.

These appeals have been answered with an alacrity which calls for his warmest thanks; and the author gladly avails himself of this opportunity of acknowledging these services, which have given a great additional value to his labours.

The work being of a technical nature, the author hopes to escape literary criticism, his main object having been to treat every subject in clear and concise language. As, however, notwithstanding his utmost care, he cannot expect to have been so fortunate as entirely to have escaped errors, ambi-« guities, or omissions, he requests of his readers the favour of the communication of any such defects, in order that those of most material import may be noticed in the Appendix to the second and ensuing volume, a great part of which is already completed.

PREFACE TO THE SECOND EDITION.

The duty that primarily devolves upon the author, in offering to the public the reprint of his first volume of the work on Turning and Mechanical Manipulation, is to express his warmest thanks for the very flattering reception the volume has met with, and which has greatly exceeded his anticipation; as the first edition was exhausted within two years of its appearance.

The author is disposed to hope that his efforts to obtain accuracy have been found successful, from the circumstance that no corrections have been suggested; and also that his descriptions have been found practical, as various amateurs previously unacquainted with some of the subjects treated of in the work, have, upon following its pages as a text-book, succeeded in their earliest attempts at various of the processes described; and amongst these in some of the more difficult, as flattening thin plates of metal, founding, soldering and others: these successes the author views as his highest encomiums.

It is the author's earnest endeavour to make his entire work keep pace with the existing state of the mechanical arts, which in this country are at most times in a state of rapid progression. This will he attempted by the introduction in the successive appendixes, of such additions and novelties as the author may consider to appertain to the portions of the work already printed.

This scheme, whilst it renders the first edition equally complete with the second, leaves the sequence of the pages unaltered, so that the index may, as intended, serve in common for the preliminary volumes; and which, instead of being limited to two, it is proposed to extend to three, as explained at length in the preface to the Second Volume, which further portion of the work is this day also laid before the public.

GENERAL SKETCH *or* THE CONTENTS OF THE WORK. VOL. L' MATERIALS, THEIR DIFFERENCES, CHOICE, AND PREPARATION; VARIOUS MODES OF WORKING THEM, GENERALLY WITHOUT CUTTING TOOLS. Introduction—Materials from the Vegetable, the Animal, and the Mineral Kingdoms.—Their uses in the Mechanical Arts depend on their structural differences, and physical characters.

The modes of severally preparing, working, and joining the materials, with the practical description of a variety of Processes, which do not, generally, require the use of Tools with cutting edges.

VOL. II. THE PRINCIPLES OF CONSTRUCTION, ACTION, AND APPLICATION, OF CUTTING TOOLS USED BY HAND; AND ALSO OF MACHINES DERIVED FROM THE HAND TOOLS. The principles and descriptions of Cutting Tools generally—namely, Chisels and Planes, Turning Tools, Boring Tools, Screw-cutting Tools, Saws, Files, Shears, and Punches. The hand tools and their modes of use are first described; and subsequently.various machines, in which the hand processes are more or less closely followed. VOL. III. ABRASIVE AND MISCELLANEOUS PROCESSES, WHICH CANNOT BE ACCOMPLISHED WITH CUTTING TOOLS. Grinding and Polishing, viewed as extremes of the same process, and as applied both to the production of form, and the embellishment of surface, in numerous cases to which, from the nature of the materials operated upon, and other causes, Cutting Tools are altogether inapplicable. Preparation and Application of Varnishes, Lackers, &c. VOL. IV. THE PRINCIPLES AND PRACTICE OF HAND OR SIMPLE TURNING. Descriptions of various Lathes;—applications of numerous Chucks, or apparatus for fixing works in the Lathe. Elementary Instructions in turning the soft and hard woods, ivory and metals, and also in Screw-cutting. With numerous Practical Examples, some plain and simple, others difficult and complex, to shew hew much may be done with hand tools alone. VOL. V. THE PRINCIPLES AND PRACTICE OF ORNAMENTAL OR COMPLEX TURNING.

Sliding Rest with Fixed Tools—Revolving Cutters, used in the Sliding Rest with the Division Plate and Overhead Motion. Various kinds of Eccentric, Oval, Spherical, Right-line and other Chucks. Ibbetaon's Geometric Chuck. Tbe Rose Engine, and analogous contrivances, Ac. With numerous Practical Examples.

VOL. VI. THE PRINCIPLES AND PRACTICE OF AMATEUR MECHANICAL ENGINEERING. Lathes with Sliding Rests for metal turning, Self-acting and Screw-cutting Lathes—Drilling Machines—Planing Enginos—Key-groove, Slotting and Paring Machines—Wheel-cutting and Shaping Engines, Ac.

With numerous Practical Examples.

"% *The First,* Second, *and lhird Volnmes of this work, are written at accompanying books, and have one Index in common, so at to constitute a general and preliminary work, the addition to which of any of the other volumes, will render the subject complete for the three classes of Amateurs referred to in the Introductory Chapter. A few additional copies of the Index have been printed for the convenience of those who may desire to bind the Index with Vols. I. and II.* TABLE OF CONTENTS OF THE FIKST VOLUME.

CHAP. I.—INTRODUCTION.

TABLE OF CONTENTS OP VOL. I.

END OF THE TABLE OF CONTENTS OF VOL. I.

The First, Second, and Third Volumes oj this work, are written as accompanying books, and have one Index in common, so as to constitute a general and preliminary work, the addition to which of any of tlie other volumes, will render the subject complete for the three classes of Amateurs referred to in the Introductory Chapter.

A few additional copies of the Index for Vols. I., II., and III., have been printed for the convenience of those who may desire to bind an Index with every volume.

VOL. I.

MATERIALS; THEIR DIFFERENCES, CHOICE, AND PREPARATION; VARIOUS MODES OF WORKING THEM, GENERALLY WITHOUT CUTTING TOOLS. CHAPTER I.

INTRODUCTION.

In offering these pages to the public, it may be expected that, in conformity with the usual custom, I should state briefly the general nature and purpose of my subject, uoticing the principal works which have already been written upon it, aud lastly, the method and arrangement adopted in the execution of my task.

It may, therefore, be premised that the practice of the Art of Turning constitutes the basis of the work, whilst the various mechanical arts associated with it, or derived directly from it, will form collateral branches of comment aud inquiry.

The importance of the lathe towards the promotion of the useful arts will be readily admitted, wheu it is considered how large a proportion of the indispensable objects we daily use, are either immediately produced by its agency, or in a greater or less degree are dependent upon this machine. Indeed it may be truly asserted, that nearly all solid objects, (particularly those of wood or metal,) in which the circle or any of its modifications can be discovered, are the offspring of the lathe, which produces, from solid materials of every descriptiou, an almost endless number and variety of forms, the origin of which can be

Vol. i. u *y* 2 PRINCIPLE OF THE LATHE, traced to that most simple, symmetrical, and best defined of all the mathematical figures, the circle.

No perfect or definite form is so easily or so accurately described as that of the circle; thus the compasses by placing one point on a sheet of paper or other material, and sweeping the pencil or style around the same, trace a line which returns into itself, and form a figure equidistant at every part from the point within, termed its center; or the proceeding may be reversed, by giving the paper a rotary motion beneath the pencil, which is kept stationary, whereby the same figure is produced.

The latter modification constitutes the principle of the lathe; the fixed center of the compasses being equivalent to the fixed axis upon which the solid material is made to revolve by some mechanical arrangement; the tracing pencil is supplanted by the cutting tool, which being held in a certain position towards the axis of rotation, cuts a circular groove in the revolving material; or if it be applied to its edge, reduces the object to a circular form.

This proceeding includes the three primary elements w hich constitute the ordinary practice of turning; namely, an immoveable axis; the revolution of the material upon that axis; and a fixed position of the cutting tool, in order that it may pare away all the parts of the body that oppose it. But the application of these elements must be modified and extended, if we desire to produce a compound form, such for instance as a vase; the first two elements, or the fixed axis and the revolution of the material, are retained, whilst the tool is moved by slow degrees along the outline or contour of the vase, both within and without, so as to remove all those parts of the material which are in excess, or project beyond the ideal line to be produced: and the continued, though temporary application of the tool, at every individual point of the vase or other object, renders every section taken at right angles to its axis, a circle. There are other less important modifications of the lathe, in which the position of the axis is changed and reudered moveable during the revolution of the work, as in oval and rose-engine turning, but these variations need only be adverted to here.

The art of turning will be admitted to be an auxiliary of great importance in the economy of mankind, as to it we are more or less immediately indebted, for nearly all the component parts of the machines and instruments, which are conducive in a ITS IMPORTANCE AND VALUE.
thousand ways to the support and clothing of the person, and the advancement of the mind.

For instance the engines which are now habitually and almost universally employed, in converting the numerous raw products of the earth to our most urgent, as well as to our most refined necessities and pleasures could scarcely exist, in the absence of the tools indispensably required for the accurate production of the circular parts, that enter so largely and in so important a manner, into their respective structures.

Again, without the lathe we could hardly possess another machine in which the circle abounds, namely, the steam-engine, which, like an obedient automaton endowed with power and endurance almost unlimited, is equally subservient, either iu converting the raw materials into their manufactured products, or in transporting them, in either state, across the ocean, or over the surface of the earth, along with the individuals, through whose energies they have been collected, transformed, and distributed.

Nor amongst our obligations to the mechanical arts, is that the least which is afforded by them iu the cause of science, the delicate apparatus for pursuing which, is due to the skill of the mechanist, whose instruments enable us to discover, and likewise to measure the planetary orbs, or to inspect in the cabinet the wonderful particles of the world we inhabit; and by means of which we find our earth to be teeming with creation, exquisite in symmetry, and beautifully adapted to the purposes of organic life; indeed, in whatever direction, and with whatever purpose the man of science may look, prospects of similar grandeur, and of equal wonder, still open in endless succession to repay the labour of research, an effort wherein the instruments, (due in great measure to the turner's skill,) are only secondary in importance to man's own mental faculties.

How largely also the circle and its many combinations enter into the elegancies and ornaments of life: more particularly in the useful and indispensable creations of taste and fancy obtained from the wheel of the manufacturer of pottery and porcelain; and more or less so iu all the arts of construction and embellishment, whether applied to the useful and agreeable purpose of ornamenting the costume of mankind; assisting towards the prosecution of the art of engraving; or in that most important of engines, the printing-machine, which disseminates, in millions of channels, the thoughts and speculations of the human mind; throughout all of which, the turner's primary element, the circle, is equally pervading and indispensable.

The important and different results to which we have cursorily referred, are in most cases greatly, and in others exclusively, indebted for their formation, to an instrument based upon the law of rotary motion, (one of the most simple though perfect yet discovered,) the turning lathe, to which and its numerous accessories and subordinates, we must largely attribute the extension of the arts, by which our comfort and well-being have been materially augmented: whilst their abstraction from our hands would deprive us of numerous sources of industrious employment, and the constructive and mechanical arts would, in all probability, eventually degenerate nearly to the low condition in which they may be still observed to be, amongst the few primitive aboriginal races yet remaining at the present day.

It will not therefore be taking too high a ground to call the lathe,—that primary machine which has conferred all these benefits upon *us,*—an engine of civilisation, and it may also be further asserted, that the extension of its employment in the higher and more important branches of manufactures and arts, especially in Great Britain, coupled with the talent, perseverance and industry of those who have developed its powers, have aided in elevating our country to its eminence among nations, by administering to its productive means, and its knowledge, and consequently, to its wealth and dominions.

I will now advert to the works that have been published on the art of turning, the honour attached to priority in which, belongs to France, the first treatise written exclusively upon the subject, being a folio volume entitled, *"L'Art de Iburner en Perfection,"* by *"le Pere Charles Plumier, (Religieux Minime")* and printed at Lyons in 1701. The author herein goes so far back as to refer the practice of the art to Tubal Cain, who is recorded in Sacred Writ to have been the first worker in metal; whilst others attribute to him the invention of wind instruments, the organ, and various machines. Plumier considered it impossible that the circular parts of such works coidd have been made otherwise than by the process of turning, which therefore he presumes to have been kuown to mankind at an extremely remote period; he also considers that the numerous circular works and objects recorded to have existed in Solomon's Temple, including the lamps and musical instruments used therein, could not have been produced otherwise than by the use of the lathe.

That account of the origin of the art which ascribes it to Daedalus, and which is quoted by Plumier and the various Encyclopaedists, appears to be derived from Felibien, (who wrote in 1690,) as will be seen by this literal extract from his pages.

"L' invention du Tour est tres ancienne; Diodore de Sicile dit que le premier qui le mil en usage estoit un neveu de Dedale nomine Talus. Pline veut que ce soil un Theodore. LLii le'cfo. *Samos; et il parte d'un Thericles qui se rendit celebre dans ces sortes d'ouvrages. "C'estoit avec cette machine qu'ils tournoient toutes sortes de vases, dont que/ques-un estoient enrichis de figures, et d'omemens en demy-bosse. Les Auteurs Grecs et Latins u!n!ofa"uusuenParent souvent, et Ciceron appelle ceux qui les per addita vitti. formoient au Tour,* Vascularis *Cestoit un proverbe Virg. Ejt.Z. J ....*
VatcvXaHos con-parmy les Anctens, de aire que les choses estoient Tfk'omi'fn Ter fates au tour Pour en exprimer la justeusse, et la delicatesse."
In conclusion the writer adverts to the great extent to which the art of turning had been practised by various persons, *gens libres,)* as a source of amusing occupation. Without pursuing these researches it may suffice to observe, that sufficient evidence exists that the art of turning has been successfully practised during a period of not less than two thousand years, although until a comparatively recent date, no description has been given of the methods pursued.

Plumier adverts to the following old authors on various subjects, in-which, amongst other matters, some brief allusions to the art are made, and which in point of date stand as follows: 1582. Besson's work, " *Theatrum Instrumentorum et Machinarum,"* has three engravings of complex lathes for screw-cutting, and oblique turning, with very slight descriptions.
1624. De Caus, *"Les Raisons des Forces Mouvantes,"* contains one engraving, and a few lines explanatory of a mode of turning the oval and of screw-cutting.
1677-83. *"Moxon's Mechanick Exercises, or the Doctrine of*
Pages 376-7.
Handy-works," published in London, in monthly parts. Vol. I. contains, *"Smithing, Joinery, Carpentry, Turning, Bricklayery, and Mechanick Dyalling,"* with a good description of the apparatus for turning. Vol. II. *"Handy-works applied to the Art of Printing."* 1690. Felibien, *"Des Principes de I'Architecture, de la Sculpture, de laPeinture, et des autres Arts qui en dependent:"* Paris. This author has devoted twelve pages to his remarks on the lathe, with a few words relative to the modes of oval turning, and to rose-engine work.

In 1719, (that is, eighteen years after Plunder's book,) a quarto volume was published at Lyons, styled *"Recueil d' Ouvrages curieux, de Mathematique, et de Mecanique; ou description du Cabinet de M. Grollicr de Serviere, par son petit-fils."* This work contains eighty plates, with etchings of his grandfather's designs for time-pieces, hydraulic machines, various bridges, military and other works, preceded by twelve plates of several of his highly-ornamental works executed in the lathe.

And lastly, in 1724-7, Leupold published at Leipzic eight folio volumes, entitled *"Theatrum Machinarum,"* &c. , which include a vast store of curious and useful matter, containing the germs and principles of many contrivances that are now commonly and abundantly used.

All these books are contained in the library of the British Museum, except that of Plunder, who appears not to have seen a rare book of more remote date than any of the above, namely, *"Panoplia Omnium,"* &c., by Hartman Schopper, printed at Frankfort-on-the-Maine in 1548, about twenty years after the Reformation: this old work contains 180 highly characteristic engravings, cut plank-ways on wood, and taken from every grade of life, civil, religious, and military, not forgetting the liberal and constructive arts, amongst which are included that of the turner, and those of a variety of artisans whose pursuits are intimately allied to our present subject. This work, which will be again referred to, shows that a great degree of perfection and subdivision in the practice of the mechanical arts existed even at that early period.

The execution of Plumier's work is honourable to its author, from the industrious care and exactness which it exhibits, more It is rather singular that not only Plumier's, but all the subsequent French and English works, written exclusively on turning (except Rich's) should be absent from that extensive and national collection of books.
especially when it is considered that it is almost the first work published upon the subject: a second edition, with extra

plates, and additional text, was published in Paris, 1749, when it also appeared in folio. It formed the basis of the article on the art of Turning, published in 1791, in *"FEncycloptdie Methodique,"* (begun in 1782,) by Diderot, DAlembert, and others, wherein forty crowded engravings of turning machinery are contained: various other French works on the same subject quickly followed.

First, the earlier edition of the " *Manuel du Tourneur,"* 2 vols, quarto, 1792-4, Paris, by L. E. Bergeron; this work is highly satisfactory, and is a record of all the material improvements introduced in the mechanism of the lathe by our continental neighbours, subsequent to the period at which Plumier wrote; and from these machines many of our modern contrivances are taken, although during the interval which has since elapsed, considerable changes have been introduced, as well in the manner of turning as in the material of the apparatus, wood being in many cases supplanted by metal, a more useful change as regards the excellence of construction, and also the strength and durability of the machinery.

A second edition of Bergeron's work, revised by his son-in-law, Hamelin Bergeron, was published in 1816; another smaller publication, entitled *"L'Art du Tourneur, par M. Paulin Desormeux,"* in 2 vols. 12mo., with an atlas, was printed in Paris in 1824; and lastly, two small volumes 16mo, with plates, entitled, *"Nouveau Manuel du Tourneur, ou Traiie complet et sirnplifii de cet Art, redigi par M. Dessables,"* the second edition of which, printed in 1839, and forming a part of the " *Encyclopedie-Roret,"* completes the list of French works devoted to the subject, the last two being in some respects compilations from Bergeron; the latter works only include the practice of hand-turning, leaving unnoticed the roseengine, the eccentric-chuck, and various apparatus described in the old books, although the" *Manuel-Roret"* contains, in an appendix, some extracts relative to the art of turning, from more recent scientific journals, and the printed transactions of various societies, with explanatory notes, by Mapod, " *Tourneur-mecanicien."*

In England, where, during the last half-century, the art has perhaps been far more extensively practised, both as a source of emolument and of amusement, we find in addition to the brief articles in the various encyclopaedias, periodicals, and a few works devoted to mechanical subjects, only the following treatises on detached portions of the art, namely: 1817. "Specimens of Eccentric Circular Turning, with Practical Illustrations for producing Corresponding Pieces in that Art. By John Holt Ibbetson, Esq." 1819. "Specimens of the Art of Ornamental Turning, in Eccentric and Concentric Patterns, with 6 copper-plate engravings; by Charles H. Rich, Esq., Southampton." 1819. "Tables; by which are exhibited at one view all the divisions of each circle on the dividing plate. By C. H. Rich, Esq." 1825. A second edition of Ibbetson's Specimens. 1S33. "A Brief Account of Ibbetson's Geometric Chuck, manufactured by Holtzapffel & Co., with a selection of 32 Specimens, illustrative of some of its powers. By J. II. Ibbetson, Esq." 1838. A third edition of Ibbetson's Specimens of Eccentric Circular Turning. "With considerable Additions, including a description and copper-plate engravings of the Compound Eccentric Chuck, constructed by the Author, and used by him in the execution of his Specimens."

The mention of the above publications by Mr. Ibbetson, enables me to particularise the services he has rendered to his fellow amateurs; and their inspection will abundantly show the great care and perseverance that he has devoted to the pursuits of turning, and the deserved eminence he has attained therein.

He has not only attended to the production of numerous highly ornamental combinations and effects, many of which are displayed in the treatises before cited; others in his " Practical View of an Invention for the better protecting Bank Notes against Forgery," editions I and 2,—1820 and 21, and in numerous communications to the Mechanics' Magazine; he has done more than this by constructing with his own hands the major part of the apparatus that he has used, many of which are original, and will be duly noticed in their appropriate places, in this work.

The best notices in our language of the *general* application of the art, are probably those contained in Rees's Cyclopaedia, under the heads of " Turning," " Lathe," and "Rose Engine." For Mr. Ibbetson's first description of his modification of the Geometric Chuck, see Mechanics' Magazine, 30th Deo,, 1826.

DIFFICULTIES OF ARRANGEMENT.

Several amateurs have undertaken the translation of Bergeron's Manuel into the English language, and others have commenced uew works, but none of these have been carried to completion. The former proceeding would have called for a re-construction of the book, which, although it abounds with a great deal of original, useful, and practical matter, is rather diffuse, and refers to apparatus that have been so far altered and superseded by others of more recent construction, and subsequent invention, that such a translation, if adapted to the present state of the art, would almost amount to a new work.

The author of these pages has been repeatedly urged, by many amateurs, to write a work upon the subject, but by no one more than by his late father, in conjunction with whom he made several beginnings; but the pressure of other business has prevented their efforts from arriving at maturity, and the delay has been materially lengthened by the difficulty of determining upon the most suitable arrangement. The first intention was to have written the book as a series of lessons; to have begun with the description of the plain or simple lathe, and so to have selected the examples, as to have successively described the more important and valuable of those instruments and methods which are now used.

The writer still pursued the same views after the loss of his father, in 1835, and the work was somewhat advanced on that plan, but ultimately abandoned, as he found the information upon each individual topic would then be scattered, difficult of reference, and introduced without any apparent order. That arrangement of the work would also have prevented him from introducing the notice of many useful contrivances, and from instituting a variety of comparisons between different methods, the insertion of which would greatly facilitate the explanations, and present a choice of proceeding.

Moreover, as the art now embraces a much more extensive and still increasing range of objects and instruments, than it did at the time when Bergeron wrote, the difficulties of arrangement that he experienced, are now proportionally increased. The author also felt some doubt as to his ability to produce, upon the first method, a work that should satisfactorily meet the wants of amateur turners generally, in reflecting upon the widely different views with which they had, even for some centuries, practised turning and the mechauical arts.

10 DIFFERENT PURSUITS OF AMATEUR TURNERS.

Many persons have followed these arts as a source of active and industrious employment, accessible at all hours, in the intervals between their other pursuits; a source of amusement that renders the amateur independent of the ordinary artisan for the supply of a great variety of works of utility for the common wants of life, including those constantly required either for the domestic establishment, or those personally experienced by its inhabitants, of every age and occupation.

Other amateurs have pursued the art of turning as a source of elegant recreation, and of inventive and skilful pastime; one closely allied to the fine arts, insomuch as its greatest success depends upon a just appreciation of sculpture and painting, and for the attainment of which the education and opportunities of the man of independent leisure eminently qualify him; whilst the embellishment of the drawing-room, cabinet, and boudoir, stimulate him to apply his knowledge and skill to that end, and in which he frequently administers at the same time to the extension and cultivation of tasteful form in ordinary manufactures.

There is also a class of amateurs who have preferred the pursuit of such branches of the art as unite, with taste and design, a certain admixture of the more exact acquirements connected with mathematical and general science, and the arts of construction; and who have devoted their time and ingenuity to the production of models, embracing a variety of objects relative to the arts of peace and war: and also to the construction of various machines and apparatus, or to the still more praiseworthy attempt of improving those already in use, or of inventing new ones; the services that have been thus rendered by men of independence and education are neither few nor slight. In all such cases the progress is more rapid and certain, when the pencil is devoted to the production of the drawing, and the tool to the formation of the rough model, as proceedings in common, prior to making the finished apparatus.

In selecting topics from the very numerous branches of the subject, the author has endeavoured to supply the more immediate wants of all classes of amateurs, and under these circumstances he has thought it best, for the convenience and choice of the general reader, to separate the practical division of the subject of Turning into three distinct and different parts, to be preceded by three general or preliminary volumes, to contain ARRANGEMENT CONSEQUENTLY ADOPTED BY THE AUTHOR. 11 miscellaneous information more or less required in the pursuit of every branch of the mechanical arts; thus dividing the entire work into six volumes—namely,

Vol. I.

Materials, Their Differences, Choice, And Preparation; Various Modes Of Working Them, Generally Without Cuttinq Tools.

Vol. II.

The Principles Of Construction, And Purposes Of Cutting Tools.

Vol. III.

Abrasive And Miscellaneous Processes.

"Vol, IV.

TriE Principles And Practice Of Hand Or Simple Turning.

Vol. V.

The Principles And Practice Of Ornamental Or Complex Turning.

Vol. VI.

The Principles And Practice Of Amateur Mechanical Engineering.

The first volume, which is now in the hands of the reader, relates principally to the materials for turning and the mechanical arts, arranged under the heads of the three great sources from which they are respectively derived; namely, the vegetable, the animal, and the mineral departments of nature; it includes also their treatment in the extended sense of the word, so far as regards their preparation for the Lathe, and their employment in various distinct branches of mechanical art, the practices of which do not in general require the use of tools with cutting edges.

The metallic materials are submitted to the greatest variety of processes, and which mainly depend on their properties of fusibility, malleability, and ductility; and consequently, the formation and qualities of alloys are considered, as also the arts of founding and soldering; those of forging works in iron and steel which are comparatively thick, and the nearly analogous treatment of thin works, or those in sheet metals; drawing tubes and wires, hardening and tempering, and a variety of correlative information is also offered, for the particulars of which the reader is referred to the Table of Contents.

The second volume, on cutting tools, is intended first to explain the general principles of cutting tools, which are few and simple; the forms and proportions of tools are however extensively modified, to adapt them to the different materials, to the various 12 CONTENTS OF THE DIFFERENT VOLUMES.
shapes to be produced, and to the convenience of the operator, or of the machine in which they are fixed. The remarks on the tools will inevitably be somewhat commingled with the account of their practical use, and the consideration of the machines with which

they are allied, as indeed it is difficult to say where the appellation of *tool* ends, and that of *machine* or *engine* begins. The tools will be treated of in different chapters, on Chisels and Planes, Turning Tools, Boring Tools, Screw Cutting Tools, Saws, Files, Shears and Punches, and in various subdivisions.

The third volume will be devoted to the description of abrasive processes; namely, those for restoring or sharpening the edges of the cutting tools; those for working upon substances to which, from their hardness or crystalline structure, the cutting tools are quite inapplicable; and also to the modes of polishing, which may be viewed as a delicate and extreme application of the abrasive process, and the final operation after the cutting tools; and lastly, to the ordinary modes of staining, lackering, varnishing, and other miscellaneous subjects.

The titles of the fourth, fifth, and sixth volumes, are, it is expected, sufficiently descriptive of their contents, which will be arranged with a similar attempt at order and classification, upon which it is unnecessary here to enlarge. From the systematic arrangement attempted throughout the six volumes, it is hoped that instead of the numerous descriptions and instructions, being indiscriminately mixed and scattered, they will assume the shape of so many brief and separate treatises, and will in a great measure condense into a few consecutive pages, the remarks offered under every head; a form that will admit of any subject being selected, and of a more easy and distinct *reference* aud *comparison,* when the reader may find it necessary; a facility that has been particularly studied.

Every one of the six volumes may be considered as a distinct work and complete in itself; this will admit of any selection being made from their number. At the same time it is to be observed that the first, second, and third are written as accompanying volumes, and will have an index in common, so as to constitute a general and preliminary work; the additioii to which, of one of the other volumes, will render the subject complete for any of the three classes of amateurs before referred to, should the entire work be deemed too extensive.

FIRST DIVISION OR PART. CHAPTER II. MATERIALS FROM THE VEGETABLE KINGDOM. SECT. I. THEIR GROWTH, STRUCTURE, AND PREPARATION.

The materials used in turning and the mechanical arts are exceedingly numerous: we obtain from the Vegetable Kingdom an extensive variety of woods of different characters, colours, and degrees of hardness, and also a few other substances.

The most costly and beautiful products of the Animal Kingdom, are the tusks of the elephant, the tortoise and pearl shells; but the horns, hoofs, and some of the bones of the ox, buffalo, aud other animals, are also extensively used for more common purposes.

From the Mineral Kingdom are obtained many substances which are used in their natural states, and also the important products of the metallic ores.

It would be altogether misplaced to attempt a minute and general description of these varied materials, as they will be found in their more appropriate places in works on natural history, physiology, mineralogy, and metallurgy; and it is the less necessary, as those which are more commonly used, are familiar to us in the buildings, machinery, implements, furniture, and ornaments, by which we are surrounded: others of less extensive supply are, in many respects, only varieties which are subject to similar usage. I shall therefore principally restrict myself to the description of those characters of the usual materials which lead the artisan to select them for his several purposes, and that also direct the choice of the tools by means of which they are respectively worked.

By far the most numerous and important of the materials from the Vegetable Kingdom are the woods, with which most parts of our globe are abundantly supplied; great numbers of them are used in their respective countries, and are known to the naturalist, although but a very inconsiderable portion of them are familiar to us in our several local practices.

14 VEGETABLE MATERIALS, THE WOODS.

The woods that are the most commonly employed in this country, are enumerated in an alphabetical list, together with the most authentic information I could obtain concerning them; in collecting which, the assistance of various kind friends has been obtained, amongst whom are numbered travellers, naturalists, merchants, and manufacturers. Various museums, collections and works, have been carefully examined, so as to include in the list the more important of the various names of the woods, their countries, general and mechanical characters, and their principal uses in the arts of construction.

The alphabetical catalogue is preceded by a tabular view, intended to classify the woods that are the most generally selected by our artisans for certain ordinary uses; it will also serve, iu a slight degree, to throw them into groups according to some of the differences between them, referrible principally to their fibrous structures, by which they are distinguished as hard or soft, elastic or non-elastic, of plain or variegated appearance, of permanent figure or the reverse.

Their other varieties, in respect to colour and scent, and the oils, resins, gums, medicinal and various other matters, they respectively contain, are questions of equal importance, but they are more connected with the chemical and economic arts, and but slightly concern this inquiry. I shall therefore, nearly restrict myself to the questions arising from the mechanical structure and treatment of the woods, which it is proposed to consider under separate heads, in the present and three following chapters.

The general understanding of the principal differences of the woods will be greatly assisted by a brief examination into their structure, whit h is now so commonly and beautifully developed b by the sections for the microscope. The figures 1, 2, 3 are drawn from thin cuttings of beech-wood, prepared by the

STRUCTURE OF THK WOODS.

15 optician for that instrument; the principal lines alone are represented, and these are magnified to about twice their

linear distances, for greater perspicuity.

Fig. 1, which represents the horizontal or transverse section of a young tree or a branch, shows the arrangement of the annual rings around the centre or pith; these rings are surrounded by an exterior covering, consisting also of several thinner layers, which it will suffice to consider collectively, in their common acceptation, or as the bark. The fibres which are seen as rays proceeding from the pith to the bark, are the medullary rays or plates.

Figs. 2 and 3 are vertical sections of an older piece of beechwood. Fig. 2 is cut through a plane, such as from *a* to *a,* in which the edges of the annual rings appear as tolerably parallel fibres running in one direction, or lengthways through the stem; the few thicker stripes are the edges of some of the medullary rays.

Fig. 3 is cut radially, or through the heart, as from *b* to *b.* In this the fibres are observed to be arranged in two sets, or to run crossways; there are first the edges of the annual rings, as in fig. 2, and secondly, the broad medullary rays or plates.

The whole of these figures, but especially the last, show the character of all the *proper* woods, namely, those possessing *two sets of fibres,* and in wbich the growth of the plant is accomplished, by the yearly addition of the external ring of the wood, and the internal ring of the bark, whence these rings are called annual rings, and the plants are said to be *exogenous,* from the growth of the *wood* being external.

In fig. 1 the medullary rays are the more distinctly drawn, in accordance with the appearance of the section, as they seem to constitute more determinate lines; whereas the annual rings consist rather of series of tubes, arranged side by side, and in contact with each other, and which could not be represented on so small a scale. At the outer part of each annual ring these tubes or pores appear to be smaller and closer; the substance is consequently more dense, from the greater proportion of the matter forming the walls of the tubes; and the inner or the softer parts of the annual rings have in general larger vessels, and therefore less density.

In many plants the wedge-form plates, intermediate between 10 STRUCTURE OK THE WOODS.

the medullary rays, only appear as au irregular cellular tissue full of small tubes or pores, without any very definite arrangement. The medullary rays constitute, however, the most characteristic part of the structure, and greatly assist in determining the difference between the varieties of the exogenous plants, as well as the wide distinction between the entire group and those shortly to be described. The medullary rays also appear, by their distinct continuity, to constitute the principal source of *combination* and strength in the substance of the woods; most of the medullary rays, in proceeding from the center to the circumference, divide into parts to fill out the increased space.

In the general way, the vertical fibres of the annual rings, and the horizontal fibres of the medullary rays, are closely and uniformly intermingled; they form collectively the substance of the wood, and they also constitute two series of minute interstices, that are viewed to be either separate cells or vessels, the majority of which proceed vertically, the others radially. In many, as the oak, sycamore, maple, and sweet chestnut, the medullary rays, when dissected, exhibit a more expanded or foliated character, and pervade the structure, not as simple radial tubes, but as broad *septa* or divisions, which resemble flattened cells or clefts amongst the general groups of pores, giving rise to the term *silver-grain,* derived from their light and glossy appearance: they vary considerably in size and number.

The beech-wood, fig. 3, has been selected as a medium example between this peculiarity and the ordinary crossings of the fibres, which in the firs and several others seem as straight as if they were lines mechanically ruled; and even in the most dense woods are in general easily made out under the microscope.

The vessels or cells running amidst the fibres are to the plant what the blood-vessels and air-cells are to the animal; a part of them convey the crude sap from the roots, or the mouths of the plant, through the external layers of the wood to the leaves, in which the sap is evaporated and prepared; the fluid afterwards returns through the bark as the elaborated sap, and combines with that in the external layers of the wood, the two constituting In the *Cismmpdos Partra,* belonging to the natural order *MenisjiermacKr,* this structure is singularly evident; the medullary rays are very thick, aud almost detached from the intermediate wedge-form plates, which are nearly solid except the few pores by which they are pierced, much like the substance of the common cane. STRUCTURE OF THE PALMS, ETC.

17 the *cambium.* The latter ultimately becomes consolidated for the production of the new annual ring that is deposited beneath the loosened bark, and which is eventually to constitute a part of the general substance or wood; the bark also receives a minute addition yearly, and the remainder of the fluid returns to the earth as an excretion.

The other order of plants grows in an entirely different manner, namely by a deposition from within, whence they are said to be *endogenous;* these include all the grasses, bamboos, palms, &c. *Endogens* are mostly hollow, and have only one set of fibres, the vertical, which appear in the transverse section, fig. 4, as irregular dots closely congregated around the margin,. and gradually more distant towards the center, until they finally disappear, and leave a central cavity, or a loose cellular structure. Fig. 5 represents the horizontal, and fig. 6 the vertical section of portions of the same, or the cocoa-nut palm, *(Cocos nucifera,)* of half their full size.

All the *endogens* are considered to commence from a circular pithy stem which is entirely solid; some, as the canes, maintain this solidity, with the exception of the tubes or pores extending throughout their length. The bamboos extend greatly in diameter, so as to become hollow, except the diaphragms at the knots; these are often used as cases for rolls of papers. The palms generally enlarge still more considerably to their extreme size, which in some cases is fifty times the diameter of the original stem, the center being soft and pithy.

Some of the palms, &c., denote each yearly increase by one of The reader is referred to the following articles in the three editions of Dr. Lindley's introduction to Botany, namely—" *Exogenous structure"* and "*Of the stem and origin of wood;"* and *aho " Exogens,"* and,' *Endogens"* by tho same author in the Penny Cyclopedia; all are replete with physiological interest.

VOL. I. O IS STRUCTURE OF THE PALMS, ETC. the rings or markings upon their stems, which are always soft in the upper part, like a green vegetable, and terminate in a cluster of broad pendent leaves, generally annual, and when they drop off they leave circular marks upon the stem, which are sometimes permanent, and indicate by their number the age of the plant. The vertical fibres above referred to, proceed from the leaves, and are considered to be analogous to their roots, and likewise to assimilate in function to the downward flow of the sap from the leaves of the *exogens:* whereas in the palms they constitute separate and detached fibres, that first proceed inwards, and then again outwards, with a very long and gradual sweep, thereby causing the fibres to be arranged in part vertically, and in part inclined, as in the figure.

The substance of the stems of the palms, is not allowed by physiological botanists to be proper wood, (which in all cases grows exteriorly, and possesses the two sets of fibres shown iu fig. 3,) whereas the *endogenous* plants have only the one set, or the vertical fibres; and although many of this tribe yield an abundance of valuable gifts to the natives of the tropical climates in which they flourish, only a portion of the lower part of the *shell* of the tree is available as wood; amongst other purposes, the smaller kinds are used by the natives as tubes for the conveyance of water, and the larger pieces as joists and beams.

The larger palms generally reach us in slabs measuring about the sixth or eighth part of the circle, as in fig. 4, the smaller sizes are sent entire; fig. 5 represents a small piece near the outside, with the fibres half size; but the different palms vary considerably in the shapes, magnitudes and distances of the fibres, and the colours and densities of the two parts.

In the vertical section, fig. 6, which is also drawn half size, the fibres look like streaks or wires embedded in a substance similar to cement or pith, whioh is devoid of fibrous structure; the inhabitants of the Isthmus of Darien pick out the fibres from some of the palms and use them as nails, they are generally pointed, and in the specimens from which the drawing was made they are as hard as rosewood, whereas the pithy substance is quite friable. Some of the smallest palms are imported into this country for walking-sticks, under the names of partridge and The leaves of the exogens are by some thought to send down similar roots or fibres, between the bark and wood for the formation of the annual ring.

Penang canes, &c. The ordinary canes and bamboos are too well known to require more than to be named.— See article Palms, in the catalogue.

To return to the more particular examination of the woods that most concern us, it will be observed that the central pith in fig. 1, happens to be of an irregular triangular shape. This, the primary portion of the plant, is in the first instance always cylindrical; it is supposed to assume its accidental form, (which is very frequently hexagonal,) from the compression to which it is subjected. The pith governs, in a considerable degree, the general figure or section, as all the series of rings will be observed, in fig. 1, page 14, to have a disposition to project at three points; but with the successive additions, the angular form is gradually lost, as it would be if we wound a ribbon upon a small triangular wire; for after a time, no material departure from the circular form would be observable.

A greater variation amongst the rings is due to the more or less favourable growth of the successive years, and to the different exposure of the tree to the sun and air, which develop that side of the plant in an additional degree; whereas the tree growing against a wall or any other obstruction, becomes remarkably stunted on that side of its axis, from being so shielded.

The growth of a tree is seldom so exactly uniform that its section is circular, or its heart central, often far from it; and as every annual ring is more consolidated, and of a deeper colour on its outer surface, they frequently serve to denote very accurately, in the woods growing in cold and temperate climates, the age of the plant, the differences of the seasons, the circumstances of its situation, and the general rapidity of its growth. "But in many hot countries the difference between the growingseason and that of rest, if any occur, is so small, that the zones are as it were confounded, and the observer finds himself incapable of distinguishing with exactness the formation of one year from that of another."

It is, however, difficult to arrive at any satisfactory conclusion respecting the qualities of woods from the appearance of their annual rings; for instance, in two specimens of larch, considered by Mr. Fincham f to be exceedingly similar, in specific gravity, Dr. Lindley's Introduction to Botany, second edition, p. 74.

t Principal builder of Her Majesty's Dockyard, Chatham.

20 CIRCULATION OP THE SAP.

strength and durability; in the one, Scotch larch, there were only three annual rings in five-eighths of an inch, whereas in Italian larch there were twenty-four layers in the same space. In some of the tropical woods the appearance of the rings can scarcely be defined, and in a specimen of the lower or butt end of teak, now before me, three annual rings alone, cover the great space of one inch and three-eighths.

The horizontal section of a tree, occasionally looks as if it were the result of two, three, or more separate shoots or stems consolidated into one; in some of the foreign woods in particular, this irregularity often gives rise to deep indentations, and most strange shapes, which become eventually surrounded by one single covering of sap; so that a stem of considerable girth may yield only an insignificant piece of wood, scarcely available for the smallest purposes of turnery, much less for cabinet work.

The circulation of the sap is considered to be limited to a few of the external layers, or those of the sap-wood, or *alburnum,* which are in a less matured state than the perfect wood, or *duramen,* beneath. The last act of the circulation, as regards the heart-wood, is supposed to be the deposition of the colouring matter, resin or gum, through the agency of the medullary rays'that proceed from the bark towards the center, and leave their contents in the layer outside the true wood perfected the year previous. We may fairly suppose by analogy, that as one ring is added each year, so one is perfected annually, and thrown out of the circulatory system.

That the circulation has ceased in the heart-wood, and that the connexion between it and the bark has become broken, is further proved by the fact, that numbers of trees may be found in tolerably vigorous growth within the bark, whereas at the heart they are decayed and rotten. In fact some of the hardest foreign woods, as king-wood, tulip-wood, and others, are rarely sound in the center, and thus indicate very clearly that This is not peculiar to the tropical woods; for example, some of the yew-trees in Hampton Court Gardens, appear to have grown in this manner from three or four separate stems, that have joined into one at a short distance above the ground. Ab an instance of the singular manner in which the separate branches of trees thus combine, I may mention that stones, pieces of metal, and other substances, are occasionally met with in the central parts of timber, from having been accidentally deposited in a,cleft, or the fork of a branch, and entirely inclosed or overgrown by the subsequent increase of the plant.

THE TREE WHEN FELLED. 21 their decay commenced whilst they were in their parent soil; and as in these, the appearance of annual rings is scarcely to be distinguished, this also appears to indicate a great term of age, enough to account for this relatively premature decay.

The quantity of sap-wood is various in different plants, and the line of division is usually most distinctly marked; in some, as boxwood, the sap-wood is very inconsiderable, and together with the bark is on the average only about the thickness of a stout card, whereas in others, as the snake-wood, it constitutes fully two-thirds of the diameter, so that a large tree yields but an inconsiderable stick of wood, of one third or fourth the external diameter.

It may be presumed that in the same variety of wood, about an average number of the layers exist as sap-wood, as in cutting up a number of pieces of the same kind, such as the black Botany-Bay wood, and others, it is found that in those measuring about two inches diameter, the piece of heart-wood is only about as large as the finger, but in pieces one, two, or three inches larger, the heart-wood is also respectively one, two, or three inches larger, or nearly to the full extent of the increase of the diameter.

The sap-wood may be therefore, in general, considered as of about an average thickness in each kind of wood: it is mostly softer, lighter, more even in colour, and more disposed to decay than the heart-wood, which prove it to be in a less matured or useful state, whether for mechanical or chemical purposes.

At the time the tree is separated from its root, its organic life ceases, and then commences the gradual evaporation of the sap, and the drying and contracting of the tubes, or tissues, previously distended by its presence.

The woods are in general felled during the cold months, when the vegetative powers of the plant are nearly dormant, and when they are the most free from sap; but none of the woods are fit for use in the state in which they are cut down, for although no distinct circulation is going on within the heart-wood, still the capillary vessels keep the trees continually moist throughout their substance, in which state they should not be employed.

If the green or wet woods are placed in confined situations, the tree or plank, first becomes stained or doated, and this

QUEEN WOOD UNFIT FOR USE.

speedily leads to its decomposition or decay—effects that are averted by careful drying with free access of air.

Other mischiefs almost as fatal as decay also occur to unseasoned woods; round blocks cut out of the entire circular stem of green wood, or the same pieces divided into quavterings, split in the direction of the medullary rays, or radially, also though less frequently upon the annual rings. Such of the round blocks as consist of the entire section contract pretty equally, and nearly retain their circular form, but those from the quarterings become oval from their unequal shrinking.

As a general observation, it may be said the woods do not alter in any material degree in respect to length. Boards and flat On this account the timbers for ships are usually cut out to their shape and dimensions for about a year before they are framed together, and they are commonly left a twelvemonth longer in the skeleton state, to complete the seasoning; as in that condition they are more favourably situated as regards exposure to the air than when they are closely covered in with the planking.

Mr. Fincham considers that the destruction of timber by the decay commonly known as dry-rot, cannot occur unless air, moisture, and heat, are all present, and that the entire exclusion of any of the three stays the mischief. By way of experiment, he bored a hole in one of the timbers of an old ship built of oak, whoso wood wos at the time perfectly sound; the admission of air, the third element, to the central part of the wood, (the two others being to a certain degree present,) caused the hole to be filled up in the course of twenty-four hours with mouldiness, a well-known vegetation, which very speedily became so compact a fungus as to admit of being withdrawn like a stick. He considers the shakes or splits in timber to predispose it to decay in damp and confined situations, from admitting the air in the same manner.

The woods differ amazingly in their resistance to decay; some perish in one or two years, whereas others are very durable, and even preserve their fragrance when they are opened after many years, or almost centuries.

Mr. Q. Loddiges says, the oak boxes, for the plants in his greenhouses, decay in two or three years, whereas he has found those of teak to last fully six or seven times as long: the situation is one of severe trial for the wood.

There are two quarto works on dry-rot; the one by Mr. M'William, 1818; the other by Mr. John Knowles, Surveyor of Her Majesty's Navy, 1821.

The process of Kyanizing is intended to prevent the re-vegetation of timber, by infusing into its pores an antiseptic salt: the corrosive sublimate is generally employed, other metallic salts are also considered to be applicable, but the general utility of the process, especially in thick timbers, or those exposed to much wet, ia still unsettled amongst practical men.

The Kyanizing is sometimes done in open tanks, at others, (by Timperley's process, Hull and Selby Railway,) in close vessels from which the air is first exhausted to the utmost, and the fluid is then admitted under a pressure of about 100 pounds on the inch.—*Sec* Minutes of Proceedings, Inst. Civ. Eng., p. 83,1841. *See* also note H. in Appendix to Vol. II., page 953, in which Payne's more recent preservative process is described.

SEASONING THE WOODS. 23 pieces contract however in width, they warp and twist, and when they are fitted as panels into loose grooves, they shrink away from that edge which happens to be the most slightly held; bnt when restrained by nails, mortices or other unyielding attachments, which do not allow them the power of contraction, they split with irresistible force, and the materials and labour thus improperly employed will render no useful service.

In general, the softest woods shrink the most in width, but no correct observations on this subject have been published. Mr. Fincham considers the rock-elm to shrink as much as any wood, namely, about half an inch in the foot, whereas the teak scarcely shrinks at all; in the "Tortoise," store-ship, when fifty years old, no openings were found to exist between the boards.

In the woods that have been partially dried, some of these effects are lessened when they are defended by paint or varnish, but they do not then cease, and with dry wood, every time a new surface is exposed to the air, even should the work have been made for many years, these perplexing alterations will in a degree recommence, even independently of the changes of the atmosphere, the fluctuations of which the woods are at all times too freely disposed to obey.

The disposition to shrink and warp from atmospheric influence, appears indeed to be never entirely subdued; some bogoak, supposed to have been buried in the island of Sheppy, not less than a thousand years, was dried for many months, and ultimately made into chairs and furniture; it was still found to shrink and cast, when divided into the small pieces required for the work.

8ECT. II. SEASONING AND PREPARING THE WOODS.

Having briefly alluded to the mischiefs consequent upon the use of woods in an improper condition, I shall proceed to describe the general modes pursued for avoiding such mischiefs by a proper course of preparation.

The woods immediately after being felled, are sometimes immersed in running water for a few days, weeks, or months, at other times they are boiled or steamed; this appears to be done under the expectation of diluting and washing out the sap, after,which it is said the drying is more rapidly and better accomplished, and also that the colours of the white woods are im24 SEASONING THE WOODS.

proved, (see article Holly in Catalogue, also Ebony,) but the ordinary course is simply to expose the logs to the air, the effect of which is assisted by the preparation of the wood into smaller pieces, approaching to the sizes and forms in which they will be ultimately used, such as square logs and beams, planks or boards of various thicknesses, short lengths or quarterings, &c.

The steins and branches of the woods of our own country, such as alder, birch, and beech, that are used by the turner, frequently require no reduction in diameter; but when they are beyond the size of the work, they are split into quarterings and stacked in heaps to dry, which latter proceeding should never be forgotten under any circumstances.

We know but little of the early treatment of the foreign woods used for cabinet-work and turning; some few of them, as mahogany and satin-wood, are imported in square logs; others, as rosewood, ebony, or coromandel, are sometimes shipped in the halves of trees, or in thick planks; but the generality of those used for turning are small, and do not require this reduction: these only reach us in billets, sometimes with the rind or bark upon them, and sometimes cleaned or trimmed.

The smaller hard woods are very much more wasteful than the timber woods; in many of the former, independently of their thick bark, the section is very far from circular, as they are often exceedingly irregular, indented, and ill-defined; others are almost constantly unsound in their growth, and either present central hollows and cavities, or cracks and radial divisions, which separate the stem into three or four irregular pieces.

Probably none of the hard woods are so defective as the black Botany-Bay wood, in which the available produce, when it is trimmed ready for the lathe, may be considered to be about one third or fourth of the original weight, sometimes still less; but unfortunately many others approach too nearly to this condition, as a very large proportion of them partake of the imperfections referred to, more especially the cracks; th,e larger hard woods are by comparison much less wasteful.

A.11 the harder woods require increased care in the seasoning, which is often badly begun by exposure to the sun or hot winds in their native climates: their greater impenetrability to the air the more disposes them to crack, and their comparative scarcity and expense, are also powerful arguments on the score of prePREPARING THE TURN-ERY-WOODS.

25 caution. It is therefore desirable to prepare them for the transition from the yard or cellar to the turning room, by removing the parts which are necessarily wasted, the more intimately to expose them to the air, some time before they are placed in the house, and they should be always kept away from the fire, or at first in a room altogether without one.

It is usual to begin by cutting the logs into pieces a few inches or upwards in length, to the general size of the work; and if possible to prepare every piece into a round block, or into two or three, when the wood is irregular, hollow, or cracked. In the latter case, a thin wedge is inserted into the principal crack, and driven down with a wooden maul; or a cleaver such as

Fig. 7.

fig. 7, which has a sharp edge, and a pole to receive the blow, is used in the same manner; these tools, or the hatchet, are likewise used in splitting up the English woods, when they are beyond the diameters required. The cleft pieces are next roughly trimmed with the hatchet, or else with the paring knife (fig. 8 over leaf) a tool of safer and more economical application in the hands of the amateur: it is a lever knife, from two and a half to three feet long, the cutting edge is near that end which terminates in a hook, the other extremity has a transverse handle; an eye-bolt for the hook to act against, is screwed into the bench or block, and a detached cutting board is fixed under the blade, to serve as the support for the wood, and for the knife to cut upon. To avoid waste of material, it is advisable, until the eye is well accustomed to the work, to score with the compasses upon each end of the rough block, as large a circle as it will allow, to serve as a guide for the knife.

The block, represented in fig. 8, is adapted to the bearers of the lathe, but any other support will serve equally well. The paring-knife is also employed for other purposes besides those of the turner: it is sometimes made with a curved edge like a Sometimes the glazier's chipping knife is used for small pieces of wood instead of the cleaver represented.—See also Appendix, Note I., page 953, vol. ii.

20 PREPARING THE TURNERY WOODS.

gouge, and is used in many shaping operations in wood, as in the manufacture of shoe-lasts, clogs, pattens, and toys.

Fig. 8.

In the absence of the paring»knife or hatchet, the work is fixed in the vice, and rounded with a coarse rasp, but this is much less expeditious; by some manufacturers the preparation both of the foreign and English woods is prosecuted still further by cutting the material into smaller pieces, rough turned and hollowed in the lathe, to the forms of boxes, or other articles for which they are specifically intended, and in fact every measure that tends to make the change of condition gradual, assists also in the economy, perfection, and permanence of the work.

Many of the timber woods are divided at the saw-pit into planks or boards, at an early stage, in order to multiply the surfaces upon which the air may act, and also to leave a less distance for its penetration: after sawing, they should never be allowed to rest in contact, as the partial admission of the air often causes stains or doating: but they are placed either perpendicularly or A paring-knife similar to the above, but working in a guide, and with an edge 12 or 14 inches long, is a most effective instrument in the hands of the toy-makers. The pieces of birch, alder, *&c.*, are boiled in a cauldron for about an hour to soften them, and whilst hot they may be worked with great expedition and perfection. The workmen pare off slices, the plankway of the grain, as large as 4 by 6 inches, almost as quickly as they can be counted: they are wedged tight in rows, like bunks, to cause them to dry flat and straight, and they seldom require any subsequent smoothing. In making the little wheels for carts, &c., say of one or two inches diameter, and one-quarter or three-eights of an inch thick, they cut them the *cross-way oj the grain,* out of cylinders previously turned and bored; the flexibility of the hot moist wood being such, that it yields to the edge of the knife without breaking transversely as might be expected.

horizontally in racks, or they are more commonly stacked in horizontal piles, with parallel slips of wood placed between at distances from about three to six or eight feet, according to the quantity of support required; the pile when carefully stacked forms a press and keeps the whole flat and straight.

Thin pieces will be sufficiently seasoned in about one year's time, but thick wood requires two or three years, before it is thoroughly fit to be removed to the warmer temperature of the house for the completion of the drying. Mahogany, cedar, rosewood, and the other large foreign woods, require to be carefully dried after they are cut into plank, as notwithstanding the length of time that sometimes intervenes between their being felled and brought into use, they still retain much of their moisture whilst they remain in the log.

In some manufactories the wood is placed for a few days before it is worked up, in a drying room heated by means of stoves, steam, or hot water, to several degrees beyond the temperature to which the finished work is likely to be subjected.

Such rooms are frequently made as air-tight as possible, which appears to be a mistake, as the wood is then surrounded by a warm but stagnant atmosphere, which retains whatever moisture it may have evaporated from the wood. Of late a plan has been more successfully practised in seasoning timber for building purposes, by the employment of heated rooms with a free circulation of air, which enters at the lower part in a hot and dry state, and escapes at the upper charged with the moisture which it freely absorbs in the heated condition. The continual ingress of hot dry air, greedy of moisture, so far expedites the drying, that it is accomplished in one-third of the time that is required in the ordinary way in the open air.f Scientifically considered, the drying is only said to be complete when the,wood ceases to lose weight from evaporation; this does not occur after twice or thrice the period usually allowed for the process of seasoning.

In many modern buildings small openings are left through the walls to the external air to allow a partial circulation amidst the beams and joists, as a preservative from decay, and for the entire completion of the seasoning.

t Price's Patent. CHAPTER III. USEFUL CHARACTERS OF WOODS. SECT. I. HARD AND SOFT WOODS, ETC.

The relative terms hard and soft, elastic or non-elastic, and the proportions of resins, gums, &c., as applied to the woods, appear to be in a great measure explained by their examination under the microscope, which develops their structure in a very satisfactory manner.

The fibres of the various woods do not appear to differ so materially in individual size or bulk, as in their densities and distances: those of the soft woods, such as willow, alder, and deal, appear slight and loose; they are placed rather wide asunder, and present considerable intervals for the softer and more spongy cellular tissue between them; whereas in oak, mahogany, ebony, and rosewood, the fibres appear rather smaller, but as if they possessed a similar quantity of matter, just as threads containing the same number of filaments are larger or smaller accordingly as they are spun. The fibres are also more closely arranged in the harder woods, the intervals between them are necessarily less, and the whole appears a more solid and compact formation.

The very different tools used by the turner for the soft woods and hard woods respectively, may have assisted in fixing these denominations as regards his art; a division that is less specifically entertained by the joiner, who uses the same tools for the hard and soft woods, excepting a trifling difference in their angles and inclinations; whereas the turner employs for the soft woods, tools with keen edges of thirty or forty degrees, applied obliquely, and as a tangent to the circle; and for the hard woods, tools of from seventy to ninety degrees upon the edge, applied as a radius, and parallel with the fibres, if so required. The tools last described answer very properly for the dense woods, in which the fibres are close and well united; but applied to the softer kinds, in which the filaments are more tender and less firmly joined, the hard-wood tools produce rough, torn, and unfinished surfaces.

In general the weight or specific gravity of the woods may be taken as a sure criterion of their hardness; for instance, the hard lignum-vitae, boxwood, ironwood, and others, are mostly so heavy as to sink in water; whereas the soft firs, poplar and willow, do not on the average exceed half tlie weight of water, and other woods are of intermediate kinds.

The density or weight of many of the woods may be increased by their mechanical compression, which may be carried to the extent of fully one third or fourth of their primary bulk, and the weight and hardness obtain a corresponding increase. This has been practised for the compression of tree-nails for ships, by driving the pins through a metal ring smaller than themselves directly into the hole in the ship's side;f ftt other times, (for railway purposes,) the woods have been passed through rollers, but this practice has been discontinued, as it is found to spread the fibres laterally, and to tear them asunder; J: an injury that does not occur when they are forced through a ring, which condenses the wood at all parts alike, without any disturbance of its fibrous structure even when tested by the microscope; The moat dense wood I have met with is in Mr. Fincham's collection; it is the Iron Bark wood from New South Wales: in appearance it resembles a close hard mahogany, but more brown than red; its specific gravity is l-426,—its strength, (compared with English oak, taken as usual at l'ooo,) is 1'557. On the other hand the lightest of the *true* woods is probably the *Cortifa,* or the *Anona paluttru,* from Brazil, in Mr. Mier's collection; the specific gravity of this is only 0 206, (whereas that of cork is 0 240,) it has only one-seventh the weight of the Iron Bark wood. The *Cortifa* resembles ash in colour and grain, except that it is paler, finer, and much softer; it is used by the natives for wooden shoes, Jtc.

The *Pita* wood, that of the *Fuurcroya gigantea,* of the Brazils, an *endogtn* almost like pith, (used by the fishermen of Rio de Janeiro, as a slow match, for lighting cigars, tc.; also like cork for lining the drawers of cabinets for insects,) and the rice paper plant of India and China, which is still lighter and more pithy, can hardly be taken into comparison.

t Mr. Annersley's Patent, 1821, for building vessels of planks only, without ribs.

J Dublin and Kingston Railway.

§ The mode at present practised by the Messrs. Ransome of Ipswich, (under their patent,) is to drive the pieces of oak into an iron ring by means of a screw press, and to expose them within the ring to a temperature of about 180 for twelve or sixteen hours before forcing them out again.

The tree-nails may be thus compressed into two-thirds their original size, and they recover three-fourths of the compression on being wetted; they are used for after compression the wood is so much harder, that it cuts very differently, and the pieces almost ring when they are struck together; fir may be thus compressed into a substance as close as pitch-pine.

In many of the more dense woods, we also find an abundance of gum or resin, which fills up many of those spaces that would be otherwise void: the gum not only makes the wood so much the heavier, but at the same time it appears to act in a mechanical manner, to mingle with the fibres as a cement, and to unite them into a stronger mass; for example, it is the turpentine that gives to the outer surface of the aunual rings of the red and yellow deals, the hard horny character, and increases the elasticity of those timbers.

Those woods which are the more completely impregnated with resin, gum, or oil, are in general also the more durable, as they are better defended from the attacks of moisture and insects.

Timbers alternately exposed to wet

and dry, are thought by Tredgold and others, to suffer from losing every time a certain portion of their soluble parts; if so, those which are naturally impregnated with substances insoluble in water may, in consequence, give out little or none of their component parts in the change from wet to dry, and on that account the better resist decay: this has been artificially imitated by forcing oil, tar, &c, through the pores of the wood from the one extremity.f

Many of the woods are very durable when constantly wet; the generality are so when always dry, although but few are suited to withstand the continual change from one to the other state; but these particulars, and many points of information respecting timber-woods that concern the general practice of the builder, or naval architect, such as their specific gravities, relative strengths, resistances to bending and compression, and other characters, are treated of in Tredgold's Elements of Carpentry, at considerable length. J railway purposes, but appear equally desirable for shipbuilding, in which the treenails fulfil an important office, and in either case their after-expansion fixes them moat securely.—*See* Minutes, Inst. Civ. Eng., 1841, p. 837.

See the treatment of the firs in Norway, article Fins, in Catalogue. t The durability of *pitch* pine, when " wet and dry," is however questioned.

J The work contains a variety of the most useful tables: the reader will likewise find a set of tables of similar experiments on American timbers, by Lieut. Deuison, Hoyal Engineers, F.K. S., &c., in the Trans. Inst. Civ. Eng., vol. ii. p. 15, and also SECT. II.—ELASTIC AND NON-ELASTIC WOODS.

The most elastic woods are those in which the annual or longitudinal fibres are the straightest, and the least interwoven with the medullary rays, and which are the least interrupted by the presence of knots; such woods are also the most easily rent, and the plainest in figure, as the lancewood, hickory and ash; whereas other woods, in which the fibres are more crossed and interlaced, are considerably tougher and more rigid; they are also less disposed to split in a straight or economical manner, as oak, beech, and mahogany, which, although moderately elastic, do not bend with the facility of those before named.

Fishing-rods, unless made of bamboo, have generally ash for the lower joint, hickory for the two middle pieces, and a strip cut out of a bamboo of three or four inches diameter as the top joint. Archery bows are another example of elastic works; the "single-piece bow " is made of one rod of hickory, lancewood, or yew tree, which last, if perfectly free from knots, is considered the most suitable wood: the "back or union bow" is made of two or sometimes three pieces glued together. The *back-piece,* or that furthest from the string, is of rectangular section, and always of lancewood or hickory; the *belly,* which is nearly of semicircular section, is made of any hardwood that can be obtained straight and clean, as ruby-wood, rose-wood, greenheart, king-wood, snake-wood, and several others: it is in a great measure a matter of taste, as the elasticity is principally due to the back-piece; the palmyra is also used for bows.

The elasticity or rather the flexibility of the woods, is greatly increased for the time, when they are heated by steaming or boiling; the process is continually employed for bending the oak two other sets of experiments on Indian timber woods, by Captain H. C. Baker, late of the Bengal Artillery, superintendent of the half-wrought timber-yard, Calcutta, at pp. 123 and 230 of the Cleanings of Science, published at Calcutta, 1829.

The union bow is considered to be " softer," that is more agreeably elastic than the single-piece bow, even when the two require the same weight to draw them to the length of the arrow. In the act of bending the bow, the back is put into tension, and the inner piece into a state of compression, and each wood is then employed in its most suitable manner. Sometimes the union bow is imitated by one solid piece of straight cocoa-wood, (of the West Indies, not that of the *cocoanut palm,)* in which case the tough fibrous sap is used for the back, and in its nature sufficiently resembles the lance-wood more generally used. and other timbers for ship-building, the lance-wood shafts for carriages, the staves of casks, and various other works.

The woods are steamed in suitable vessels, and are screwed or wedged, at short intervals throughout their length, in contact with rigid patterns or moulds, and whilst under this restraint they are allowed to become perfectly cold; the pieces are then released. These bent works suffer very little departure from the forms thus given, and they possess the great advantage of the grain being parallel with the curve, which adds materially to their strength, saves much cost of material and time in the preparation, and gives in fact a new character to the timber.

The inner and outer plankings of ships are steamed or boiled before they are applied; they are brought into contact with the ribs by temporary screw-bolts which are ultimately replaced by the copper bolts inserted through the three thicknesses and riveted: or they are secured by oak or locust tree-nails, which are caulked at each end.

Boiling and steaming are likewise employed for softening the woods, to facilitate the cutting as well as bending of them.f

When the two sets of fibres meet in confused angular directions, they produce the tough cross-grained woods, such as See the description of Mr. Wiliam Hookey's apparatus for bending ships' timbers, rewarded by the Society of Arts, and described in their Trans, vol. 32, p. *HI.*

Preference is now given to the "Steam Kiln" over the "Water Kiln," and the time allowed is one hour for every inch of the thickness of the timber; it loses much extractive matter in the process, which is never attempted a second time, as the wood then becomes brittle.

Colonel O. A. Lloyd devised an ingenious and economical mode of bending the timbers to constitute the ribs of a teak-bridge which he built in the Mauritius. Every rib was about 180 ft. long, and of 8 ft. rise, and consisted of five thicknesses of wood of various lengths and widths. The wood had been cut down about a month; it was well steamed and brought into contact with a strong mould, by means of an iron chain attached to a hook at the one extremity of the mould and passed under a roller fixed at the other; the chain was drawn tight by a powerful capstan. Whilst under restraint the neighbouring pieces were pinned together by tree-nails, after which a further portion of the rib was proceeded with: the seasoning of tho timber was also effected by the process. t Thus in Taylor's Patent Machinery for making casks, the blocks intended for the staves are cut out of white Canada oak to the size of thirty inches by five, and smaller. They are well steamed, and then sliced into pieces one half or five-eighths inch thick, at the rate of 200 in each minute, by a process far more rapid and economical than sawing; the instrument being a revolving iron plate of 12 or 14 feet diameter, with two radial knives, arranged somewhat like the irons of an ordinary plane or spokeshave. lignum-vitae, elm, &c., and, like the diagonal braces in carpentry and shipping, they deprive the mass of elasticity, and dispose it rather to break than to bend, especially when the pieces are thin, and the fibres crop out on both sides of the same; the confusion of the fibres is, at the same time, a fertile source of beauty in appearance to most woods.

Elm is perhaps the toughest of the European woods; it is considered to bear the driving of bolts and nails better than any other, and it is on this account, and also for its great durability under water, constantly employed for the keels of ships, for boat-building, and a variety of works requiring great strength and exposure to wet.

A similar rigidity is also found to exist in the crooked and knotted limbs of trees from the confusion amongst the fibres, and such gnarled pieces of timber, especially those of oak, were in former days particularly valued for the knees of ships: of later years they have been in a great measure superseded by iron knees, which can be more accurately and effectively moulded at the forge to suit their respective places, and they cause a very great saving in the available room of the vessel.

The lignum-vitae is a most peculiar wood, as its fibres seem arranged in moderately thick layers, crossing each other obliquely, often at as great an angle as thirty degrees with the axis of the tree; when the wood is split, it almost appears as if the one layer of annual fibres grew after the manner of an ordinary screw, and the succeeding layer wound the other way so as to cross them like a left-hand screw. The interlacement of the fibres in, lignum-vitae is so rigid and decided, although irregular, that it exceeds all other woods in resistance to splitting, which cannot be effected with economy; the wood is consequently always prepared with the saw. It is used for works that have to sustain great pressure and rough usage, several examples of which are given under the head Lignum-vitjs, in the Catalogue already referred to.

CHAPTER IV.

ORNAMENTAL CHARACTERS OF WOODS.

SECT I.—FIBRE OR GRAIN, KNOTS, ETC.

The ornamental figure or grain of many of the woods, appears to depend as much or more upon the particular directions and mixings of the fibres, as upon their differences of colour. I will first consider the effect of the fibre, assisted only by the slight variation of tint, observable between the inner and outer surfaces of the annual layers, and the lighter or more silky character of the medullary rays.

If the tree consisted of a series of truly cylindrical rings, like the tubes of a telescope, the horizontal section would exhibit circles; the vertical, parallel straight lines; and the oblique section would present parts of ovals; but nature rarely works with such formality, and but few trees are either exactly circular or straight, and therefore although the three natural sections have a general disposition to the figures described, every little bend and twist in the tree disturbs the regularity of the fibres, and adds to the variety and ornament of the wood.

The horizontal section, or that parallel with the earth, only displays the annual rings and medullary rays, as in fig. 1, p. li, and this division of the wood is principally employed by the turner, as it is particularly appropriate to his works, the strength and shrinking being alike at all parts of the circumference, in the blocks and slices cut out of the entire tree, and tolerably so in those works turned out of the quarterings or parts of the transverse pieces.

But as the cut is made intermediate between the horizontal line, and the one parallel with the axis, the figure gradually slides into that of the ordinary plank, magnified portions of which are shown in figs. 2 and 3: and these are almost invariably selected for carpentry, &c.

The oblique slices of the woods possess neither the uniformity of grain of the one section, nor the strength of the other, and it would be likewise a most wasteful method of cutting up the timber; it is therefore only resorted to for thin veneers, when some particular figure or arrangement of the fibres has to be obtained for the purposes of ornamental cabinet-work.

The perpendicular cut through the heart of the tree is not only the hardest but the most diversified, because therein occurs the greatest mixture and variety of the fibres, the first and the last of which, in point of age, are then presented in the same plank; but of course the density and diversity lessen as the board is cut further away from the axis. In general the radial cut is also more ornamental than the tangental, as in the former the medullary rays produce the principal effect, because they are then displayed in broader masses, and are considered to contain the greater proportion of the colouring matter of the wood.

The section through the heart displays likewise the origin o» most of the branches, which arise first as knots, in or near the central pith, and then work outwards in directions corresponding with the arms of the trees, some of which, as in the cypress and oak, grow out nearly horizontally, and others, as in the poplar, shoot up almost perpendicularly.

Those parts of wood described as curls, are the result of the confused filling in of the space between the forks, or the springings

Fig. 9. B

A of the branches. Fig. 9 represents the section of a piece of yew-tree, which shows remarkably well the direction of the main stem A B, the origin of the branch C, and likewise the 36. BRANCHES AND KNOTS.

formation of the curl between B and C; fig. 10 is the end view of the stem at A. In many woods, mahogany especially, the curls are particularly large, handsome and variegated, and are generally produced as explained.

It would appear as if the germs of the primary branches were set at a very early period of the growth of the central stem, and gave rise to the knots, many of which however fail to penetrate to the exterior so as to produce branches, but are covered over, by the more vigorous deposition of the annual rings. All these knots and branches, act as so many disturbances and interruptions to the uniformity of the principal zones of fibres, which appear to divide to make way for the passage of the off-shoots, each of which possesses in its axis a filament of the pith, so that the branch resembles the general trunk in all respects, except in bulk, and again from the principal branches smaller ones continually arise, ending at last in the most minute twigs, each of which is distinctly continuous with the central pith of the main stem, and fulfils its individual share in causing the diversity of figure in the wood.

The knots are commonly harder than the general substance, and that more particularly in the softer woods; the knots of the deals, for example, begin near the axis of the tree, and at first show the mingling of the general fibres with those of the knot, much the same as in the origin of the branch of the yew in fig. 9, but after a little while it appears as if the branch, from elongating so much more rapidly than the deposition of the annual rings upon the main stem, soon shot through and became entirely detached, and the future rings of the trunk were bent and turned slightly aside when they encountered the knot, but without uniting with it in any respect.

This may explain why the smooth cylindrical knots of the outer boards of white deal, pine, &c., so frequently drop out when exposed on both sides in thin boards; whereas the turpentine in the red and yellow deals may serve the part of a cement, and retain these kinds the more firmly.

The elliptical form of the knots in the plank, is mostly due to the oblique direction in which they are cut, and their hardness, (equal to that of many of the tropical hard woods,) to the close grouping of the annual rings and fibres of which they are themselves composed. These are compressed by the surrounding wood of the parent stem, at the time of deposition; whereas the principal layers of the stem of the tree are opposed alone by the loosened and yielding bark, and only obtain the ordinary density.

The knots of large trees are sometimes of considerable size; I have portions of one of those of the Norfolk Island pine, *(Auracaria excelsa,)* which attained the enormous size of about four feet long, and four to six inches diameter. In substance it is thoroughly compact and solid, of a semi-transparent hazelbrown, and it may be cut almost as well as ivory, and with the same tools, either into screws, or with eccentric or drilled work, &c.; it is an exceedingly appropriate material for ornamental turning.

It is by some supposed that the root of a tree is divided into about as many parts or subdivisions as there are branches, and that, speaking generally, the roots spread around the trunk under ground, to about the same distance as the branches wave above; the little germs or knots from which they proceed being in the one case distributed throughout the length of the stem of the tree, and in the other crowded together in the shorter portion buried in the earth.

If this be true, we have a sufficient reason for the beautiful but gnarled character of the roots of trees when they are cut up for the arts; many a block of the root of the walnut-tree, thus made up of small knots and curls, and that was first intended for the stock of a fowling-piece, has been cut into veneers and arranged in angular pieces to form the circular picture of a table, and few pictures of this natural kind will be found more beautiful. The roots of many trees also display very pretty markings; some are cut into veneers, and those of the olive-tree, and others, are much used on the Continent for making snuff-boxes.

The tops of the pollard trees, such as the red oak, elm, ash, and other trees, owe their beauty to a similar crowding together of the little germs, whence have originated the numerous shoots which proceeded from them after they had been lopped. The burrs or excrescences of the yew, and some other trees, appear to arise from a similar cause, apparently the unsuccessful attempts I am indebted to Maj. Brown for my specimens of this knot. and the information concerning it; a part of a knot of the same species, with some of the sur rounding wood, is in the model-room of the Admiralty, Somerset House.

at the formation of branches from one individual spot, from this may arise those bosses or wens, which almost appear as the result of disease, and exhibit internally crowds of knots, with fibres surrounding them in the most fantastic shapes. Sometimes the burrs occur of immense size, so as to yield a large and thick slab of highly ornamental wood of most confused and irregular growth: such pieces are highly prized, and are cut into thin veneers to be used in cabinet work.

It appears extremely clear likewise, that the beautiful East Indian wood, called both Kiabooca and Amboyna, is, in like manner, the excrescence of a large timber tree. Its character is very similar to the burr of the yew-tree, but its knots are commonly smaller, closer, and the grain or fibre is more silky. The Kiabooca has also been supposed to be cut from around the base of the cocoanut palm, a surmise that is hardly to be maintained although the latter may resemble it, as the Kiabooca is imported alone from the East Indies, whereas the cocoa-nut palm is common and abundant both in the eastern and western hemispheres. *(See* Kiabooca in the Catalogue.)

The bird's-eye maple shows in the finished work the peculiar appearance of small dots or ridges, or of little conical projections with a small hollow in the centre (to compare the trivial with the grand, like the summits of mountains, or the craters of volcanoes,) but without any resemblance to knots, which are the apparent cause of ornament in woods of somewhat similar character, as the burrs of the yew and kiabooca, and the Russian maple (or birch tree): this led me to seek a different cause for its formation.

On examination, I found the stem of the American bird's-eye maple, stripped of its bark, presented little pits or hollows of irregular form, some as if made with a conical punch, others ill-defined and flattened like the impression of a hob-nail; suspecting these indentations to arise from internal spines or points in the bark, a piece of the latter was stripped off from another block, when the surmise was verified by their appearance. The I have a beautiful specimen of a Burr, found occasionally upon the teak, which is fully equal in beauty to the Amboyna, but a smaller figure; I owe it to the kindness of Dr. Horsfield of the India-House.

Mr. G. Loddiges considers the burrs may occur upon almost all *old* trees, and that they result from the last attempt of the plant to maintain life, by the reparation of any injury it may have received.

Bird's-eye Maple, Etc. no layers of the wood being moulded upon these spines, each of their fibres is abruptly curved at the respective places, and when cut through by the plane, they give, in the *tangental* slice, the appearance of projections, the same as in some rose-engine patterns, and the more recent medallic glyptographic or stereographic engravings, in which the closer approximation of the lines at their curvatures, causes those parts to be more black, (or shaded,) and produces upon the plane surfaces, the appearances of waves and ridges, or of the subject of the medal.

The short lines observed throughout the maple wood, between the dots or eyes, are the edges of the medullary rays, and the same piece of wood when examined upon the radial section, exhibits the ordinary silver grain, such as we find in the sycamore, (to which family the maple tree belongs,) with a very few of the dots, and those displayed in a far less ornamental manner.

The piece examined measured eight inches wide, and five and a half inches radially, and was apparently the produce of a tree of about sixteen inches diameter; the effect of the internal spines of the bark was observable entirely across the same, that is through each of the 130 zones of which it consisted. The curvature of the fibres was in general rather greater towards the center, which is to be accounted for by the successive annual depositions upon the bark, detracting in a small degree from the height or magnitude of the spines within the same, upon which the several deposits of wood were formed. Other woods also exhibit spines, which may be intended for the better attachment of the bark to the stem, but from their comparative minuteness, they produce no such effect on the wood as that which exists, I believe exclusively, in the bird's-eye maple.

This led me to conclude, that in woods the figures of which resemble the undulations, or the ripple marks on the sands, that frequently occur in satinwood and sycamore, less frequently in box-wood, and also in mahogany, ash, elm, and other woods, to be due to a cause explained by fig. 11, namely, a serpentine or *ffuilloche* form in the grain: and on inspection, the fibres of all such pieces will be found to be wavy, on the face, at right angles to that on which the ripple is observed, if not on both faces. Those parts of the wood which happen to receive the light, appear the brightest, and form the ascending sides of the ripple, 40 SERPENTINE GRAIN; SILVER GRAIN. just as some of the medallic engravings appear in cameo or in intaglio, according to the direction in which the light falls upon them.

The woods possessing this wavy character, generally split with an undulating fracture, the ridges being commonly at right angles to the axis of the tree, or square across the board; but in a specimen of an Indian red wood, the native name of which is *Caliatour,* the ridges are inclined at a considerable angle, presenting a very peculiar appearance, seen as usual on the polished surface.

In those woods which possess in abundance the *septa* or *silver grain,* described by the botanist as the medullary plates or rays, the representations of which as regards the beech tree are given in fig. 3, p. 14, another source of ornament exists; namely, a peculiar damask or dappled effect, somewhat analogous to that artificially produced on damask linens, moreens, silks, and other fabrics, the patterns on which result from certain masses of the threads on the face of the cloth running lengthways, and other groups crossways. This effect is observable in a remarkable degree in the more central planks of oak, especially the light-coloured wood from Norway, and the neighbourhood of the Rhine, called wainscot and Dutch oak, &c., aud also in many other woods, although in a less degree.

In the oak plank, the principal streaks or lines are the edges of the annual rings, which show, as usual, parallel lines more or less waved from the curvature of the tree, or the neighbouring knots and branches; and the damask pencillings, or broad curly veins and stripes, are caused by groups of the medullary rays or *septa,* which undulate in layers from the margin to the center of the tree, and creep in betwixt the longitudinal fibres, above some of them and below others. The plane of the joiner, here and there, intersects portions of these groups, exactly on a level Dr. Royle favoured me with this curious specimen.
with their general surface, whereas their recent companions are partly removed

in shavings, and the remainder dip beneath the edges of the annual rings, which break their continuity; this will be seen when the *septa* are purposely cut through by the joiner's plane.

Upon inspecting the ends of the most handsome and showy pieces of wainscot oak and similar woods, it will be found that the surface of the board is only at a *small* angle with the lines of the medullary rays, so that *many* of the latter "crop out" upon the surface of the work: the medullary plates being seldom flat, their edges assume all kinds of curvatures and elongations from their oblique intersections. All these peculiarities of the grain have to be taken into account in cutting up woods, when the most showy character is a matter of consideration.

The same circumstances occur in a less degree in all the woods containing the silver grain, as the oriental plane-tree, or lace-wood, sycamore, beech, and many others, but the figures become gradually smaller; until at last, in some of the foreign hard woods, they are only distinguishable on close inspection under the magnifier. Some of the foreign hard woods show lines very nearly parallel, and at right angles to the axis of the tree, as if they were chatters or utters arising from the vibration of the plane-iron. The medullary rays cause much of the beauty in all the showy woods, notwithstanding that the rays may be less defined than in the woods cited.

In many of the handsomely figured woods, some of the effects attributed to colour would, as in damask, be more properly called those of light and shade, as they vary with the point of view selected for the moment. The end grain of mahogany, the surfaces of the table-cloth, and of the motherof-pearl shell, are respectively of nearly uniform colour, but the figures of the wood and the damask, arise from the various ways in which they reflect the light.

Had the fibres of all these substances been arranged with the uniformity and exactitude of a piece of plain cloth, they would have shown an even uninterrupted colour, but fortunately for The *Cuticaem branco,* from Carvalho da Terra, Brazils, and *Cuticaem vermo,* brought over by Mr. Morney, (Admiralty Museum,) show the silver grain very prettily; the first in peculiar straight radial stripes, the other in small close patches. The *Rewa-rewa, (Knightia cjccelsa,)* from New Zealand, is of similar kind; all would be found handsome light-coloured furniture woods.

the beautiful and picturesque such is not the case; most fibres are arranged by nature in irregular curved lines, and therefore almost every intersection through them, by the hand of man, partially removes some and exposes others, with boundless variety of figure.

If further proof were wanted, that it is only the irregular arrangement that causes the damask or variegated effect, I might observe that the plain and uniform silk, when passed in two thicknesses face to face, between smooth rollers, comes out with the watered pattern; the respective fibres mutually emboss each other, and with the loss of their former regular character they cease to reflect the uniform tint.

To so boundless an extent do the interferences of tints, fibres, curls, knots, &c., exist, that the cabinet-maker scarcely seeks to match any pieces of ornamental wood for the object he may be constructing. He covers the nest of drawers, or the table, with the neighbouring veneers from the same block, the proximity of the sections causing but a gradual and unobserved difference in the respective portions; as it would be in vain to attempt to find two different pieces of handsomely figured wood exactly alike.

SECT. II.—VARIATIONS OF COLOUR.

The figures of the woods depend also upon the colour as well as on the fibre; in some the tint is nearly uniform, but others partake of several shades of the same hue, or of two or three different colours, when a still greater change in their appearance results.

In the horizontal sections of such woods, the stripes wind partly round the center as if the tree had clothed itself at different parts with coats of varied colours with something like caprice: tulip-wood, king-wood, zebra-wood, rose-wood, and many others, show this very distinctly; and in the ordinary plank these markings get drawn out into stripes, bands, and The brilliaut prismatic colours of the pearl are attributed to the decomposition and reflection of the light by the numerous minute grooves or strife, a more vivid effect of the same general kind.

A beautiful artificial example of the same description was produced by Sir John Barton, then comptroller of the Koyal Mint; he engraved with the diamond, the surfaces of hard steel dies in lines as fine as 2000 in the inch, arranged in hexagons, &c. The gold buttons struck from these dies display the brilliant play of iridescent colours of the originals.

patches, and show mottled, dappled, or wavy figures of the most beautiful or grotesque characters, upon which it would be needless to enlarge, as a glance at the display of the upholsterer will convey more information than any description, even when assisted by coloured figures.

Those woods which are variegated both in grain and colour, such as Amboyna, king-wood, some mahogany, maple, partridge, rose-wood, satin-wood, snake-wood, tulip-wood, zebra-wood, and others, are more generally employed for objects with *smooth* surfaces, such as cabinet-work, vases, and turned ornaments, as the beauties of their colours and figures are thereby the best displayed. Every little detail in the object causes a diversion in the forms of the stripes and marks existing in the wood: these terminate abruptly round the mouldings which have sharp edges, and upon the flowing lines they are undulated with infinite variety into curves of all kinds, which often terminate in fringes from the accidental intersections of the stripes in the woods.

The elegant works in marquetry, in which the effect of flowers, ornamental devices, or pictures, is attempted by the combination of pieces of naturally coloured woods, are invariably applied to smooth surfaces. In the same manner the beautifully tesselated wood floors, abundant in the buildings of one or two centuries back, which exhibit geometrical combinations of the various ornamental woods, (an art that has been recently pursued in miniature by the Tunbridge turners in their Mosaic works,) are other instances, that in such cases the plain smooth surface is the most appropriate to display the effect and variety of the colours, for such of the last works as are turned into mouldings fail to give us the same pleasure.

Even-tinted woods are best suited to the work of the eccentric chuck, the revolving cutters, and other instruments to be explained; in which works, the *carving* is the principal source of ornament: the variation of the wood, in grain or colour, when it occurs, together with the cutting of the surface, is rather a source of confusion than otherwise, and prevents the effect either Attempts have been made to stain some of our European woods during their growth, by inserting certain portions of their roots in vessels filled with colouring matters, but I am not aware with what success. It is not however to be expected, that such a mode would be either so effective or permanent, as that produced by the natural absorption during the entire period of the life of the plant, an experiment of too lengthened and speculative a character to be readily undertaken.

44 PLAIN WOODS PROPER FOR, ECCENTRIC TURNING. of the material, or of the work executed upon it, from being" thoroughly appreciated.

The transverse section, or end grain of the plain woods, is the most proper for eccentric turning, as all the fibres are then under the same circumstances; many of the woods will not admit of being worked with such patterns, the plankway of the grain: and of all the woods the Black Botany-Bay wood, or the black African wood, by which name soever it may be called, is most certainly the best for eccentric turning; next to it, and nearly its equal, is the cocoa-wood (from the West Indies, not the cocoa-nut palm); several others may also be used, but the choice should always fall on those which are of *uniform* tint, and sufficiently hard and close to receive a polished surface from the *tool,* as such works admit of no subsequent improvement.

Contrary to the rule that holds good with regard to most substances, the colours of the generality of the woods become considerably *darker* by exposure to the light; tulip-wood is, I believe, the only one that fades. The tints are also rendered considerably darker from being covered with oil or lacker, and although the latter checks their assuming the deepest hues, it does not entirely prevent the subsequent change. The yellow colour of the ordinary varnishes greatly interferes also with the tints of the light woods, for which the whitest possible kinds should be selected. When it is required to give to wood that has been recently worked, the appearance of that which has become dark from age, as iu repairing any accident in furniture, it is generally effected by washing it with lime-water; or iu extreme cases, by laying on the lime a3 water-colour, and allowing it to remain for a few minutes, hours, or days, according to circumstances. In many cases the colours of the woods are heightened or modified, by applying colouring matters either before or with the varnish; and in this manner handsome birch-wood is sometimes converted into factitious mahogany, by a process of colouring rather than dyeing, that often escapes detection.

The bog-oak is by some considered to assume its black colour from the small portion of iron contained in the bog or moss, Specimens of woods for cabinets should be left in their natural state, or at most they should be polished by friction only; or if varnished, then upon the one side alone. Their colours are best preserved when they are excluded from the light, either in drawers or in glass cases, covered with some thick blind.

CARVED AND MOULDED W0RK9. combining with the gallic acid of the wood, and forming a natural stain, similar to writing ink. Much of the oak timber of the Royal George that was accidentally sunk at Spithead, in 1782, and which has been recently extricated by Col. Pasley's submarine explosions, is only blackened on its outer surface, and the most so in the neighbourhood of the pieces of *iron;* the inside of the thick pieces, is in general of nearly its original colour and soundness. Some specimens of cam-wood have maintained their original beautiful red and orange colours, although the inscription says that they were " washed on shore at Kay Haven in October, 1840, with part of the wreck of the Royal Tar, lost near the Needles twenty years ago, when all the crew perished."

The recent remarks on colour equally apply to the works of statuary, carving and modelling generally: the materials for which are either selected of one uniform colour, or they are so painted. Then only is the full effect of the artist's skill apparent at the first glance; otherwise it frequently happens either that the eye is offended by the interference of the accidental markings, or fails to appreciate the general form or design, without a degree of investigation and effort, that detracts from the gratification which would be otherwise *immediately* experienced on looking at such carved works.

This leads me to advert to modes sometimes practised to produce the effect of carving; thus, in the Manuel de Tourneur,f a minute description will be found of the mode of making embossed wooden boxes, which are pressed into metallic moulds, engraved with any particular device. The wood is first turned to the appropriate shape, and then forced by a powerful screwpress into the heated mould, (which is made just hot enough to avoid materially discolouring the wood,) it is allowed to remain in that situation until it is cold; this method however only applies to subjects in small relief, and is principally employed on knotty pieces of box-wood and olive-wood of irregular curly grain.

The following method may be used for bolder designs, more resembling ordinary carving: the fine sawdust of any particular wood it is required to imitate, is mixed with glue or other cementitious matter, and squeezed into metallic moulds, but in the latter case the peculiar characteristic of the wood, namely

its fibrous structure, is entirely lost, and the eye only views the Received from the hands of H. Hardman, Esq.
t Second Edition, vol. ii., pp. 441-51. work as a piece of cement or composition, which might he more efficiently produced from other materials, and afterwards coloured.

Each of these processes partakes rather of the proceeding of the manufacturer than of the amateur; extensive preparations, such as very exact moulds consisting of several parts, a powerful press, and other apparatus, are required, and the results are so proverbially alike, from being "formed in the same mould," that they lose the interest attached to original works, in the same manner that engravings are less valued than the original paintings from which they are copied.

Another method of working in wood may be noticed, which is at any rate free from the objections recently advanced: I will transcribe its brief description

"Raised figures on wood, such as are employed in picture frames and other articles of ornamental cabinet-work, are produced by means of carving, or by casting the pattern in Paris plaster or other composition, and cementing or otherwise fixing it on the surface of the wood. The former mode is expensive, the latter is inapplicable on many occasions.

"The invention of Mr. Straker may be used either by itself or in aid of carving; and depends on the fact that if a depression be made by a blunt instrument on the surface of wood, such depressed part will again rise to its original level by subsequent immersion in water."

"The wood to be ornamented having first been worked out to its proposed shape, is in a state to receive the drawing of the pattern; this being put in, a blunt steel tool, or burnisher, or die, is to be applied successively to all those parts of the pattern intended to be in relief, and at the same time is to be driven very cautiously, without breaking the grain of the wood, till the depth of the depression is equal to the subsequent prominence of the figures. The ground is then to be reduced by planing or filing to the level of the depressed part; after which, Ihe piece of wood being placed in water, either hot or cold, the parts previously depressed will rise to their former height, and will thus form an embossed pattern, which may be finished by the usual operations of carving." See Appendix, Note A, page 459 of this volume, and also Appendix, Notes J.K.L. , of Vol. II., pages 954-956, for recent and more available modes of carving by machinery.
See the Section ou Tortoiseahell. t Trans. Soc. of Arts, voL vliii, p. 52.

CHAPTER V. PERMANENCE OF FORM, AND COMBINATION OF THE WOODS.

SECT. I. SHRINKING AND WARPING.

The permanence of the form and dimensions of the woods requires particular consideration, even more than their comparative degrees of ornament, especially as concerns those works which consist of various parts, for unless they are combined with a due regard to the strength of the pieces in different directions, and to the manner and degree in which they are likely to be influenced by the atmosphere, the works will split or warp, and may probably be rendered entirely useless.

The piece of dried wood is materially smaller than in its first or wet state, and as it is at all times liable to re-absorb moisture from a damp atmosphere, and to give it off to a dry one, even after having been thoroughly seasoned, the alterations of size again occur, although in a less degree.

The change in the direction of the *length* of the fibres is in general very inconsiderable. It is so little in those of straight grain, that a rod split out of clean fir or deal is sometimes employed as the pendulum of a clock, for which use it is only inferior to some of the compensating pendulums; whereas a piece of the same wood taken diametrically out of the center of a tree, or the crossway of the grain, forms an excellent hygrometer, and indicates by its change of length the comparative degree of Good box-wood and lance-wood are approved by the Tithe Commissioners as materials for the verified scales to be employed in laying down the plans for the recent Parliamentary survey, as being next in accuracy to those of metal; whereas scales of ivory are entirely rejected by them, owing to their material *variation in length* under hygrometrical influence. See their printed papers.

Mr. Fincham says he has found a remarkable variation in the New Zealand pine the Kowrie or Cowrie, corrupted into Cowdie, which expands so much as to cause the strips constituting the inside mouldings of ships to expand and buckle, probably from the comparative moisture of our atmosphere: and Colonel Lloyd says he found the teak timbers used by him in constructing a large room in the Mauritius, to have shrunk three-quarters of an inch in thirty-eight feet, although this wood is by many considered to shrink sideways least of all others.
moisture of the atmosphere. The important difference in the general circumstances of the woods, in the two directions of the grain, I propose to notice, first as regards the purposes of turning, and afterwards those of joinery-work, which will render it necessary to revert to the wood in its original, or unseasoned state.

The turner commonly employs the transverse section of the wood, and we may suppose the annual rings then exhibited, to consist of circular rows of fibres of uniform size, each of which, for the sake of explanation, I will suppose to be the onehundredth of an inch in diameter.

When the log of green wood is exposed to a dry atmosphere, the outer fibres contract both at the sides and ends, whereas those within, are in a measure shielded from the immediate effect of the atmosphere, and nearly retain their original dimensions. Supposing all the outside fibres to be reduced to the one hundred and tenth, or the one hundred and twentieth of an inch, as the external series can no longer fill out the original extent of the annual ring, the same as they did before they were dried; they divide, not singly, but into groups, as the unyielding center, or the incompressible mass within the arch, causes the parts of which the latter is composed to separate, and the divisions occur in preference at the natural indentations of the margin, which appear to indicate the places where the splits are likely to commence.

The ends being the most exposed to the air are the first attacked, and there the splits are principally radial with occa sional diversions concentric with the layers of fibres, as in fig. 12, and on the side of the log, the splits become gradually extended in the direction of its length. The air penetrates the cracks, and extends both cause and effect, and an exposure of a few weeks, days, or even one day, to a hot dry atmosphere, will, sometimes, spoil the entire log, and the more rapidly the harder the wood, from its smaller penetrability to the air. This effect is in part stayed by covering the ends of the wood with grease, wax, glue or paper, to defend them, but the best plan is to transfer the pieces very gradually from the one atmosphere to the other, to expose them equally to the air at all parts, and to avoid the influence of the sun and hot dry air.

The horizontal slice or block of the entire tree, is the most proper for the works of the lathe, as it is presented by nature the most nearly prepared to our hand, and its appearance, strength, grain, and shrinking, are the most uniform. The annual rings, if any be visible, are, as in fig. 13, nearly concentric with the object, the fibres around the circumference are alike, and the contraction occurs without causing any sensible departure from the circular form. Although thin transverse slices are necessarily weak from the inconsiderable length of the fibres of w hich they are composed, (equal only in length to the *thickmess* of the plate,) they are strengthened in the generality of turned works by the margin, such as we find in the rim of a snuff-box, which supports the bottom like the hoop of a drum or tambarine.

The entire circular section is therefore most appropriate for turning, next to it the quartering, fig. 14, should be chosen, but its appearance is less favourable; and a worse effect happens, as the shrinking causes a sensible departure from the circle, the contraction being invariably greater upon the circular arcs of fibres, than the radial lines or medullary rays. If such works VOL. I. E 50 SHRINKING AND WARPING be turned before the materials are thoroughly prepared, they will become considerably oval; so much so, that a manufacturer who is in the habit of working up large quantities of pear-tree, informs me that hollowed pieces rough turned to the circle, alter so much and so unequally in the drying, that works of three inches will sometimes shrink half an inch more on the one diameter than the other, and become quite oval; it is therefore necessary to leave them half an inch larger than the intended size. Even in woods that were comparatively dry, a small difference may in general be detected by the callipers, when they have been turned some time, from their unequal contraction.

In pieces cut lengthways, such as fig. 15, circumstances arc still less favourable; there being no perceptible contraction in the length of the fibres, the whole of the shrinking takes place laterally, at right angles to them, and the work becomes oval to the full extent of the contraction that occurs in the fibres.

The plank-wood is almost solely employed for large discs which would be too weak if cut out transversely; and in some cases for objects made of those ornamental woods which are best displayed in that section, as the tulip, rose, king, zebra, partridge, and satin woods. Specimens of oak from ancient buildings are sometimes thus worked, but in all such cases the wood should be exceedingly well dried beforehand; otherwise in addition to the inconvenience arising from the greater departure from the circle, the pieces will warp and twist, an effect that more generally concerns the joiner's art, and to the consideration of which we will now proceed.

When the green wood is cut up into planks, boards, and veneers, the splitting which occurs in the transverse section is less to be feared than distortion or warping, from the unequal contraction of the fibres. Thick planks are partially stayed from splitting and opening, by cleets nailed upon each end; boards are left unprotected, and veneers are protected from accidental violence by slips of cloth glued upon each end.

One plank only in each tree can be exactly diametrical, the others are parallel therewith, and, as shown in fig. 12, the two sides of all the boards, but that from the centre, are differently circumstanced as regards the arrangement of the fibres, and con tract differently. It will be generally found that the boards exposed to similar conditions on both sides, become, from the simple effect of drying, convex on the side towards the center of the tree; this will be explained by a reference to the diagram, fig. 16, which shows that the longest continuous line of fibres is concentric with the axis of the tree. Thus let *a, b, c, d, e, f,* represent the section of a board, the line *b, e,* of which is supposed to contain five fibres, and the arc *d, b,f,* thirty: therefore supposing every fibre to shrink alike in general dimensions, the contraction on the arc, will be six times that upon the short radial line, and the new margin of the board will be the dotted line which proceeds from *g* to *h,* the departure of which from the original straight line will be five times as much at *d* as at *e.*

This is not imaginary, as it is in all cases borne out by observation, where the pieces are exposed to similar circumstances on both sides. "When a true flat board is wanted, it is a common practice to saw the wide plank in two or four pieces, to change sides with them alternately, and glue them together again(as in fig. 17, so that the pieces, 1, 3, 5, may present the sides

Fig. 17 1 2 3 4 5 6 towards the axis of the tree, and 2, 4, 6, those towards its circumference; the curvature from shrinking will then become a serpentine line consisting of six arcs, instead of one continuous circular sweep.

When the opposite sides of a board are exposed to *unequal* conditions, the moisture will swell the fibres on the one side and make that convex, and in the opposite manner that exposed to the dry air or heat will contract and become concave; from these circumstances, when several pieces of wood are placed around the room or before the fire, *"to air,"* the sides should be continually changed, that hoth may have equal treatment, so as to lessen the tendency to curvature. To remedy the defect when it may have occurred, the joiner exposes the convex side to the fire, but it is obviously better to be sparing of these sudden changes.

Any unequal treatment of the two sides is almost sure to curl the board; if, for instance, we paste a sheet of paper upon one side of a board, it will in the first instance swell the surface and make it convex; as the paper dries it contracts, it forces the wood to accompany it, and the papered side becomes hollow; when two equal papers are pasted on opposite sides, this change does not generally occur. A similar effect is often observed when a veneer' is glued on a piece qf wood; hence it is usual to swell the surface on which the veneer is to be laid, by wetting it with a sponge dipped in thin size, so as to make it moderately round; in this case, the wetted surface of the board, and the glued surface of the veneer, are expanded nearly alike by the moisture, and in drying they also contract alike, so that under favourable management the board recovers its true flat figure.

The woods are much less disposed to become curved in the direction of their *length,* than crossways; but another evil equally or more untractable is now met with, as the general figure of the board is more or less disposed to twist and warp, so that when it is laid upon a flat surface it touches only at the two diagonal corners, and is said to be " *in winding."* This error is the less experienced in the straight-grained pines and mahogany, which are therefore selected for works in which constancy of figure is a matter of primary importance, as in models for the foundry, and objects exposed to great vicissitudes of climate.

The warping may arise from the curved direction of the fibres in respect to the length of the plank, and also from the *spiral* direction in which many trees grow; in some, for example, the furrows of the bark are frequently twisted as much as fifteen or twenty degrees from the perpendicular, and sometimes even thirty and forty. The woods themselves when split through the center of the tree differ materially; they sometimes present a tolerably flat surface, at others they are much in winding or twisted, a further corroboration of the "spiral growth;" we cannot be therefore much surprised that the planks cut out from such woods, should in a degree pursue the paths thus early impressed upon them.

Boxwood is often very much twisted in tins manner. I have a block, the diameter of which is nine inches; its surface is split at five parts, with spiral grooves, at an angle of nearly thirty degrees with the axis; these make exactly *one complete revolution,* or one turn of a screw in the length of the piece, which is just three feet.

On the other hand, the *Alerce,* a pine growing in the island of Chiloe in South America, to the diameter of about four feet, and whose wood resembles the cedar of Lebanon in colour, is so remarkably straight in the grain, that it is the custom of the country to *split* it into planks about eight feet long and seven inches wide, which are almost as true as if they were cut with the saw, although of course not quite so smooth.

To correct the errors of winding and curvature in length, the joiner in working upon rigid pieces, first planes off the higher points so as to produce the true form by reduction. But when the objects are long and thin, they are corrected by the hands, just as we should straighten a cane, or a walking stick, except that the one angle of the board is rested upon the bench or floor, the other is held in the hand, and the pressure is applied between them.

Broad thin pieces are sometimes warmed on both sides before the fire to lessen their rigidity; they are then fixed between two stout flat boards by means of several hand-screws, and allowed to remain until they are quite cold; this is just the reverse of the mode of bending timber for ship-building and other purposes, but applied in a less elaborate manner.

In concluding this division of the subject, I may observe that the shrinking and contracting of the straight-grained woods, especially deal and mahogany, cause but little distortion of their general shape after they have been properly dried; but the diversity of grain, a principal cause of beauty of figure in the ornamental woods, is at the same time a source of confusion in their shrinking, which being called on to pursue many paths, (which are parallel with the fibres, however tortuous,) gives rise to a greater disturbance from the original shape, or in extreme cases, even causes them to split where the contraction is restrained by the peculiarity of growth.

In the handsome furniture woods the economy of manufacture corrects this evil, as from their great value they are cut into very thin slices or veneers, and glued upon a stout fabric of straight-grained wood, commonly inferior mahogany, cedar, or deal, by which the opposite characters, of beauty of appearance and permanence of form, are combined at a moderate expense; these processes will be explained.

SECT. II.— COMBINING DIFFERENT PIECES OF WOOD.

In combining several pieces of wood for works in carpentry and cabinet-making, the different circumstances of the plank as respects its length and width should be always borne in mind. Provision must be made that the shrinking and swelling are as little restrained as possible, otherwise the pieces may split and warp with an irresistible force: and the principal reliance for permanence or standing, should be placed on those pieces, (or lines of the work,) cut out the lengthway of the plank, which are, as before explained, much less disposed to break or become crooked, than the crossway sections: these particulars will be more distinctly shown by one or two illustrations.

Fig. 18. Let *a, b, c, d,* represent the flat sur face of aboard: *e,f,* the edge of the same, and *g, h,* the end; no contraction will occur upon the line *e,f,* or the length, and in the general way, that line will remain pretty straight and rigid; but the whole of the shrinking will take place on *g,* A, the width, which is slender, flexible, and disposed to become curved from any unequal exposure to the air; the *d* four marginal lines of *a, b, c, d,* are not

L-/ likely to alter materially in respect to *9""* -each other, but they will remain toler ably parallel and square, if originally so formed.

A dove-tailed box consists of six such pieces, the four *sides* of which, A, B, C, D, fig. 19, are interlaced at the angles by the dove-tails, so that the flexible lines, as *g, h,* on B, are connected with, and strengthened by, the strong lines, as *c, d* on A, and so on: the whole collectively form a very rigid frame, the more especially when the bottom piece is fixed to the sides by glue or screws, as it entirely removes from them the small power of racking upon the four angles, (by a motion like that of the jointed parallel rule,) which might happen if the dovetails, shown on a larger scale in fig. 20, were loosely fitted.

Fig. 19. Fig. 20.

When the grain of the four sides, A, B, C, D, runs in the *same* direction, or parallel with the edges of the box or drawer, as shown by the shade lines on A and B, and the pieces are equally wet or dry, they will contract or expand equally, and without any mischief or derangement happening to the work; to ensure this condition, the four sides are usually cut out of the same plank. But if the pieces had the grain in different directions, as C and D, and the two were nailed together, D would entirely prevent the contraction or expansion of C, and the latter would be probably be split or cast, from being restrained. When admissible, it is therefore usual to avoid *fixing* together those pieces, in which the grain runs respectively lengthways and crossways, especially where apprehension exists of the occurrence of swelling or shrinking.

A wide board, fig. 21, composed of the slips, A, B, C, D, E, (reversed as in diagram, fig. 17, page 51,) is rendered still more permanent, aud very much stronger, when its ends are confined by two clamps, such as G, H, (one only seen;) the shade lines represent the direction of the grain. The group of pieces, A to E, contract in width upon the line A, E, and upon it they are also flexible, whereas the clamp G, H, is strong and incapable of contraction in that direction, and therefore unless the wood is thoroughly dry the two parts should be connected in a mauner that will allow for the alteration of the one alone. This is effected by the tongue and groove fitting as represented; the end piece, G, II, is sometimes only fastened by a little glue in the center of its length, but in cabinet-work, where the seasoning of the wood is generally better attended to, it is glued throughout.

If the clamp G, H, were fixed by tenons, (one of which, *ij,* is shown detached in fig. 22,) the contraction of the part of the board between the tenons might cause it to split, the distance between the mortises in G, II, being unalterable: or the swelling of the board might cause it to bulge, and become rounding; or the entire frame would twist and warp, as the expansion of the center might be more powerful than the resistance to change in the two clamps, and force them to bend.

It is therefore obvious that if any question exist as to the entire and complete dryness of the wood, the use of clamps is hazardous; although in their absence, the shrinking might tear away the wood from the plain glue joint, even if it extended entirely across, without causing any further mischief, but more generally the shrinking would split the solid board.

Another mode of clamping is represented at K; it is there placed edgeways, and attached by an undercut or dovetailed groove, slightly taper in its length, and is fixed by a little glue at the larger end, which holds the two in firm contact: each of these modes, and some others, are frequently employed for the large drawing boards required by architects and engineers for the drawings, made with squares and instruments.

From a similar motive, the thin bottom of a drawer is grooved into the two sides and front, and only fixed to the back of the drawer by a few small screws or brads, so that it may swell or shrink without splitting, which might result were it confined all around its margin. It is more usual, however, to glue thin slips along the *sides* of large drawers, as in fig. 23, which strengthen the sides, and being grooved to receive the bottom, allow it to shrink without interfering either with the front or back of the drawer.

In an ordinary door with two or more panels, all the marginal pieces run lengthways of the grain: the two sides, called the *stiles,* extend the whole height, and receive the transverse pieces or *rails,* now mortised through the stiles, and wedged tight, but without risk of splitting, on account of their small width; every panel is fitted into a groove within four edges of the frame. The width of the panel should be a trifle less than the extreme width of the grooves, and even the mouldings, when they are not worked in the solid, are fixed to the frame alone, and not to the panel, that they may not interfere with its alterations; therefore in every direction we have the frame-work in its strongest aud most permanent position as to grain, and the panel is unrestrained from alteration in width if so disposed.

This system of combination is carried to a great extent in the tops of mahogany billiard tables, which consist of numerous panels about 8 inches square, the frames of which are 3£ in. wide and 1 in. thick; the panels are ploughed and tongued, so as to be level on the upper side, and from their small size the individual contraction of the separate pieces is insignificant, and consequently the general figure of the table is comparatively certain. Of late years, I am told , that slate, a material uninfluenced by the atmosphere, has been almost exclusively used; the top of a full sized table, of 12 by 0 feet, consists of four slabs one inch thick, ground on their lower, and planed by machinery on their upper surfaces: the iron tables are almost abandoned for several reasons. Large thin slates, from their permanence of form, are sometimes used by engineers and others for drawing upon, and also in carpentry for the panels of superior doors.

SECT. HI.—ON GLUEING VARIOUS WORKS IN WOOD.

Glue is the cement used for joining different pieces of wood; it is a common jelly, made from the scraps that are pared off the hides of animals before they are subjected to the tan-pit for conversion into leather. The inferior kinds of glue are often contaminated with a considerable portion of the lime used for removing the hair from the skins, but the better sorts are transparent, By Mr. Thurston, of Catherine-street.
especially the thin cakes of the Salisbury glue, which are of a clear amber colour.

In preparing the glue for use, it is most usually broken into small pieces, and soaked for about twelve hours in as much water as will cover it; it is then melted in a glue-kettle, which is a double vessel or water bath, the inner one for the glue, the outer for the water, in order that the temperature applied may never *exceed* that of boiling water. The glue is allowed at first to simmer gently for one or two hours, and if needful it is thinned by the addition of hot water, until it runs from the brush in a fine stream; it should be kept free from dust and dirt by a cover, in which a notch is made for the brush. Sometimes the glue is covered with water, and boiled without being soaked.

Glue is considered to act in a twofold manner, first by simple adhesion, and secondly by excluding the air, so as to bring into action the pressure of the atmosphere. The latter however alone, is an insufficient explanation, as the strength of a well-made glue joint is frequently greater than the known pressure of the atmosphere: indeed it often exceeds the strength of the solid wood, as the fracture does not at all times occur through the joint, and when it does, it almost invariably tears out some of the fibres of the wood: mahogany and deal are considered to hold the glue better than any other woods.!" It is a great mistake to depend upon the quantity or thickness of the glue, as that joint holds the best in which the neighbouring pieces of wood are brought the most closely into contact; they should first be well wetted with the glue, and then pressed together in various ways to exclude as much of it as possible, as will be explained.

The works in turnery do not in general require much recourse to glue, as the parts are more usually connected by screws cut upon the edges of the materials themselves; but when glue is used by the turner the mode of proceeding is so completely similar to that practised in joinery works, that no separate instructions appear to be called for, especially as those parts in which glue is required, as for example in Tunbridge ware, partake somewhat of the nature of joinery work.

When glue is applied to the end grain of the wood, it is rapidly absorbed in the pores; it is therefore usual first to glue the end wood rather plentifully, and to allow it to soak in to fill the grain, and then to repeat the process until the usual quantity will remain upon the face of the work; but it never holds so well upon the endway as the lengthway of the fibres.

In glueing the edges of two boards together, they are first planed very straight, true, and square; they are then carefully examined as to accuracy, and marked, to show which way they are intended to be placed. The one piece is fixed upright in the chaps of the bench, the other is laid obliquely against it, and the glue-brush is then run along the angle formed between their edges, which are then placed in contact, and rubbed hard together lengthways, to force out as much of the glue as possible. When the joint begins to feel stiff under the hand, the two parts are brought into their intended position and left to dry; or as the bench cannot in general be spared so long, the work is cautiously removed from it, and rested in contact with a slip of wood placed against the wall, at a small inclination from the perpendicular. Two men are required in glueing the joints of long boards.

In glueing a thin slip of wood on the edge of a board, as for a moulding, it is rubbed down very close and firm, and if it show any disposition to spring up at the ends, it is retained by placing thereon heavy weights, which should remain until the work is cold: but it is a better plan to glue on a wide piece, and then to saw off the part exceeding that which is required.

Many works require screw-clamps and other contrivances, to retain the respective parts iu contact whilst the glue is drying; in others the fittings by which the pieces are attached together, supply the needful pressure. For instance, in glueing the dovetails of a box, or a drawer, such as fig. 19, page 55, the dovetails, if properly fitted, hold the sides together in the requisite manner, and the following is the order of proceeding.

The dove-tail pins, on the end B, fig. 19, are first sparingly glued, that piece is then fixed in the chaps of the bench, glue upwards, and the side A, held horizontally, is driven down upon B by blows of a hammer, which are given upon a waste piece of wood, smooth upon its lower face, and placed over the dove-tail pins, which should a little exceed the thickness of the wood, so that when their superfluous length is finally planed off, they may make a good clean joint. When the pins of the dove-tails come flush with the face/ the driving block is placed *beside* them to 00 BLUEING A BOX. A FRAME FOR A PANEL.
allow the pins to rise above the surface. The second end, D, is then glued the same as B, it is also fixed in the bench, and A is driven down upon it as before; this unites the three sides of the square. The other pins on the ends B and D are then glued, and the first side, A, is placed downwards on the bench, upon two slips of wood placed close under the dove-tails, that it may stand solid, and the remaining side, D, is driven down upon them to complete the connexion of the four sides.

The box is then measured with a square, to ascertain if it have accidentally become rhomboidal, or *out of square,* which should be immediately corrected by pressure in the direction of the longer diagonal; lastly, the superfluous glue is scraped off whilst it is still soft with a chisel, and a sponge dipped in the hot water of the glue-kettle is occasionally used, to remove the last portion of glue from the work.

The general method pursued in glueing the angles of the frame for a panel, is somewhat similar, although modified, to meet the different structure of the joints. The tenons are made quite parallel both ways, but the mortises are a little bevelled or made longer outside, to admit the small wedges by which the tenons are fastened: and the stiles are made somewhat longer than when finished, to prevent the mortises-from being broken out in driving the wedges, which are mostly cut out of the waste pieces sawn off from the tenons in forming their *shoulders* or *haunches*. These details are seen in fig. 22, p. 56.

In glueing the frame for a single panel which is fitted into a groove, the whole of the frame is put together before commencing the glueing, and the stiles are knocked off one at a time, by which the misplacement of the pieces is avoided. The tenons are glued, and a little glue is thrust into the two mortises with a thin piece of wood; when the stiles have been driven down close, the joint is completed by the insertion of a wedge on each side of the tenon; their points are dipped in the glue, and they are driven in like nails, so as to fill out the mortises, after which the tenons cannot be withdrawn: sometimes the wedges are driven into saw-kerfs, previously made near the sides of the tenons; the other stile is then knocked off, glued, and fixed in the same manner. Occasionally all four tenons are glued at the same time, and the two stiles are pressed together by screw-clamps, stretching across the frame just within the tenons; the wedges are lastly driven in, before the removal of the clamps, and the door if square and true is left to dry.

In many other cases also, the respective pieces are pressed together by screws variously contrived; the boards employed to save the work from being disfigured by the screws are planed flat, and are warmed before the fire, to supply heat to keep the glue fluid until the work is screwed up, and the warmth afterwards assists in drying the glue: such heated boards are named *cauls*, and they are particularly needed in laying down large veneers, which process is thus accomplished.

The surfaces of the table or panel, and both sides of the veneer, are scratched over with a tool called a toothing-plane, which has a perpendicular iron full of small grooves, so that it always retains a notched or serrated edge; this makes the roughness on the respective pieces, called the *tooth* or *key*, for the hold of the glue. A caul of the size of the table is made ready; and several pairs of clamps, each consisting of two strong wooden bars, placed edgeways and planed a little convex or rounding on their inner edges, and connected at their extremities with iron screw-bolts and nuts, are adjusted to the proper opening; the table is warmed on its face, and the veneer and caul are both made very hot.

All being ready, the table is brushed over quickly with thin glue or size, the veneer is glued and laid on the table, then the hot caul, and lastly the clamping bars, which are screwed down as quickly as possible, at distances of three or four inches asunder, until they lie exactly flat. The slender veneer is thereby made to touch the table at every point, and almost the whole of the glue is squeezed out, as the heat of the caul is readily communicated through the thin veneer to the glue and retains it in a state of fluidity for the short space of time required for screwing down, when several active men are engaged in the process. The table is kept under restraint until entirely If the clamps were straight, their pressure would be only exerted at the sides of the table, but being curved to the extent of one inch in three or four feet, their pressure is first exerted in the center, and gradually extends over their entire length, when they are so far strained as to make the rounded edge bear flat upon the table and caul respectively.
cold, generally for the whole night at least, and the drying is not considered complete under two or three days.

When the objects to be glued are curved, the cauls, or moulds, must be made of the counterpart curve, so as to fit them; for example, in glueing the sounding board upon the body of a harp, which may be compared to the half of a cone, a trough or caul is used of a corresponding curvature, and furnished all along the edge with a series of screws to bring the work into the closest possible contact.

In glueing the veneers of maple, oak, and other woods upon curved mouldings, such as those for picture frames, the cauls or counterpart moulds, are made to fit the work exactly. The moulding is usually made in long pieces and polished, previously to being mitred or joined together to the sizes required.

In works that are curved in their length, as the circular fronts of drawers, and many of the foundry patterns that are worked to a long sweep, the pieces that receive the pressure of the screws used in fixing the work together "whilst it is under glue," are made in narrow slips, and pierced with a small hole at each end; they are then strung together like a necklace, but with two strings. This flexible caul can be used for all curves; the strings prevent the derangement of the pieces whilst they are being fixed, or their loss when they are not in use.

I have mentioned these cases to explain the general methods, and to urge the necessity of thin glue, of a proper degree of warmth to prevent it from being chilled, and of a pressure that may cause the greatest possible exclusion of glue from the joint. But for the comparatively small purposes of the amateur, four or six hand-screws, or ordinary clamps, or the screw-chaps of the bench, aided by a string to bind around many of the curvilinear and other works, will generally suffice.

As however the amateur may occasionally require to glue down a piece of veneer, I will, in conclusion, describe the method of " laying it with the hammer," which requires none of the In some of the large manufactories for cabinet-work, the premises are heated by steam-pipes, in which case they have frequently a close stove in every workshop heated many degrees beyond the general temperature, for giving the final seasoning to the wood, for heating the cauls, and for warming the glue, which is then done by opening a small steam-pipe into the outer vessel of the glue-pot. The arrangement is extremely clean, safe from fire, and the degree of the heat is very much under control.

apparatus just described, but the *veneering hammer* alone. This is either made of iron with a very wide and thin pane, or more generally of a piece of wood from three to four inches square, with a round handle projecting from the center; the one edge of the hammer head is sawn down for the insertion of a piece of sheet iron or steel, that projects about one quarter of an inch, the edge of which is made very straight, smooth, and round; and the opposite side of the square wooden head of the veneering hammer is rounded, to avoid its hurting the hand.

The table and both sides of the veneer having been toothed, the surface of the table is warmed, and the *outer* face of the veneer and the surface of the table are wetted with very thin glue, or with a stiff size. The inner face of the veneer is next glued; it is held for a few moments before a blazing fire of shavings to render the glue very fluid, it is turned quickly down upon the table, and if large is rubbed down by the outstretched hands of several men; the principal part of the remainder of the glue is then forced out by the veneering hammer, the edge of which is placed in the center of the table, the workman leans with his whole weight upon the hammer, by means of one hand, and with the other he wriggles the tool by its handle, and draws it towards the edge of the table, continuing to bear heavily upon it all the time.

The pressure being applied upon so narrow an edge, and which is gradually traversed or scraped over the entire surface, squeezes out the glue before it, as in a wave, and forces it out at the edge; having proceeded along one line, the workman returns to the center, and wriggles the tool along another part close by the side of the former; and in fact as many men are generally engaged upon the surface of the table as the shop will supply, or that can cluster around it. The veneer is from time to time wetted with the hot size, which keeps up the warmth of the glue, and relieves the friction of the hammers, which might otherwise tear the face of the wood.

The wet and warmth also render the veneer more pliable, an prevent it from cracking and curling up at the edges, as should the glue become chilled the veneer would break from the sudden bending to which it might be subjected, by the pressure of the hammer just behind the wave of glue, which latter would be then too stiff to work out freely, owing to its gradual loss of VENEERING A TABLE WITH THE HAMMER.

fluidity; the operation must therefore be conducted with all possible expedition.

The concluding process is to tap the surface all over with the back of the hammer, and the dull hollow sound will immediately indicate where the contact is incomplete, and here the application of the hammer must be repeated; sometimes when the glue is too far set in these spots, the inner vessel of the glue-pot or heated irons, are laid on to restore the warmth. By some, the table is at the conclusion laid flat on the floor, veneer downwards and covered over with shavings, to prevent the too sudden access of air. Of course the difficulty of the process increases with the magnitude of the work; the mode is more laborious and less certain than that previously described, although it is constantly resorted to for the smaller pieces arid strips of veneer, even where the foregoing means are at hand.

The former chapters were in type before the Author was aware of the existence of two excellent papers, by A. Aikin, Esq., F.G.S., &c., "On Timber," and "On Ornamental Woods," read before the Society of Arts in 1831 *(see* their Trans. Vol. L., Part ii. p. 140-170). The Author is very happy to find, that so far as the present pages treat of parallel parts of this extensive subject, they are in general confirmed by Mr. Aikin, although the construction of the two papers is entirely different.

Mr. Aikin adverts in a very interesting manner to circumstances relative to the growth of the tree in its native forest, and the process of seasoning, 4c., in which M. Duhamel's great work, *Sur tExploitation des Bois,* is referred to; and also to the luxurious employment of ornamental woods amoDgst the Romans as derived from the Natural History of Pliny the Elder. (Plin. Hist. Nat. xiii. 29—xvi. 24-34.)

"By far the most costly wood was procured from a tree called citrus, a native of that part of Mauritania which is adjacent to Mount Atlas. In leaf, odour, and trunk, it resembles the female wild cypress. The valuable part is a tuber or warty excrescence, which, when found on the root and under ground, is more esteemed than when growing on the trunk or branches. When cut and polished it presents various figures, of which the most esteemed are curling veins, or concentric spots like eyes, the former being called tiger-wood, the latter panther-wood."—" Tables of this material appear to have been first brought into fashion by Cicero, who is said to have given lor a single one a million of sesterces, *i. e.* 8072*/.*"—Others of these *tolid* tables were sold at greater prices, and one as high as 11,800£

"In the time of Pliny the art of veneering was a recent invention; and he descants in his usual antithetical way, on thus converting the cheaper into the most valuable woods, by plating them with these latter; and of the ingenuity of cutting a tree into thin slices, and thus selling it several times over. The woods employed for this purpose were the citrus, the terebinth, various kinds of maple, box, palm, holly, ilex, the root of elder and poplar. The middle part of a tree, he observes, shows the largest and most curling veins, while the rings and spots are chiefly found near the root. The veneers, or plates, were secured, as at present, by strong glue."—Pages 162-4.

CHAPTER VI. CATALOGUE OF THE WOODS COMMONLY USED IN THIS COUNTRY. SOURCES FROM WHENCE IT WAS COLLECTED.

In presenting this descriptive catalogue of woods to the reader, it becomes the author's first and pleasing duty, to acknowledge the valuable assistance he has received from numerous kind friends, of various pursuits, acquirements, and occupations; to most of whom he has submitted the manuscript and rough proofs of thc catalogue, in their various stages through the press, for confirmation or correction, and which has led to the attainment of numerous valuable additions, or he may say, the major part of its contents.

Amongst those to whom he is thus indebted, he has to mention, with gratitude, the following naturalists and travellers, &c.: namely, Arthur Aikin, Esq. , late Secretary to the Society of Arts, London; John Fincham, Esq., Principal Builder in Her Majesty's Dockyard, Chatham; Colonel G. A. Lloyd, Her Majesty's Surveyor-General of the Mauritius; G. Loddiges, Esq.; John Macueil, Esq., Civil Engineer; John Miers, Esq., long resident in the Brazils; and also W. Wilson Saunders, Esq., Colonel Sir James Sutherland, and Colonel Sykes, all three of the East India Company's Service. The author is likewise indebted, in a similar manner, to the following wood-merchants, manufacturers, and others, Messrs. Bolter, Cox, Edwards, Fauntleroy, Jaques, Russell, Saunders, Seddons, Shadbolt, &c., and in a less degree to numerous others.

The extensive botanical notes interspersed (in a smaller type), throughout the list of woods, are from the pen of Dr. Royle, to whom he submitted the early proofs of the catalogue, with the request that he would examine the botanical names so far as he had been able to collect them. The unlooked-for and careful manner in which the professor has executed this request, both from his personal knowledge, and also by a very laborious comparison of the scattered remarks in various works on botany and VOL. L natural history, contained in his select library, will be duly appreciated by those interested in the natural history of the subject, or in the search for the woods themselves in their various localities, whether for the purpose of science or commerce; aud from the mode adopted, the one or the other part of the catalogue may be separately consulted.

To attain the means of comparing the descriptions with the woods themselves, the author has procured a quantity of most of the woods, those employed in turning especially, from which he has cut his own specimens, (these have been kindly augmented by several of the friends before named), aud he has been fortunate in having purchased a very fine cabinet of seven hundred specimens, collected by a German naturalist, and arranged with both the Linnean and German names; all of which specimens are open to the inspection of those who may feel interested therein.

Still further to test the descriptions in the catalogue, he has also carefully examined a variety of museums and collections, from which scrutiny it would have been an easy task to have extended this list in a considerable degree, by the introduction of the names, localities, and descriptions of a variety of wellauthenticated specimens of woods, apparently useful; but he has purposely endeavoured to keep himself within the strict limits called for by this work, in noticing those woods only which are used in England, and that may in general be procured there.

For the use of those who may desire to follow this interesting Bubject with other views, the names of the several museums that have been kindly laid open to his inspection, and a slight notice of their contents, are subjoined in a note.

Many of the remarks on the Timber Woods are derived from that excellent work before named," Tredgold's Elements of Carpentry:" all the French books on turning, enumerated in the introduction, have been consulted, besides those referred to in the various notes, and some others; and, in fact, the author has spared no pains to obtain the most authentic information within his reach, but upon a subject, pronounced by those who have paid attention to it, to be so boundless and confessedly difficult, it is necessary to ask a lenient judgment, and the kind notice and communication of any inaccuracies that may inadvertently exist, notwithstanding his efforts to the contrary.

It is indeed a matter of great and real regret, that upon a subject of general importance, there should in many respects be such a scarcity of exact and *available* information. The true names and localities of some of the most familiar woods, are either unknown or enveloped in considerable doubt; in many cases we have only the commercial names of the woods, and a vague notion of their localities; in others we have authenticity as to their locality and their *native* names; and, lastly, we have also very extensive lists and descriptions of woods in botanical works, and in the writings of travellers, but these three nomenclatures are often incompatible, and admit of surmise only, rather than strict and satisfactory comparison, which drawback was strongly experienced by Dr. Royle in collecting the notes attached to the catalogue.

This deficiency arises from the little attention that has been given to the scientific part of the subject by naturalists and travellers, and from the arbitrary manner in which the commercial names are fixed, often from some faint and fancied resemblance , sometimes from the port whence the woods are shipped, or rather from that whence the vessel "cleared out," or obtained her official papers; as it frequently happens, that the woods are picked up at different points along the coast, the names even of which places cannot be ascertained, much less those of the inland districts or territories in which the woods actually grew.

Naturalists and travellers, and also merchants residing abroad, would therefore confer a great benefit, not only on science, but likewise on the arts, by correcting our knowledge on these points. This might be done by transmitting along with specimens collected on the spot, the exact particulars of their locality, and of the soil; their relative abundance, native names, and uses f. Iu The Komans had their tiger and panther woods, namely the pieces of citrus, marked with *ttripes* or *spots (see* note, p. 64;) the moderns have partridge, snake, porcupine, zebra, and tulip woods, and others. *Ste* the Catalogue.

t The specimens should be stamped with numbers, as a mode preferable to affixing labels, and it should be noted whether the tree from which it was cut were of superior, average, or inferior quality, and also its size. It would be still better to collect three or four samples from different trees, and the transverse sections especially those with the bark would be highly characteristic.

The trouble of preparing the notes to accompany the specimens, would be greatly diminished by the employment of a tabular form, on the model of that adopted at Lloyd's Registry, described in the note, psge 69.

cases of doubt as to their true botanical names (by which alone their identity can be ensured for future years), then some of the leaves, fruits, flowers, &c. , should, if possible, be preserved, by which their species might be afterwards exactly determined by those possessed of the requisite knowledge of the vegetable kingdom.

This would also be important in a commercial point of view, as numerous woods, of which small quantities, perhaps one single importation, have been received, might be again procured, whereas they are now unattainable, from the absence of these particulars.

Latitude exerts a general influence in the distribution of the woods, but it must be remembered that alone it is insufficient to limit the locality; it must be viewed in connexion with the elevation of the land; for even under the equator, as we ascend the mountains, the products of the temperate and even the frigid zones are met with, as Nature appears to set no bounds to her liberality and munificence.

MUSEUMS, Etc., CONSULTED.

The Admiralty Muskem, Somerset House, which is principally due to the superintendence of Sir William Symonds, the Surveyor-General of the Navy, is very rich in specimens of woods. It contains the foundation of a fine collection with their foliage, acorns, cones, and other seed-vessels, &c.; at present the oaks and firs are the most complete: there are also, from Brazil, 56 specimens; from Australia, 13; and from New Zealand, 40; all with native names and foliage. And the following woods, with native names, from various contributors.

N. America, 30, Capt. C. Perry, U.S. N. Brazil, 152, Mr..

Cuba, 168,, Tyrie, Esq. Malabar, 25.

Jamaica, 100, Capt. T. M. C. Symonds, Java, 83.

R.N.. Australia, 25, Sir Thomas Mitchell.

Brazil, 140, S. Homey, Esq., Engineers. Norfolk Island, 16, Leslet, Esq.

This fine museum also includes, amongst others not specified, sets of specimens from the different Government dock-yards, of the timbers used respectively therein. Many of the specimens are worked into cubes and blocks of aim lar size, and their several weights are marked upon them.

There are also 84 pieces from the " Gibraltar" of 80 guns, launched in 1751, and recently broken up: these are intended to show the durability of the woods. East India House.

Indian woods, 117 kinds, in the form of books, about half with their native names.

Indian woods, from Dr. Roxburgh; large pieces of the principal kinds.

Indian and Himalayan woods, from Dr. Wallich; 437 specimens.

Java woods, 100 kinds, presented by Dr. Horsfield.

Asiatic Societt.

Ceylon woods, 255 specimens, with their native names, and alphabetical catalogue.

United Service Museum. Travancore, 110, with native names, Lieut.-Col. J. M. Frith, Madras Artil., C.B. New Zealand and New South Wales, 30, R. Cunningham, Esq.,Bot. Gard., Sydney. Ceylon, 31, names in the native character, from Captain Chapman, B.A. Jamaica, 80, names principally English, from Lieut. J. Grignon, 37th Regt. Jamaica, 31 large handsome polished specimens, Capt.Ethelred Hawkins,22ndRegt. Society Of Arts.

Indian woods, a duplicate set of Dr. Wallich's collection, namely, 457 specimens, enumerated in the Trans, of the Society, Vol. 48, Part 2, pp. 439—479.

India, various parts, Cape of Good Hope, Pitcairn's Islands, &&, 452 specimens, Capt. H. C. Baker, Bengal Art. *Ice. See* Trans. Vol 50, Part 2, pp. 173—189. Lloyd's Reoistby Op Shipping. 160 specimens of ship-building woods, oaks the most numerous, next firs, pines, and elms. They are accompanied by a list which contains seven columns, respectively, headed "Stamped number on Specimen, Name of Wood, Place of Growth, Soil, Durable or otherwise, Purpose for which used, Remarks."

Private Collections Of Specimens.

Mr. Fincham's contains most of those woods in the subjoined list, generally in two sections, with their specific gravities and relative degrees of strength. Also from Nova Scotia, 8; Rio Janeiro, 11; Isle of France, 34; Malabar, 19; Ceylon, 59; New South Wales, 14; Van Diemen's Land, 6; New Zealand, 17; all with native names, brought over direct by the captains of Government ships.

G. Loddiges, Esq., F.L.S., F.H.S., F.Z S., &c., has a fine cabinet. Of the woods of Europe, 100; Jamaica, 100; Brazils, 250; Chili, 45; Sierra Leone, 20; East Indies, 25; South Seas, 33; all with native names; and 25 from China, marked in that character. Also about 100 commercial and dye woods, and not less than 1000 from all parts of the globe not yet prepared for his cabinet.

J. Miers, Esq., F.L.S., &c., has 75 Brazilian specimens, collected by himself on the spot.

W. Wilson Saunders, Esq., F.L.S., &o.: Brazilian, 70; Grecian, 17; British, 70; various localities, 65.

Mexico. Dr. Coulter, M.D., M.R.I.A. , Hon. Fel. Col. Phys., Hon. Fel. Roy. Dub. Soc. &c., has collected 800 specimens in Mexico, 788 with the leaf, flower, and sometimes the fruit. They have been presented by him to Trinity College, Dublin. These I have not seen.

Isthmus of Panama. *See* Colonel Q. A. Lloyd's Notes and Catalogue of Woods, Trans. Royal Geog. Soc., VoL I., p. 71.

Ship-building Woods used in our Government Yards. Oaks.—English. Adriatic. Italian. Sussex. New Forest. Canada, white and red. Pollard. Istrian. Live oak. African. And also Teak.
Pines.—Yellow. Red. Virginian Nil Red. Pitch-pine. Riga.
Firs—Norway and American Spruce fir. Dautzic and Adriatic fir.
Larches.—Hackmetack. Polish. Scotch. Italian, 1. 2. 3. Athol. Cowdie, or New Zealand Larch.
Cedars —Cuba. Lebanus. New South Wales and Pencil cedar.
Elms. —English and Wych elm.
Miscellaneous Woods, used in small quantities.—Rock Elm. English and American ash. Birch, black and white. Beech. Hornbeam. Hickory.
Mahogany. Lime-tree. Poon-wood, and Lignum vitae, &c.

TABULAE VIEW OF THE WOODS COMMONLY USED IN THIS COUNTRY.

For Buildino. *Ship-building.*
Cedars.
Deals.
Kline.
Firs.
Larches.
Locust.
Oaks.
Teak, &c. &c.
Wet works, as piles, foundations, tc.
Alder.
Beech.
Elm.
Oak.
Plane-tree.
White Cedar.
House-carpentry.
Deals.
Oak.
Pines.
Sweet Chesnut.
For Tornert. For Furniture.
For Machinery And' Mill-Work.
Frames, tic.
Uh.
Beech.
Birch.
Deals.
Elm,
Mahogany.
Oak.
Pines.
Hollers, itc.
Box.
Lignum-vitse,
Mahogany.
Teeth of Wheels, Jkc.
Crabtree.
Hornbeam.
Locust.
Foundry patterns.
Alder.
Deal.
Mahogany.
Pine.
Common woods for toys: softest.
Alder.
Aps.
Beech) „
Birch (smaU-
Sallow.
Willow.
Best woods for Tunbridge ware.
Holly)
H. Chesnut
a *I* W00O8 oycamore)
Apple-tree 1
Pear-tree brown
Plum treeJ wood8
Hardest English woods.
Beech, large.
Box.
Emi.
Oak.
Walnut.
Common Furniture and inside works.
Beech.
Birch.
Cedars.
Cherry-tree.
Deal.
Pines.
Best Furniture.
Amboyna.
Black Ebony.
Cherry-tree.
CoromandeL
Mahogany.
Maple.
Oak, various kinds.
Rose-wood.
Satin-wood.
Sandal wood.
Sweet Chesnut.
Sweet Cedar.
Tulip-wood.
Walnut.
Zebra-wood.
Miscellaneous Properties.
Elasticity.
Ash.
Hazel.
Hickory.
Lancewood.
S. Chesnut, small.
Snakewood.
Yew.
Inelasticity and toughness.
Beech.
Elm.
Lignum-vitie.
Oak
Walnut.
Earn grain, proper for Carving.
Lime-tree.
Pear-tree.
Pine.
Durability in dry' works.
Cedar.
Oak.
Poplar.
Sweet Chesnut.
Yellow Deal.
Colouring Matter.
Red Dyes. Brazil. Braziletto. Cam-wood. Log-wood. Nicaragua. Red Sanders. Sapan-wood.
Green Dye.
Green Ebony.
Yellow Dyes.
Fustic.
Zante.
Seenl.
Camphor-wood.
Cedar.
Rose-wood.
Sandal-wood.
Satin-wood.
Sassafras.

DESCRIPTIVE CATALOGUE or THB CHAKACTEES AND USES OF THE WOODS COMMONLY EMPLOYED IN THIS COUNTRY FOB TBI MECHANICAL AND ORNAMENTAL ARTS.

N. B.—The botanical name?, the notes

printed in a smaller type, and the articles marked with an asterisk, have been added by Dr. Royle, M.D., F.R.S., L. S., & G.S., Ac., Ac., of tks East India House, Professor of Materia Medica and Therapeutics, King's College, London.

ABELE. *Set* Poplab. ACACIA, true. The. 1TMTM *proximo, Mordi,* A. GuiHard's MSS., called in Cuba, *Sabicu,* and in England *Savico* and *Savacn,* is a heavy durable wood of the red mahogany character, but rather darker and plainer; it is highly esteemed in ship-building. In the Admiralty Museum the leaves, 4c. of this tree are to be seen, and also specimens of the original timbers of the "Gibraltar," of 80 guns, Launched in 1751, some of the wood of which is now in such perfect condition that Sir W. Symonds intends to use the old keel of the "Gibraltar," (made of Savico,) for that of a new frigate.

The true acacias are found in warm parts of the world, and yield valuable theugh usually small timber, which is remarkable for being hard and tough, as *Acacia tortuom,* called Cashaw tree in the West ladies. On the west coast of Africa, *Acacia vcrtk* has very hard white wood, as well as other species. -/ *viclanoxylon.* Black wattle tree aud Slack wood,and *A. decurrem,* Orcen wattle, occur in New Holland.

In India *Acacia arabica* and *fanusiana* commonly called *bubod A. tpeciata,* and. (. *tundra,* yield timber valued for different purposes. Many of theao trees exude gum, aud their bark is employed iu tanning leather.

ACACIA, false, the common acacia or locust-tree. See Locust-trek. AFRICAN BLACKWOOD. *See* Black Botant-bat Wood. ALDER, *(Alnut glutinosa,)* Europe and Asia. There are other species in N. America and the Himalayas. The oommon alder se dom exceeds *iO* ft. iu height, is very durable under water, and was used for the piles of the Rialto at Venice, the buildings at Ravenna, tc.: the wood is also much used for pipes, pumps, and sluices. The colour of alder is reddish-yellow of different shades, and ALDER—*continued.* nearly uniform; the wood is soft, and the smaller trees are much used for inferior turnery, as tooth-powder boxes, common toys, brushes, and bobbins, and occasionally for foundry patterns. The roots and knots are sometimes beautifully veined, and used in cabinet-work. The charcoal of the alder is employed in the manufacture of gunpowder. ALOES-WOOD. *See* Calembeq. ALMOND-TREE, *(Amygdalut community* is very strongly recommended by Desormeaux, as being hard, heavy, oily or resinous, and somewhat pliable; he says the wood towards the root so much resembles *lignum-vita,* as to render it difficult to distinguish between them. It is sometimes called false *lignum-vita,* and is used for similar purposes; as handles, the teeth and, bearings of wheels, pulleys, &c., and any work exposed to blows or rough usage. It is met with in the South of Europe, Syria, Barbary, 4c. The wood of the bitter almond, grown in exposed rocky situations, is preferred. AMBOYNA-WOOD. *See* Kiabooca-wood.

ANGICA-WOOD. See Canoioa-wood.

APS. *See* Poplar.

APPLE-TREE, *(Pyrm Malm).* The woods of the apple trees, especially of the uncultivated, are in general pretty hard and close, and of red-brown tints, mostly lighter than the hazel-nut. The butt of the tree only is used; it is generally very straight and free from knots up to the crown, whence the branches spring. The apple tree splits very well, and is one of the best woods for standing when it is properly seasoned: it is very much used in Tunbridge turnery, for bottle-cases, &c.: it is a clean-working wood, and being harder than chesnut, sycamore, or lime-tree, is better adapted than they are for screwed work, but is inferior in that respect to pear-tree, which is tougher. The millwright uses the crab-tree for the teeth of mortise-wheels. APRICOT-TREE, *(Armeniaca xulgarit,)* a native of Armenia, is mentioned in all of the French works on turning, beginning with Bergeron, (1792,) who says, the wood of the apricot-tree is very rarely met with sound, but that it is agreeably veined, and better suited to turning than carpentry. He elsewhere very justly adds, that we are naturally prejudiced in favour of those trees, from which we derive agreeable fruits, and expect the respective woods to be either handsome in appearance, or agreeable in scent, but in each of which expectations we are commonly disappointed: this applies generally to the orange and lemon trees, and we may add, to the quince, pomegranate and coffee trees, the vine, and many others occasionally met with, rather as objects of curiosity, than as materials applicable to the arts. ARBOR VITE. The different species of *Thuja,* are called *Arbor viia,* and are chiefly found in North America and China. *T. occidentals,* or American *A rbor vita,* attains a height of from 40 to 50 feet, and has reddish-coloured, CATALOGUE OF THE WOODS. 73 ARBOR *YITM—continued.* somewhat odorous, very light, soft and fine-grained wood. It is softer than white pine, and much used in house carpentry, and also for fences.

The Chinese *Arbor vita* or *T. orientaiit,* is smaller, but the wood is harder. *T. articulata,* a native of the north coast of AfriOi, is the *Alerce* of the Moors, and was employed in the woodwork of the mosque, now the cathedral, of Cordova. The plant is now called *Calliiris quadrivalvis.* ASH, *(Fraxinw exceUa* ;) Europe and North of Asia; mean size, 38 ft. long by 23 in. diam., sometimes much larger. The young wood is brownish-white with a shade of green; the old, oak brown with darker veins. Some specimens from Hungary with a zigzag grain, and some of the pollards, are very handsome for furniture.

Ash is superior to any other British timber for its toughness and elasticity; it is excellent for works exposed to sudden shocks and strains, as the frames of machines, wheel-carriages, agricultural implements, the felloes of wheels, and the inside work of furniture, &c. The wood is split into pieces for the springs of bleachers' rubbing boards, which are sometimes 40 fyet long; also for handspikes, billiard cues, hammer handles, rails for chairs, and numerous similar works, which are much stronger when they follow the natural fibre of the wood.

Ash is too flexible and insufficiently durable for building purposes; the young branches serve for hoops for ships' masts, tubs, churns, &c.

Several species are found In North America: of these it is theught that the White Ash, or *Fraxinus americana,* comes the nearest in quality of wood to thc common ash. *F. JUnriounda* and *tan-thexyloidet* are two ashes found in the Himalayas.

Fraxinut ornut produces manna. *Fraxinus excelea* produces a manna somewhat similar.

Ash, the Moontain Ash, or Quicken or Rowan tree, *Pyrut Aucuparia,* grows in almost every soil or situation, has fine-grained hard wood, which may be stained of any colour, and takes a high polish, and is applied to the samo purposes as the wood of the beam aud service trees. See S«bvice-ires. ASPEN. *See* Poplar. BARBERRY WOOD, *(Berberis vulgaris,)* is of small size, generally about 4 in. diam.; the rind is yellow, and about half an inch thick: the wood re. sembles elder, and is tolerably straight and tenacious.

BAR-WOOD, Africa. Two kinds are imported from Angola and Gaboon respectively, in split pieces 4 to 5 ft. long, 10 to 12 in. wide, and 2 to 3 in. thick. It is used as a red dye-wood, the wood is dark-red, but the dye rather pale; it is also used for violin bows, ramrods, and turning. BIT-TREE. The sweet bay-tree, *(Laurus nobilti,)* a native of Italy and Greece, grows to the height of 30 feet, and is an aromatic wood. It is the laurel that was used by the ancients for their military crowns. BEECH. Only one species, *(Fagiu sylvatica,)* is common to Europe; in England the Buckinghamshire and Sussex beech are esteemed the best. Mean dimensions of the tree, 44 ft. long and 27 in. diam. The colour, (whitishbrown,) is influenced by the soil, and is described as white, brown, and black. *(Tredgold.)*

Beech is used for piles in wet foundations, but not for building; it is excel74 ALPHABETICAL AND DESCRIPTIVE BEECH—*continued.* lent from its uniform texture and closeness for in door works, as the frames of machines, common bedsteads and furniture; it is very much used for planes, tools, lathe chucks, the keys and cogs of machinery, Shoe-lasts, pattens, toys, brushes, handles, &c. Carved moulds for the composition ornaments of picture-frames, and for pastry, and large wooden types for printing, are commonly made of beech: the wood is often attacked by worms, when stationary as in framings, but tools kept in use are not thus injured.

Beech is stained to imitate rose-wood and ebony, and it is considered to be almost chemically free from foreign matters; for example, the glassblowers use the wood almost exclusively in *welding,* or fusing on, the handles of glass jugs, which process fails when the smallest portion of sulphur, &c., is present: oak is next in estimation for the purpose.

The wfaito beech of North America, *Fagiui sylvestru,* is by some theught to be Identical with the common beech, but the wood Is little valued in America; the bark bowever is employed in tanning.

BEEF-WOOD. Bed-coloured woods, are sometimes thus named, but it is generally applied to the Botany-Bay oak—which see. BIRCH-WOOD, a forest tree common to Europe and North America; the finest is imported from Canada, St. John's, and Pictou. It is an excellent wood for the turner, being light-coloured, compact, and easily worked: it hi in general softer and darker than beech, and unlike it in grain.

Birch-wood is not very durable, it is considerably used in furniture; some of the wood is almost as handsomely figured as Honduras mahogany, and when coloured and varnished, is not easily distinguished from it. The bark of the birch-tree is remarkable for being harder and more durable than the wood itself, amongst the Northern nations it is used for tiles for roofs, for shoes, hats, &c., and in Canada for boats. The Russians employ the tan of one of the birch-trees to impart the scent to Russia leather, which is thereby rendered remarkably durable. The inner bark is used for making the Russia mats.

The English birch is much smaller than the foreign, and lighter in colour; it is chiefly used for common turnery. Some of the Russian birch, (called Russian maple,) is very beautiful and of a full yellow colour.

Betula alba is the common birch of Europo, and he most common tree throughout the Russian empire. Tbo Russian maple of commerce is tbought to be the wood of the birch. *Betula lenta,* mahegany birch and mountain mahegany, of America, has close-grained, reddish-brown timber, which ie variegated and well adapted to cabinet-work. It is imported in considerable quantities into England under the name of American birch. *Betula execha;* tall, also called yellow birch, has wood much like the last, and *B. nigra,* or black, ie also much esteemed. *B. papyiarta,* paper, or canoe birch, is employed by the North American Indians in constructing their portable canoes. *B. Bhe-jputtra* is a Himalayan species, of which the bark is used for writing upon and for making the snakes of heokahs. BITTER-NUT WOOD, a native of America, is a large timber wood measuring 3O inches when squared, plain and soft in the grain, something like walnut. *Juglani amara,* white or swamp nickory or Bitter-nut, and *J. aqnatica,* or water bitter-nut hickory, are probably the trees which yield this wood. BLACK BOTANY-BAY WOOD, called also African Black-wood, is perhaps BLACK BOTANY-BAY WOOD—*continued.* the hardest, and also the most wasteful of all the woods; the billets arc vcry knotty and crookcd, and covered with a thick rind of the colour and hardness of boxwood; the section of the heart-wood is very irregular, and mostly either indented from without, or hollow and unsound from within; many of the pieces have the irregular scrawling growth that is observed in the wood of the vine. The largest stem of Black Botany-Bay wood I have ever seen, measured transversely eleven inches the longest and seven aud a half the shortest way, but it would only produce a circular block of five inches, and this is fully two or three times the ordinary size.

The wood, when fresh cut, is of a bluish-black, with dark-grey streaks, but soon changes to an intense jet black;

of the few sound pieces that are obtained, the largest may perhaps be five inches, but the majority less than two inches diameter. It is most admirably suited to excentric turning, as the wood is particularly hard, close, free from pores, but not destructive to the tools, from which, when they are in proper condition, it receives a brilliant polish. It is also considered to be particularly free from any matter that will cause rust, on which account it is greatly esteemed for the handles of surgeons' instruments.

The exact locality of this wood has long been a matter of great uncertainty. It has been considered to be a species of African ebony, but its character is quite different and peculiar; I have however recently heard from two independent sources, that it comes from the Mauritius, or Isle of France. Col. Lloyd says the wood is there called *Cocobolo prieto;* that it is not the growth of the Mauritius, but of Madagascar, to the interior of which island Europeans are not admitted; and that it is brought in the same vessels that bring over the bullocks, for the supply of food. The stonemasons of the country use splinters of it as a pencil for marking the lines upon their work; it makes a dark blue streak not readily washed off by rain.

I have only met with one specimen of this wood in the numerous collections I have searched, namely, in Mr. Fincham's: he assures me that his specimen grew in Botany Bay, and was brought direct from thence with several others, by Captain Woodroffe, R.N. As I have recently purchased a large quantity imported from the Mauritius, it is probable that this wood, in common with many others, may have several localities.

It would be very desirable for the amateur turner that the wood should be selected on the spot, and the better pieces alone sent, as a large proportion is scarcely worth the expense of shipment, but the fine pieces exceed all other woods for excentricturned works.

BLUE-GUM WOOD. *Set* Gcm-wood.

BOTANY-BAY OAK, sometimes called Beef-wood, is from New South Wales; it is shipped in round logs, from 9 to 14 in. diam. In general colour it resembles a full red mahogany with darker red veins; the grain is more like the evergreen oak than the other European varieties, as the veins are small, slightly curled, and closely distributed throughout the whole surface. 76 ALPHABETICAL AND DESCRIPTIVE BOTANY-BAY OAK—*continued.*

It is used in veneer for the backs of brushes, Tunbridge ware,and turnery; some specimens are very pretty.

The trees called oaks in New South Wales do not belong to the genus Qurrwu.like the European, Xorth American, and Himalayan oaks. There, the tree called

Forest Oik, is *Caauariua tnrulosa;* Swamp Oak, is *C. palniosa:* He Oak is *C. equuitifolia;* while *C. strictn.* is called She Oak, and also Beef-wood.

BOXWOOD, *(Bums tempenirens,)* is distinguished as Turkey and European boxwood. The former is imported from Constantinople, Smyrna, and the

Black Sea, in logs felled with the hatchet, that measure from 2 to 6 ft. long, and 2J to 14 in. diam. The wood is yellow inclining to orange; it has a thin rind with numerous small knots and wens; some of it is much twisted, and such pieces do not stand well when worked; on the whole, however it is an excellent, sound, and useful wood.

Boxwood is much used for clarionets, flutes, and a great variety of turned works; it makes excellent lathe-chucks, and is selected by the woodengraver to the exclusion of all other woods. It is also used for carpenters' rules, and drawing scales; although lance-wood, satinwood and elder, are sometimes substituted for it. Boxwood is particularly free from gritty matter, and on that account its sawdust is much used for cleaning jewellery; it is frequently mentioned by the Roman authors as a wood in great esteem at the period in which they wrote.

Some of the boxwood is as handsomely mottled as fine satin-wood; but it differs much in colour, apparently according to the age and season at which it is cut, as only a small portion of the Turkey boxwood is of the full yellow so much admired.

European boxwood is imported from Leghorn, Portugal, &c. The English boxwood is plentiful at Boxhill in Surrey, and in Gloucestershire; it is more curly in growth, softer and paler than the Turkey boxwood; its usual diameters are from 1 to 5 in.; it is used for common turnery, and is preferred by brass finishers for their lathe-chucks, as it is tougher than the foreign box, and bears rougher usage. It is of very slow growth, as in the space of 20 to 25 years it will only attain a diameter of 1J to 2 inches. A similar wood, imported from America under the name of Tugmutton, was formerly much used for making ladies' fans.

Murraya (Mackay li. fr. Tavoy.) Specimen 275 of Dr. Wallich's, and 118 of Captain Baker's Collection of Indian woods, and *Oaripe apugnt braco* of ; Mr. — *'b* from the Brazils, (Admiralty,) seem fully equal to boxwood, in most respects. *Bnxtui tempervii ent,* or common evergreen box, is found througheut Europe, attaining a height sometimes of from 15 to 20 feet. Turkey box is yielded by *Baxnt balearwa,* which is found in Minorca, Sardinia, and Corsica, and also in both European aud Asiatic Turkey, and large quantities of it are imported from Constantinople into England.

A new species has lately been introduced from tho Himalayas, *Buxut emarginatus,* of Dr. Wall:ch: this is found of considerable size and thickness, aud the wood appears as good and compact as that of the boxwood in use in Europe. Royle, 11 lust. Himal. Bot. p. 327. On actual comparison the Himalayan boxwood is found to be softer than the common kinds, but is like them in other resjects; as maybe seen in the woodcut, figs. 9 aud 10, which havo been eugraved upon a piece of the wood of the Himalayan *Buxut tmarginatas.* CATALOGUE OK THE WOODS. 77 BRAZILWOOD, called also Pernambuco, was supposed by Dr. Bancroft to have been known as a red dyewood before the discovery of the Brazils, which country, he says, was so named by Europeans from its abounding in this wood. 1 he best kind is from Pernambuco, where it is called *Pao da rainha,* or queen's wood, and by the natives *1 birapitanrja;* it is also found in the West Indies generally, and is often called Pernambuco-wood. The tree is large, crooked, and knotty, and the bark is so thick, that the wood only equals the third or fourth of the entire diameter; the leaves are of a beautiful red, and exhale an agreeable odour. The *Pao da rainha* 'grows to the diameter of 15 or 16 inches, the *Pao Brazil,* an inferior kind, to 50 or 60 in. Brazil-wood is a royal monopoly, and the best quality has the imperial brand mark at the end; it is shipped in trimmed sticks, from 1 to 4 in. diam. and 3 to 8 ft. long, and its colour becomes darker by exposure to the air. Its principal use is for dyeing; the best pieces are selected for violin bows and turning.

Ccualpina tchinata, the *Ibirapitonga* of Piso, yields the Brazil-wood of commerce. Do Condolle inquires whether it is not rather a species of *Quilandina. C.* crista, a native of the West Indies, is coiled *Breeillet,* because its wood is reddishcolourtd like Brazil-wood. *C. Sapan* is a native chiefly of the Asiatic Isles and of the Malayan Peninsula; its wood is like Bi azil-wood, and well known in cotnmerco as Snp;m-wood.

BRAZILETTO is quite unlike the Brazil wood; its colour is ruddy orange, sometimes with streaks; it is imported from Jamaica in sawn logs from 2 to 6 ft. long and 2 to 8 in. diam. with the bark, (which is of the ordinary thickness,) left on them; and also from New Providence, in small cleaned aticks. Braziletto is thought to be an inferior species of Brazil-wood; it is principally used for dyeing, also for turnery and violin bows.

It is considered to bo botauically allied to the above, and is called *Catalpina brazilunsti,* a native of the West Indies, and also fouud in Brazil.

BULLET-WOOD, from the Virgin Isles, West Indies, is the produce of a large tree, with a white sap; the wood is greenish-hazel, close and hard. It is used in the country for building purposes, and resembles the Greenheart.

A specimen (at Lloyd's Registry, &c.) of the Booley or Bully-tree, from the Quarawtive River, South America, appeared an excellent hard wood, very dense, and of a plain deep purple rod.

The name of Bullct-wood is perhaps taken from the *Boil de balle* or Bullet-wood of the French, *Guarea triehilioides,* which in Jamaica is called musk or alligatorwood. Bullet is perhaps a change from Bully-wood, which is that of the bullytreo, called also Nasebcrry bullet-tree, or *Achrai Sapota* of botanists, described as one of the best Umber-trees. The buliy-treo of Guiana is also an Achras. The bastard bully-trees of Jamaica are species of *Bumelia.*

BULLET-WOOD, another species so called, is supposed to come from Berbice; its colour is hazel-brown, of an even tint without veins; it is a very close, hard and good wood, well adapted to general and to excentric turning, but is not common.

The latter agrees pretty closely with a wood described by Dr. Bancroft as Bow-wood, or *Watceba,* of Guiana.

Different specimens marked Naseberry bullet-wood, and one of an iron-wood, were exceedingly near to the above, if not identical with it, and the Bull Hoof and Bread Nut Heart, all from Jamaica, approached more distantly.—(United Service and Admiralty Collections.) 78 ALPHABETICAL AND DESCRIPTIVE BUTTON-WOOD TREE. *See* Planb Tbbk.

CABBAGE-WOOD. *See* Pabtbidqk-wood.

CALAMANDER, *Dioupyros hirsutu. See* Cobomandkl.

CALEMBERRI. *See* Cobohandel.

CALEMBEG. A wood similar to Sandal-wood in grain, and similarly, but lesa powerfully, scented; its colour is olive-green, with darker shades. It appears entitled to the name of Green Sandal-wood.

Calembeg, or Calambac, sometimes called Aloes-wood, is the Agallochnm of the ancients, and the Agila or Eple-wood of the moderns. It is produced in Sism aud Silbot by *Aqailuria AgaUocha.* V. Boyle, Illustr. p. 171.

CAMPEACHY LOGWOOD. *See* Logwood.

CAMPHOR-WOOD, is imported from China, the East Indies and Brazils, in logs, and planks of large size; it is a coarse and soft wood, of a dirty greyish yellow colour, sometimes with broad iron-grey streaks, and is frequently spongy, and difficult to work. It is principally used in England for cabinet-work and turnery, on account of its scent.

The Campher-tree of Sumatra is *Dryobalanopt Camphera,* of which the wood is hard, compact, and brownish-coloured; there is a genuine specimen in the museum of King's College, London. The fragrant light-coloured soft wood of which the trunks end boxes from China are made, is supposed to bo that of the Campher-tree of Japan, *Laurus Camphera,*now *Camphora officinalis.* Oueormore of the tribo of Laurels yield the *Sirwabali* wood of Guiana, which is light, fragrant, aud mnch used in the building of boats.

CAM-WOOD, an African dyewood, is shipped from Rokella, Sierra Leone, 4c. in short logs, pieces, roots, and splinters. When first opened, it is tinted with red and orange; the dust is very pungent, like snuff; it would be a beautiful wood if it retained its original colours, but it changes to dark red, inclining to brown. Cam-wood is the best and hardest of the red dyewoods; it is very fine and close in the grain, and suitable to ornamental and excentric turning.

Cam-wood is yielded by a leguminous plant, which has been introduced into, and flowered in this country, and has boon described and figured by-Air. G. Loddiges, iu his botanical cabinct, vol. iv. t. 367, undor the name of *Baphia niivla.* CANARY-WOOD from the Brazils, Para, &c.; known at the Isthmus of Darien as *A marillo.* It is imported in round logs from 9 to 14 in. diam., and sometimes in squared pieces. The wood is of a light orange colour, and generally sound; it is straight and close in the grain, and very proper for cabinet work, marquetry, and turnery; is similar, if not the same, to a wood called Vantatico and Vigniatico, corrupted from *Vinhatico,* a Portuguese name for several yellow woods, besides that imported from the Brazils under the same name.

Zaunu *indica,* or Royal Bay. is a native of the Canary Isles. The wood is of a yellow colour, not heavy, but well suited to furniture; it is called *Vigniatico* in the island of Madeira, and is probably what is imported into England under the name of Madeira Mahegany; it is less brown than mahegany.

CANGICA WOOD, from the Brazils, also called in England Angica, is of the rosewood character, but of a lighter and more yellow brown, less abrupt and more fringed, sometimes straight in grain and plain in figure. It is imported in trimmed logs from 6 to 10 in. diam., and is used for cabinetwork and turning.

CEDAR. The name Cedar lias been given to trees of very different natural orders, and has occasioned much confusion.

The cedar of Lebanon, or great cedar, *(Pinui Cedrus,)* is a cone-bearing resinous tree, and one of the pines. It is tall and majestic, and grows to a great size; the mean dimensions of its trunk are 50 feet high and 39 inches diameter. The wood is of a rich yellowish brown, straight-grained, and it has a peculiar odour. The tree is famous in Scripture for its size and durability (Ezekiel, xxxi. 3, 5, 8;) it was used in the construction of Solomon's temple at Jerusalem, and many Grecian temples and statues. A few fine trees are said still to remain on Mount Lebanon; but the wood was also procured in the time of Vitruvius, from other parts of Syria, and from Crete, Africa, &c.—*Tredgold.*

The Pencil cedar is the *Juniperus vtrginiana;* it is also of the same natural order as the pine-tree. It is imported from North America, in pieces from 6 to 10 inches square. The grain of the wood is remarkably regular and soft, on which account principally, it is used for the manufacture of pencils, and from its agreeable scent, for the inside work of small cabinets; from the same reason it is made into matches, for the drawing-room.

Another species is the *Juniperut bermudiana;* it is a much harder and heavier wood than the pencil cedar, with a similar smell and appearance. It was formerly much used in ship-building: many of the timbers of the Spanish ships taken in the last war were of the Bermuda cedar.

"Up to this time there are great quantities of the finest cedar growing in the British island of Bermuda, and the best ships and schooners arc always built of it; it is imperishable."—*Col. G. A. Lloyd.*

The cedar known to cabinet-makers by the name of Havannah cedar, is the wood of the *Cedrela odorata* of Linnaius, and belongs to the same natural order as mahogany, which it resembles, although it is softer and paler, and without any variety of colour. It is imported in considerable quantities from the island of Cuba, and is excellent for the insides of drawers and wardrobes: all the cigar-boxes from Havannah are made of this kind of cedar; the wood is brittle and porous. Some kinds of the Havannah cedar are not proper for cabinet-work, as the gum oozes out and makes the surface of the work very sticky and unpleasant.

There is another kind more red in colour, called red cedar; there are also white cedars common to America: one kind is called prickly cedar, from its being covered with spines: this is very like the white hemlock, and grows to *i* ft. diam. and 60 to 70 ft. high, and is much used for railway works.

Another sort, from New South Wales, is the wood of the *Cedrela Toona;* it is somewhat similar to the Havannah, but more red in colour, and of a coarser grain; it sometimes measures 4 feet diameter. This kind Ib also found in the East Indies; it is in common use in joinery-work. Most of the cedars have been used for ship-building.

The Himalayan cedar *(Junipa-ut exceUa)* is harder and less odoriferous

C E D AR—*cant in ued.* than the Pencil cedar, but is an excellent light wood between pencil cedar and deal in general character.—*See* Dr. Wallich's Collection, 202.)

The cedar of Lebanon is usually called *Pinut Celrut,* but sometimes *Cednu* iitanwt,-the lofty *Deodara,* a native of the Hirnulayas, with fragrant and almost imperishable wood, and ofteu called the Indian cedar, is sometimes referred to the genus *Pinut.* and somotimes to that of *Cednu* or *LarU,* with the specific DNme of *Deodara.*

The wood of several of the Conifiana Is hewever called cedar. The wood of *Juni.p?rnt vinjiniuna* is callei Red or Pencil Cedar, and that of *J. bermudiana* is called Bermuda Cedar; of *J. barbadensu.* is called Barbadoes Cedar; while the Juniper of the North of Spain, and South of Franco, and of the Levant, is called */. oxycedrus;* the White Cedar of North America, a less valuable wood than the red cedar, is yielded by *Capreesut TKyoula,* and the cedar-wood of Japan, according to Thunbcrg, is a species of cypress.

The name cedar is hewever applied to a number of woods In our different colonies, whiohare in no way related to tho *Conifera:* thus the cedar of Guiana lb the wood of *Icica oliiuima,* whito wood or white cedar of Jamaica is *Bijivtnis ieucoxy'on,* and bastard cedar Is *Ouaxuma ulmifolia.* In New South Wales again the term white cedar is applied to *Melia Azetlerach,* aud red cedar to that of *Ftindereia australu,* as well as to the wood of the Toou-trce, or *Cedrela Toona.* CHERRY-TREE, is a hard, close-grained wood, of a pale red brown, that grows to the size of 20 or 24 inches, but it is more usually of half that size. When stained with lime, and oiled or varnished, it closely resembles mahogany; it is much used for common and best furniture and chairs, and is one of the best brown woods of the Tunbridge turners. The wood of the black-heart cherry-tree is considered to be the best. "The Spanish American cherry-tree is very elastic, and is used for felucca masts."—*Cut. 0. A. Lloyd. Ceratut avium* is the wild cherry. *C. dnracini* is the heart cherry or Bigarreau. The wood of C. Mahalob is much used by the French, and is called *boude Saintt Lucie.* CHESNUT, *(Castanea raca,)* is common to Europe; mean size 44 ft. high, 37 in. diameter; is very long-lived and durable. The sweet, or Spanish chesnut, is very much like oak, and is sometimes mistaken for it; it was formerly much used iu house carpentry and furniture. The young wood is very elastic, and is used for the rings of ships' masts, the hoops for tubs, chums, 4c.

, but the old wood is considered to be rather brittle.—*Sec* Housii Chesni/t.

The ediblo or sweet chesnut is tho *Castanea raca,* but the Horse Chesnut (which see) belongs to a very different genus. The wood formerly much used in heuse. binldiug aud carpentry, and which, famed for its durability, has been mistaken for chesnut, is now considered to be that of an oak, *Quercut tamitijl'tra.* COCOA-WOOD, or Cocus, is imported from the W. Indies in logs from 2 to 8 in.
diameter, sawn to the length of 3 to 6 ft.
, tolerably free from knots, with a thick yellow sap: the heart, which is rarely sound, is of a light yellow brown, streaked, when first cut with hazel and darker brown, but it changes to deep brown, sometimes almost black. Cocoa-wood is much used for turnery of all kinds, and for flutes; it is excellent for excentric turning, and in that respect is next to the African black-wood.

An apparent variety of cocoa-wood from 2 to 6 or 7 inches diameter, with a large proportion of hard sap of the colour of beechwood, and heart wood of a chesnut brown colour, is used for tree-nails and pins for ship-work and purposes similar to lignum-vitse, to which it bears some resemblance, although it is much smaller, has a Tough bark, the sap ia more red, COCOA-WOOD—*toHt-inued.* and the heart darker and more handsomely coloured when first opened than lignum-vitas; it is intermediate between it and cocoa-wood. Another but inferior wood, exactly agrees with the ordinary cocoa-wood, but that the heart is in wavy rings, alternately hard and soft.

Cocoa-wood has no connection with the Cocoa-nut, which is the fruit of a palm-tree common to the East and West Indies, the *Cocoa nucifcra;* neither can it have any relation to the other endogenous trees which produce the Coquilk nut, the *Atlulia fuuifera* according to Martius, and *Cocos lapidea* »f Givrtner, or of the *Cacao Theobroma,* or the Chocolate-nut tree.

It is really singular that the exact localities and the botanical name of the cocoa-wood that is so much used, should be uncertaiu: it appears to come from a country producing sugar, being often imported Ob *dunnage,* or the stowage upon which the sugar hogsheads are packed: it is also known as I!rown Ebony, but the *Amcrininuin Ebcnua* of Jamaica seems dissimilar.

I have scarcely found any specimens of it in the various collections recently examined. The piece in Mr. G. Loddiges' collection from Rio Janeiro, with Portuguese names,) was marked Cocoa, by which it is generally designated in this country, as cocus-wood is the name given by the wholesale merchant. The cogwood of the West Indies, used for the cogs of wheels and building purposes, is a similar but lighter-coloured wood of larger size.

In Mr. Tyrie's collection of Cubanel woods, in Sir W. Symonds's museum, there are nine woods of about the same density and general character as cocoa-wood; they are arranged the lightest first, with their Spanish names; the figures denote the apparent diameters of the trees from which the specimens were cut: No. 108, *Acacio real,* hazel brown, slightly veined, (8 inches); No. 141, *Xacaco,* very like cocoa but much lighter (3 in.); No. 144, *Gateado,* more veined, ruddy cast (4 in.); No. 5, *Yayti,* slightly darker than last, with greenish cast; No. 12, *Almiqul,* chesnut-brown, only more ruddy, very rich tint (4 in.); No. 133, *Oerilio,* the complexion of tolerably dark walnut, sap is paler than cocoa (5 in.); No. 42, *China,* very near to cocoa in colour, specimen had a very small heart and much sap, No. 101, *Granadilla,* greenish cast (3 in.); No. 72, *ifabno,* rather darker than cocoa, the heart apparently 15 or 18 in. diam., one inch of sap left on the specimen; Nos. 108, 12, and 72, appear to be desirable woods.

The Cocus wood of commerce is not easy to trace to any of the trees of the West
Indies, the cocoa plum is *Chrysobah'. nut leaco,* which forms only a shrub;
Voccoloba uvi/ei-ai or mangrove grape tree, grows largo and yields a beautiful wood for cabinet-work, but which is light and of a white colour. In appearance and description it comes noar to the Greenheart or *tnnrnt rhlonxylon,* which is also called Cogwood.

COCOA-NUT TREE and COCOA-NUT. *Set* Palms, and Scpti.emknt.

COCUS. *See* Cocoa-wood.

COFFEE-TREE *(Coffea arabica).* The wood is of a light greenish-brown or duskyyellow, with a bark externally resembling boxwood, but thicker and darker. The specimen I have is nearly as close-grained and hard as boxwood; it has Vol. L COFFEE-TREE—*continued.* no smell, and but little taste. The tree does not grow more than a few feet high, and it is cut down in the plantations to five or six feet, and is not therefore useful in manufactures.

The tree called Kentucky coffee-tree, or hardy *bondue,* is very different from the common coffee; it forms a large tree called *Qymnocladui canadensit;* the wood is compact, of a rosy hue; and used by cabinet-makers. CORAL-WOOD, says Bergeron, is so named from its colour. When first cut it is yellow, but soon changes to a fine red or superb coral; it is hard, and receives a fine polish: he also speaks of a damasked coral-wood. It is difficult to associate these with the red woods; they are perhaps, from the descriptions, nearest to the cam-wood from Africa.

The coral-treo, so culled from the colour of its flowers, *itErythrina CoroUodautnn:* but the *hois U corail* of the French, is the wood of *Aiienaix-thora* /aro»i *ina,* which is hard, reddish coloured, and sometimes confounded with rod Sanders wood.

COQUILLA NUT. *Sec* Supplement page 111. COROMANDEL, or Calamander, the produce of Ceylon, and the coast of India, is shipped in logs and planks from Bombay and Madras. The figure is between that of rose wood and zebra-wood; the colour of the ground is usually of a red hazel brown, described also as chocolate brown, with black stripes and marks. It is said to be so hard as almost to require grinding rather than cutting; this is not exactly true, as the veneer saws cut it without particular difficulty, it is a very handsome furniture wood and turns well: it is considered to be a variety of ebony.

Mr. Laird says there are three varieties of Coromandel; the *Calamander or Coromandel,* which is the darkest, and the most commonly seen in this country, the *Calembcrri,* which is lighter coloured and striped, and the *Omander,* the ground of which is as light as English yew, but of a redder cast, with a few slight veins and marks of darker tints. He says, the wood is scarce and almost or quite limited to Ceylon; that it grows between the clefts of rocks—this renders it difficult to extract the roots, which are the most beautiful parts of the trees.

The Calamander-wood tree is *Vlotpynt hiriuta,* and Kadum Beriya is *D. Ebe,ie-sler,* according to Moore's Catalogue of Ceylon Plants, and therefore of the same gonns as the true ebony.

COROMANDEL, falsely so called, has a black ground, and is either striped, mottled, or dappled, with light yellow, orange, or red; it is a description of accidental or imperfect East Indian black ebony. Some of the pieces are very handsome; it is used for similar purposes to the true coromandel, from which, however, it is entirely different, and generally inferior, although it is considered a variety of the same group. COROSOS, or Ivory Nut. *Sec* Supplement page 112.

COWDIE. *See* Pines.

CRAB-TREE, the wild Apple-tree; principally used by millwrights for the teeth of wheels. *Sec* Apple-tree. CATALOGUE OF THE WOODS. 83 CYPRESS-TREE. Of this there are many varieties; the principal are the *Cu*2irts«i *sempervirens,* and the white cypress or white cedar of North America, the *Cuprtssts Thyoidcs;* the latter is much need as a timber wood, it is an immense tree, and is considered to be more durable even than the cedar of Lebanon. The *Cupressm semperrirens* is said to have been much used by the ancients; by the Egyptians for the cases for some of their mummies, by the Athenians for coffins, and for tho original doors of St. Peter's at Rome, which, on being replaced after six hundred years by gates of brass, were found to be perfectly free from symptoms of decay, and within, to have retained part of the original odour of the wood.—*Trcd'jold.* It is probable that the wood of *Thuja articulata,* (see Arbor vitas,) was also used by the ancients, and has sometimes been mistaken for that of Cypress. DEAL. SccPi.vES. DOQ-WOOD, a small underwood, which is so remarkably-free from silex, that little splinters of the wood are used by the watchmaker for cleaning out the pivot-holes of watches, and by the optician for removing the dust from small deep-seated lenses; dog-wood is also used for butchers' skewers, and tooth-picks.

The charcoal of the black dog-wood is employed in the manufacture of the best sporting gunpowder, alder and willow charcoal for the government powder.—*Wilkinson's Engines of War,* 1841.

Cornut tanguinta is the wild cornel or common dog-wood, *C. mas.* Is the male dogwood or Cornelian cherry, whilo *C. fiorida* is an American species; others are fonnd in the Himalayas. The name dog-wood is applied in Jamaica to *Pit?idia Erithrina.* EAST INDIAN BLACK-WOOD, *(Dalbergia lalifolia,)* called Black-wood tree by the English and *Sit Sdl* by the natives of India, on the Malabar coast, where it grows to an immense size. The wood of the trunk and largo branohes is extensively used for making furniture; it is heavy, sinking in water, close grained, of a greenish or greenish black colour, with lighter coloured veins running in various directions, and takes a fine polish. EBONY is described as of several colours, as yellow, red, green, and black. The existence of yellow and red ebonies appears questionable. The black ebony is the kind always referred to when the name is mentioned alone; in fact, "as black as ebony," is an old proverb. The wood is surrounded by a white sap 3 or 4 inches thick. The green ebony is an entirely different tree, with a thin smooth bark, growing in the West Indies.

Three kinds are imported; No. 1, from the Mauritius, in round sticks like scaffold poles, they seldom exceed 14 in. diameter; No. 2, the East Indian which grows in Ceylon, the East India islands, and on the continent of India, this is mostly shipped from Madras and Bombay in logs from 6 to 20 and sometimes even 28 in. diameter, and also in planks; and No. 3, the African ebony, shipped from the Cape of Good Hope in billets, the general sizes of which are from 3 to 6 ft. long, 3 to 6 in. wide, and 2 to 4 in. thick, these are rent out of the trees, and are thence often called billet-wood.

No. 1, the Mauritius, is the blackest and finest in the grain, as well as the hardest and most beautiful of the three, but also the most costly and EBONY—*continued.* unsound; No. 2, the East Indian, is less wasteful, but of an inferior grain and colour to the above; and No. 3, the African, is the least wasteful, as all the refuse is left behind, and all that is imported is useable, but it is the most porous, and the worst in point of colour.

They are all used for cabinet, mosaic, and turnery works; also for flutes, the handles of doors, knives, and surgeons' instruments, and many other purposes. Piano-forte keys are generally made of the East Indian variety.

The African stands the best, and is the only sort used for sextants.

Colonel Lloyd says, the Mauritius ebony when first cut is beautifully sound, but that it splits like all other woods from neglectful exposure to the sun. The workmen who Mac it, immerse it in water as soon as it is felled for 6 to 18 months, it is taken out and the two ends are secured from splitting by iron rings and wedges. He considers the Mauritius ebony to be the finest, next the Madagascar, and afterwards the Ceylon.

In Mr. Fincham's collection there is a specimen of White Ebony from the Isle of France; it resembles boxwood in most of its characters. There is also a specimen of Young Ebony, of similar description, or rather more like ash in general tint, intermixed with light iron-grey streaks or stains, as if the black were in course of deposition. And in Capt. Baker's collection at the Society of Arts, there are nine different specimens of *Diospyrm,* two only of which are black, the remainder, more or less like the above.

The black ebony is also met with in South America, but much less generally than in Asia and Africa.

The ebony of Mauritius is yielded by *Diospyrus Ebenut,* that of Ceylon is *D. Elmoster,* while the ebony-tree of the Coromandel coast is *D. melanoxylon,* other species as *D. tomentosa* and *D. Roylei,* yield ebony on the continent of India. The tree yielding the African ebony is not ascertained. A kind of ebony is produced by *Amcrimnum EUnutt* in the West Indies, and called Jamaica ebony.

Mountain Ebont. The different species of *Bauhinia* are so called: *B. porrteta* grows on the hills in Jamaica, and has wood which is hard and veined with black.

See Gbeen Ebony and Coromandel. ELDER, *(Sambucus nigra).* The branches of the elder contain a very light kind of pith, which is used when dried for electrical purposes. The surrounding wood is peculiarly strong and elastic. The trunk-wood is tough and closegrained; it is frequently used for common carpenters' rules and inferior turnery work, for weavers' shuttles, (many of which are also made of boxwood,) for fishermen's netting pins, shoemakers'"pegs, &c. ELM *(Ulmut),* a European timber tree, of which there are five species; mean size, *H* ft. long, 82 in. diameter. The heart wood is red brown, darker than oak, the sap yellowish or brownish white, with pores inclining to red; the wood is porous, cross-grained, and shrinks and twists much in drying. Elm is not liable to split, and bears the driving of nails or bolts better than any other timber, and it is exceedingly durable when constantly wet; it is therefore much used for the keels of vessels, and for wet foundations, waterworks, piles, pumps, and boards for coffins; from its toughness, elm is selected for the naves of wheels, shells for tackle-blocks, and sometimes for the gunwales

CATALOGUE OK THE WOODS.

So

K LM—*continued.* of ships, and also for many purposes of common turnery, as it bears very rough usage without splitting.

Wtch Elm. This sometimes grows to the height of 70 feet, and the diameter of 3 J feet; the branches are principally at the top, the wood is lighter and more yellow in colour than the above, also straighter and finer in the grain. It is tough, similar to young sweet chesnut for bending, and is much used by coachmakers, and by shipwrights for jolly-boats.

Rock Elm appears very like the last; it is extensively used for boat buildujg, and sometimes for archery bows, as it is considered to bend very well.

Plmus eamislrit is the common small-leaved elm, *U. efitua* is the spreading, branched, *U. glabra* is the smooth-leaved, and *U. montana* the Wych elm. *Ulmu Americana,* or the American elm, is used for the same purposes as the European species, theugh the wood is inferior in quality. *17. fulra* and *alata* are other American species, and several species are found in the Himalayas. FIRS AND PINES. *See* Pines. FUSTIC, is the wood of a species of Mulberry, *(Morut tinctoria,)* growing in most parts of South America, the United States, and West Indies. It is a large and handsome tree; it is shipped in trimmed logs from 2 to *i* ft. long, 3 to 8 in. diameter; the colour of the wood is a greenish yellow, it is principally used for dyeing greens and yellows, and also in mosaic cabinet-work and turning. *Sec* Zante, or Young Fustic. GHKNADILLO, Granillo, or Grenada Cocus, from the West Indies, is apparently a lighter description of the common cocoa or cocus-wood, but changes ultimately to as dark a colour, although more slowly. It is frequently imported without the sap.

The tree yielding this has not boen ascertained, the *boit de Grtnaditle* of the French is also called red ebony by their cabinet-makers.

GREEN EBONY, from Jamaica, and the West Indies generally. It is cut in lengths of 3 to 6 ft., has a bark much like cocus, but thinner and smoother, the heart wood is of a brownish green, like the green fig. It is used for round rulers, turnery, and marquetry-work, and it cleaves remarkably well. The dust is very pungent, and changes to red when the hands are washed with soap and water. The wood is very much used for dyeing, and it contains so much resinous matter, that the negroes in the West Indies employ it in fishing as a torch. The candle-woods of the West Indies obtain their name probably from the same circumstance, they are allied to the rosewoods, but are of lighter yellow colours.

The ebony of Jamaica is *Amerimnum Ebenut* and has been mentioned under Ebony. The wood is described as being of a fine greenish brown colour, hard, durable, and capable of taking a fine polish; *B. leucoxylon* of South America yields *te boit d'ibi-nt vert.* GREEN-HEART; from Jamaica, Demerara, and the Brazils, bears a general resemblance to cocoa-wood both in size and bark, but the latter has a redder tint. Qreenheart when first cut is of a light green brown, and striped, but it changes to the colour of *Lignum-vita,* and is by some

ALPHABETICAL AND DESClUPTIVE

GHEENHEAKT— *continued.* ponsidered to be pernicious. It is used for turnery and other works, but its texture is coarse, and it will not cleave at all profitably.

Gheenheart used in ship-building is entirely different from the above, and runs into several varieties.

Dr. Bancroft describes Greenheart, or the *Skjnera* tree, to be in size like tho locust-tree, say 60 or 70 feet high: there are two species, the black and the yellow, differing only in the colour of their bark and wood. He says there is also a purple-heart wood, of a bright crimson colour, but which changes to purple, and is esteemed more valuable than the preceding.— *Dr. Bancroft's Guiana,* p. 68-9.

These descriptions exactly agree with Mr. Finchain's specimens described as Greenheart, and black and brown Greenheart; they are large heavy woods and of olivo green even tints, varying from very pale to dark. These, as well *m* the Purple-heart woods, are used for ship-building, but more particularly in their native countries; they appear excellent also for the lathe.

The greenheart of Jamaica and Guuma, i s the *Latum Chloroziihn* of botanists; it is aUo called Cogwood ia the former, and *Sipieri* in the latter locality.

UU.U-WOUD, or blue Gum-wood, is the produce of Now South Wales, it is sent over iu large logs and planks, the colour is similar to that of dark Spanish mahogany, with a blue, sometimes a purple-grey cast: it is used in shipbuilding. There is also a variety of a redder tint called red Gum wood, which is used for ramrods, both are also employed by the turner. *Encalyptuspiperita* is tho bluo guin-treo of Now South. Wuloa, whilo red gum-tite is another species, probably *E. rainifera.*

HACKMETACK LARCH. &e Pines.

HARE-WOOD. *See* Sycamore.

HAWTHORN, *(Cratagus oxyacantlta,)* has hard wood of a whitish colour, with a tinge of yellow; the grain is line, and tho wood takes a good polish, but being small and difficult to work it is not much used. HAZEL, a small underwood, but little used for turning, except for a few toys. It is very elastic, and is used, as well as the ground ash, for the rods of blacksmiths' chisels, hoops of casks, &c. Its botanical name is *Conjlius A idlana.,* HICKORY, or White Walnut, *(Jwjlans alba,)* is a native of America; it is a large tree, sometimes exceeding 3 ft. diameter. The wood of young trees is exceedingly tough and flexible, and makes excellent handspikes, and other works requiring elasticity. The bark of hickory is recommended by Dr. Bancroft as yellow dye. HOLLY, *Hex aqulfulium,)* is a very clean, fine-grained wood, the whitest and most costly of those used by the Tunbridge ware manufacturer, who employs it for a variety of his best works, especially those which are to be painted in water colours. It is closer in texture than any other of our English woods, and does not readily absorb foreign matters, for which reason it is used for painted screens, the squares of draft-boards, and for tho stringings or lines of HOLLY—*continued.* cabinet-work, both in the white state and when dyed black, also for some of the inside works of pianofortes, harps, for calico-printers' blocks, &c. When larger wood than holly is required, the horse-chesnut is employed, but the latter is much softer.

The holly requires very particular care in its treatment: immediately it is felled it is prepared into pieces of the form ultimately required, as planks, veneers, or round blocks for turning. The veneers are hung up separately to dry, as resting in contact even for two or three hours would stain them; the round blocks are boiled in plain water for two or three hours, and on removal from the copper they are thrown in a heap and closely covered up with sacking to exclude the air, which would otherwise cause, them to split. The heap is gradually exposed as it dries; at the end of about four weeks the pieces look greenish, and are covered with mildew sometimes as thickly as one-sixteenth of an inch; this is brushed off at intervals of three or four weeks, and in about six months the wood is fit for use.

Holly is a remarkably tough clean wood, and is used for chucks; but this troublesome preparation to whiten the wood, (and which is not generally practised on other woods,) is not then required, although a good boil hastens the extraction of the sap, and the subsequent seasoning of the wood. Birdlime is prepared from the middle bark of tho Holly.

There is an American species of this genus, the *Ilex opuca,* opaque-leaved or American helly, of which the wood is em ployed in turnery and cabinet-making; there aro other species in tho Himalayas.

HORNBEAM, *(Carpinut £ctulus,)* sometimes also called yoke-elm, is a very tough and stringy European wood, which is used by millwrights for the cogs of wheels, also for skittles, plumbers' dressers or mallets, and a variety of things required to bear rough usage. Hornbeam is sometimes used for planes, it turns very well, and is occasionally imported from America. HORSE-CHESNUT, *(Jtecuhu hippocaslum,)* bas no relation to the Spanish or sweet chesnut, which latter is more nearly allied to the oaks. The horse chesnut is one of the white woods of the Tunbridge turner; it is close and soft, even in the grain, and is much used for brush backs, it turns very well in the lathe, and is a useful wood. It is softer than holly, but is preferable to it for large painted and varnished works, on account of its greatly superior size.

HORSE-FLESH WOOD, one of the Mangroves, which see.

INDIAN BLACK-WOOD. Sec East Indian Black-wood.

IRON-WOOD, is imported from the Brazils, the East and West Indies, and other countries, in square and round logs, 6 to 9 in. and upwards through. Its colours are very dark browns and reds, sometimes streaked, and generally straight grained.

A specimen in Mr. Fincham's collection, from the Isle of France, wa as light in colour as pencil cedar, but of a yellower brown; and in the same cabinet a piece called Iron Fark, from New South Wales, had the density of 88 ALPHABETICAL AND DESCRIPTIVE IRON-WOOD—*continued.* s. g. 1-426, and the strength of 1557, (English oak being called 1000,) in appearance it resembled plain brown Spanish mahogany, and it seemed to be not only the heaviest but the most solid of the woods; Mr. Fincham considers that the Australian woods, taken on the whole, are the most dense with which we are acquainted.

The iron-woods are commonly employed by the natives of uncivilised countries for their several sharp-edged clubs and offensive weapons; in England they are principally used for ramrods, walking-sticks, for turning, aud various purposes requiring great hardness and durability: the more red varieties are frequently called beef-wood.

Iron-wood is a term applied to a great variety of woods, in consequence of tbeir hardness, and almost every country has an iron-wood of its own. *Miiua ferrm,* which has received its specific name from the hardness of its wood, is a nativeof the peninsula of India and of the islands.

Melrosidtrot vera is called truo ironwood: the Chinese are said to make their rudders aud anchers of it, and among the Japmose it is so scarce and valuable that it is only allowed to be manufactured for the service of their king. The iron-wood of southern China is *Baryzt/lum ri'fum*; of the leland of Bourbon *Stadmannia Sidcroxylon,* and of the Cane of Good Hope *Sideroxylou melaiwphleum,* which latter is very hard, close-grained, and sink? in water. The iron-wood of Guiana is *Robitaa Panaewo* (of Aublet); that of Jamaica is *Fagara Ptcrota,* and also *Erythroxyl-nm artolatum,* which is also called red-wood, *JBgipkila martinicensii* and *Co-coLoba latifolia,* are other West Indian trees, to the woods of which the name of iron-wood bos been applied.

Ostrva virginica, called American bop hernbeam, has wood exceedingly hard and heavy, whence it is generally called iron-wood in America, and in some places lover-wood. JAK-WOOD, is the wood of *Artocarpus integrifolia,* or the entire-leaf bread-fruit tree, a native of India, is imported in logs from 3 to 5 feet diameter, and also in planks; the grain is cross and crooked, and often contains sand. The wood is yellow when first cut, but changes to a dull red or mahogany colour. It is very much used in India for almost every purpose of house carpentry and furniture, and in England for cabinet-work, marquetry, and turning, and also for brush-backs. The jak-wood is very abundant, and its fruit is commonly eaten by the natives, and also sometimes by Europeans at dessert, with salt and water, like olives. The jak-wood is sometimes misnamed orange-wood from its colour, and also jack-wood, /(laci-wood and *Kutkul. See* Baker's Papers. JACARANDA, the Portuguese and continental name for Rose-wood, which see. JUNIPER-WOOD. The wood of all the species is more or less aromatic, and very durable; they are found in the cold and temperate parts of the world. Some have already been mentioned under the head of Cedar. The common juniper, *Junipcrus communis,* has wood which is aromatic, finely veined, and of a yellowish brown colour; *J. txctlea,* lofty or Himalayan cedar, is found on those mountains, as well as in Siberia and North America. KIABOOCA-WOOD, or Amboyxa Wood, imported from Sincapore, appears to be the excrescence or burr of some large tree; it is sawn off in slabs from 2 to 4 ft. long, 4 to 24 in. wide, and 2 to 8 in. thick; it resembles the bun of the yew-tree, is tolerably hard, and full of small curls and knots, the KIABOOCA WOOD—*continued.* colour is from orange to chesnut-brown, and sometimes red brown. It is a very ornamental wood, that is also much esteemed in China and India, where it is made into small boxes and writing-desks, and other ornamental works, the same as by ourselves.

The Kiabooca is said by Prof. Reinwardt, of Leyden, to be the burr of the *Pterospermum indicum;* by others that of *Plerocarptts draco,* from the Moluccas, the island of Borneo, Amboyna, &c. The native name appears from Mr. Wilson Saunders' specimen, to be *Seri-oulcut,* the wood itself is of the, same colour as the burr, or rather lighter, and in grain resembles plain mahogany.

"The root of the cocoa-nut tree is so similar, when dry and seasoned, to the 'bird's-eye' part of the wood here termed kiabooca, that I can perceive no difference; the cocoa has a tortuous and silky fracture, almost like indurated asbestos."— *Col. 0. A. Lloyd.*

The comparison of the palm wood with the kiabooca, renders the question uncertaiu, as amongst the multitudes of ordinary curly woody fibres, that one cannot account for in a palm, there are a few places with soft friable matter much resembling its cement.

KING-WOOD, called also Violet-wood, is imported from the Brazils, in trimmed logs from 2 to 7 in. diameter, generally pipy, or hollow in the heart. It is beautifully streaked in violet tints of different intensities, finer in the grain than rose-wood, and is principally used in turning and small cabinet-work; being generally too unsound for upholstery. It is perhaps one of the most beautiful of the hard woods in appearance.

The specimen in my German cabinet is marked *Spartium Arbor. Trifol. Ivjn. violaceo barrillieri.* It is also marked Guiana-wood, and King-wood. The description is sufficiently distinct, but the arbitrary nature of many of the names renders it difficult to be traced in books. KOURIE. See Pines. LABURNUM, *Cytitut Laburnum,)* possesses poisonous seeds, and a small dark greenish brown wood, that is sometimes used in ornamental cabinet-work and marquetry. Mr. Aikin says: " In the Laburnum there is this peculiarity, which I have not observed in any other wood, namely, that the medullary plates, which are large and very distinct, are white, whereas the fibres arc a dark brown; a circumstance that gives quite an extraordinary appearance to this wood."—Page 160 of Vol. 50, Trans. Soc. of Arts. The Alpine laburnum, with blackish wood, is *Cytimt alpinut.* LANCE-WOOD is imported in long poles from 8 to 6 in. diameter from Cuba and Jamaica; it has a thin rind, externally similar to that of cocoa-wood, it is called one of the rough-coated woods, and has a bark distinct from the sapwood, but together they are very thin. Lance-wood is of a paler yellow than box, and rends easily; it is selected for elastic works, such as gig shafts, archery bows and springs; these are bent by boiling or steaming, LANCE-WOOD—*continued.* lance-wood is also used for surveyors' rods, billiard cues, and for ordinary rules which are described as being made of box-wood.

In Captain Baker's Papers an Indian Lance-wood is called *Menaban.*

The lance-wood of Jamaica is *Oualteria vlrpata,* formerly *Umria lanaolala.* Thtt of Guiana Is on *Anonartoue* plant, and probably thc saine species.

LAKCH. &e Pines. LETTER-WOOD. *See* Snake-wood.

LEMON-TREE. *See* Orange-tree.

LEOPARD-WOOD. *See* Palms.

Lignum-vitje, or *Ouaiacum,* is a very hard and heavy wood. It is shipped from Cuba, Jamaica, St. Domingo, and New Providence, in logs from 2J to 36 in. diameter, and is one of the heaviest of the woods. CoL Lloyd says that it grows in the Isthmus of Darien to the size of 5 or 6 ft., and is there called *Quallacan,* and that it is one of the most abundant woods of the country. When first cut, it is soft and easily worked, but it becomes much harder on exposure to the air. The wood is gross-grained, covered with a smooth yellow sap-like box, almost as hard as the wood, which is of a dull brownish green, and contains a large quantity of the gum guaiacum, which is extracted for the purposes of medicine. Lignum-vitte is much used in machinery, &c. for rollers, presses, mills, pestles and mortars, sheaves for ship-blocks, skittle-balls, and a great variety of other works requiring hardness and strength. It was employed by the Spaniards for making guncarriages and wheels.

The fibrous structure of this wood is very remarkable: the fibres cross each other sometimes as obliquely as at an angle of 30 degrees with the axis, as if one group of the annual layers wound to the right, the next to the left, and Bo on, but without much apparent exactitude.

The wood can hardly be split, it is therefore divided with the saw; and when thin pieces, such as old sheaves, are broken asunder, they exhibit a fracture more like that of a mineral than an ordinary wood. The chips, and even the corners of solid blocks, may be lighted in the candle and will burn freely from the quantity of gum they contain, which is most abundant in the heart wood.

The Bahama lignum-vite has a very large proportion of sap-wood, pieces of 8 or 10 inches diameter have heart wood that scarcely exceeds 1 or 2 inches diameter. One variety of cocoa-wood and also the almond-wood are.
somewhat similar to lignum-vittc.

Thero are two species, *Quaiacum ojlicinale* and *G. mnchira,* both of which probably yield the lignum-vito of commerce This uamo is also sometimes applied to the wood of *Arbor vita.* LIME-TREE, called also the Linden-tree, *Tilia,* is common to Europe, and attains considerable size. The wood is very light-coloured, fine and close in the grain, and when properly seasoned, it is not liable to split or warp. It is nearly or quite as soft as deal, and is used in the construction of piano LIME-TREE— *contin ued.* fortes, harps, and other musical instruments, and for the cutting-boards for curriers, shoemakers, &c. as it does not draw or bias the knife in any direction of the grain, nor injure its edge; it turns very cleanly; this wood has recently been used for the frames of the best japanned chairs inlaid with mother-of-pearl. Lime-tree is particularly suitable for carving, from its even texture and freedom from knots: the works of Gibbons, at Windsor Castle and St. Paul's, London, are of lime-tree, which wood, as well as boxwood, was eulogised by Virgil, Georgics, book ii. ver. 449.

The lime-tree, *Tilia esropai,* is usually divided into several species; as *T. intermedia, mkrophylla, rubra,* and *platyphyUa.* LOCUST-TREE. The locust-tree of North America is *Robinia pseudacacia.* The wood is greenish yellow, with a slight tinge of red in the pores, it is used like oak. Locust is much esteemed for tree-nails for ships, and for posts, stakes, pales, itc., as it is very tough and durable; it works similarly to ash, and is very good for turning.

"It grows most abundantly in the Southern States; but it is pretty generally diffused throughout the whole country. It sometimes exceeds four feet in diameter and seventy feet in height. The locust is one of the very few trees planted by the Americans."

"According to Mr. Browne, there are no less than 140 species of forest trees indigenous to the United States which exceed thirty feet in height. In France there are about thirty, and in Great Britain nearly the same number."—*Stqilunton 'a Civil Engineering of North America*, p. 183.

The locust-tree of the West Indies and Guiana is *Ilymcnea, Courbaril,* (Semiri,) a tree from 60 to 80 feet in height, and 5 or 6 feet in diameter: the colour of the wood of West Indian locust-tree is light reddish-brown, with darker veins, and the mean size 36 inches. The wood in its native country is used for mill rollers and cogs of wheels. Another tree called Honey Locust, *Qleditechia triacanthvs,* of which the wood splits with great ease, is coarse grained, and but little used. LOGWOOD, called also Campeachy logwood, is from the bay of that name, and from Jamaica, Honduras, &c. It is scarcely used for turning, and is a dark purple red dye-wood, that is consumed in large quantities: its botanical name is *Uctmatoxylon campechianum.* MAHOGANY, the *Swittenia Mahogoni,* is a native of the West Indies and the country round the Bay of Honduras. It is said to be of rapid growth, and so large that its trunk often exceeds 40 feet in length and 6 feet in diameter. This wood was first brought to London in the year 17144; its Spanish name is *Cu&ba.*

Spanish Mahogany is imported from Cuba, Jamaica, Hispaniola, St. Domingo, and some other of the West India islands, and the Spanish Main, in logs from about 20 to 26 in. square, and 10 ft. long. It is close-grained, hard, sometimes strongly figured, and generally of a darker colour than Honduras mahogany; but its pores frequently appear as if chalk had been rubbed into them.

92 ALPHABETICAL AND DESCRIPTIVE
MAHOGANY—*continued.*

Honduras Mahogany is imported in logs of larger size than the above, that is, from 2 to 4 ft. square, and 12 to 18 feet in length: sometimes planks have been obtained 6 or 7 ft. wide. Honduras mahogany is generally lighter than the Spanish, and also more open and irregular in the grain; many of the pieces are of a fine golden colour, with showy veins and figures. The worst kinds are those the most filled with grey specks, from which the Spanish mahogany, (except the Cuba,) is comparatively free.

Specimens of the leaves, and of the handsome seed-vessels of the mahogany tree, are in Sir W. Symonds's museum.

oth Spanish and Honduras mahegany are supposed to bo produced by the same tree, *Sicietcnia mahegoni* of botanists, but some suppose that the Honduras Is the wood of a different species, (V. Don, Syst, 1. p. 683.) but Long, in his history of Jamaica, says, "What grows on rocky grounds is of small diameter but of closer grain, heavier weight, and more beautifully veined; what is produced in low and rich moist land is larger iu dimensions, more light and porous, and of a palo complexion. This constitutes the difference between the Jamaica wood iuid that which is collected from the coast of Cuba and the

Spanish Main; tbc former is mostly found on rocky eminences, the latter is cut in swampy soils near the sea-coast." African Mahogany, *(ihoictenia senegt-densis,)* from Gambia, is a more recent importation; it twists much more than either of the above, and is decidedly inferior to tbem in all respects except hardness. It is a good wood for mangles, curriers' tables, and other uses where a hard and cheap wood of great size is required: it admits of being turned equally as well as the others.

African mabogany is the wood of *Kkaya itwgalmtit,* a genus very closely allied to the Swictcnia.

Mahogany shrinks but little in drying, and twists and warps less than any other wood; on which account it is used for founders' patterns, and other works in which permanence of form is of primary importance. For the same reason, and from its comparative size, abundance, soundness, and beauty, it is the most useful of the furniture woods, and it holds the glue the best of all. Mahogany is also used for a variety of turned works, apart from upholstery and cabinet-work. The Spanish mahogany is in general by far the best, although some of the Honduras nearly approaches it, except in hardness and weight. The African is by no means so useful or valuable as either of the above, especially as it alters very much in drying.

There are two other species of *Svnetmia,* besides the Mahogany tree, which are natives of the East Indies: the one, a large tree of which the wood is of a dull red colour, and remarkably hard and hcavy; the other is only a middle-sized tree, the wood of which is close-grained, heavy, and durable, of a deep yellow colour, and much resembles boxwood; but neither of these species is in common use in this country.—*Tredgold.*

The first of these trees was formerly referred to Swietenia, but is now. *Soymidv febrifuga,* the second Is probably *ChloroxyUm Suutenia,* which is the Satin-wood of India and Ceylon. A third species much admired for its light colour, close grain, and being elegantly veined, is the *Chiknma* of the natives, and *CkiI'rauia tabularit* of botanists: the wood is much employed in making furniture and cabinet-work. The wood of the Toon tree, *Ctdrela Toona,* is sometimes called Indian Mahegany.

MANCHINEEL, a large tree of the West Indies and South America; the wood possesses some of the general characters of mahogany and is similarly used but it is much less common. The wood is described as being yellow brown, beautifully clouded, and very close, hard, and durable. It is said the Indians poison their arrows with its juice, and that the wood-cutters make a fire around it before felling it, to cause the poisonous sap to run out, to avoid injuring their eyes.

This has been accurately described in Bancroft's Guiana, p. 86-7, and Colonel Lloyd says of it: "The juice of this tree is a most deadly poison; it bears a little apple appearing so like the English fruit, and so tempting, that many new comers have been poisoned by eating it. The tree is poisonous while green; sleeping under it has the most deadly effect, and I have myself been blistered most severely by passing under one in a shower of rain, when some of the drops have fallen on me 5 its effects are like molten lead." *Mpomane Mancinella,* is the Mauchineel-tree of the West Indies. *Camcraria latifolia* Is called bastard Mainhined.

MANGROVE. Native woods of the shores of the tropics, bearing this name, and those of Mango, Mangle, *Maniglier* (Fr.) &c., differ very much in kind: some that I have examined bear the appearance of very indifferent ash and elm, others of good useful woods of the same kind, some are dark coloured, and many of them have the red mahogany character.

One of the latter kind known to our cabinet-makers, has less of the brown and more of the red tint than mahogany, it becomes darker on exposure, but not in general as much so as mahogany. This mangrove is straightgrained, hard, and elastic, and stands almost better than Spanish mahogany, and it is therefore preferred for straight-edges and squares.

A specimen in Mr. Loddiges's collection, named *Jthizophora decandru,* and another *Mangle rermelho,* both from Rio Janeiro, much resembled the beautiful wood last described, as likewise the *Savaeoa,* (see Acacia,) although in grain it is somewhat coarser.

"The timbers are very much valued for ship-building, and a large quantity comes from Crab Island and Porto Rico. "—*Col. G. A. Lloyd.*

The mangrove-tree is *Rhizophora Mangh,* of which, the weod is employed in making staves for sugar hegsheads. Growing in the same situations with it are two trees to which the name mangrove is also applied: the *Coiwcarpui racemot«,* is called white mangrove by Sloane, and *Avkeaniv iomeniosa,* olive mangrove. *Coceoloba uviftrra,* sea-side grape, also grows in the same situations, and is a large tree of which the wood it of a reddish colour.

MAPLE, is considered to be allied to the Sycamore, which is sometimes called the great maple, *(Acer Pseudo-platanut,)* or the plane-tree. The English, or common maple, is of this kind; its colour is pale yellow brown, and it is only used for ordinary Tunbridge-ware, such as boxes, butter prints, &c.

The American, especially that from Prince Edward's Island, is very beautiful, and distinguished as bird's-eye maple and mottled maple. The latter is principally used for picture-frames; the former is full of small knots that give rise to its name: the grain varies accordingly as the saw MAPLE—*continued.* has divided the eyea transversely or longitudinally, as pieces cut out in circular sweeps, such as chair backs, sometimes exhibit both the bird's-eye and mottled figures at different parts. Much sugar is made in America from this variety of maple. The common maple, *(A cer eampatrit,)* is very much used in America for house-carpentry and furniture.

The so-called Russian maple is considered to be the wood of the birchtree; it is marked in a manner similar to the American maple, but is unlike it, inasmuch as there are little stripes that appear to connect the eyes, which in the American are quite distinct, and arise from a different cause, which is explained at page 38. All but the first are much used in handsome cabinet-work, and their diversities of grain are very beautifully shown in turned works. Some of the Russian birch is beautifully yellow.

Acer comptrtre is the common maplo, and *A. ilatanoid?.i* the plonatus-like or Norway maple, while *A. pseudo-ptataniis* is the great maple, sycamore, or mock plane-troe. *A. taccharlnum* is the sugar maple, and its wood is often called bird'seyo maple. *A. rnbrinn, cir-cinatum, striatum,* and *eriocarpum,* are other American species of which the timber is employed and more or less valued. *Acer oblo, "gum, eultratum, cau-datum, sterculiaceuui,* and ri7/ou;n, aro Himalayan species of which the timbers may be employed for the same purpos-es. MARACAYBO is a furniture wood of moderate size, as hard as good ma-hogany, and in appearance between it and tulip-wood. It ia sometimes called Mara caybo cedar, but it has no resem-blance to the cedar, although it may grow in the vioinity of the Bay of Mara-caybo. A wood in Mr. Morney's col-lection from Rio Janeiro, marked *Mara-cauha,* is similar to the above. MEDLAR-TREE, *(Mapilus germanica,)* the wood is white, soft, and being small is not much used, except for walking-sticks. MICO-COULIER. *See* Xettlr-tree. MORA-WOOD. Specimens of the Mora-tree were brought home by Mr. Schomburgk, and have been described by Mr. Bentham under the head *Mora exedsa;* the tree is 100 feet high, and abundant, the wood is close-grained like teak, and superior to oak, esteemed for ship-building, and likewise fitted for knees from the branches growing crooked; in colour it resembles moderately red mahogany. MOSATAHIBA. *See* Mustaiba. MULBER-RY-TREE, *(Morns,)* consists of about twenty varieties, of which the yellow fustic ia one that is imported in consid-erable quantities from Rio de

Janeiro. Bergeron very strongly recom-mends the white mulberry, which he de-scribes as similar to elm, but very close in the grain, and suitable for furniture. He says the white is greatly superior to the black mulberry.

Morut niora is the black, and *Moi'us al-ba* the white mulberry; there aro sever-al other species of which the wood Is esteemed for its toughness, as of *Morvi parvifolia* in India, for hardness and tenacity. *See* Fustic. MDSTAIBA, from the Brazils and Rio Janeiro, is imported in logs about 7 by 10 in., also in planks; it is generally of an inferior rosewood character, but harder, and is sometimes equally good; the veins are of a chesnut brown, running into black. In its grain it resembles some of the iron-woods and MUSTAIBA—*continued.* black partridge-wood, it has fewer resinous veins than the rose-woods.

Mosatahiba, as well as lignum-vitaj, co-coa-wood, &c., is used at Sheffield for the handles of glaziers' and other knives; some of the better kinds are very good for turning, as the wood is close, sound, and heavy.

I have copied the Portuguese name for this wood from Mr. Morney's and Mr. 's specimens in Sir W. Symonds' collection; it is known in

England as Mosatahiba.

NETTLE-TREE, *(Celtis auatralk,) Mico-coulieur* of the French, has wood that is compact, between oak and box for den-sity, and takes a high polish; it is de-scribed in the French works as a heavy, dark, close wood, without bark, very durable and free from flaws. It is said to be used for flutes, and for carving; it is also called dot's *de Perpignan.* NICARAGUA-WOOD, a native of South America, is imported from the Bay of Nicaragua, and also from St. Lucia, Rio de la Hache, Mexico, &c., in rough groovy logs without sap, that measure from 2 to 9 inches through, and 2 to 3 feet long.

Another sort, from Lima, Jamaica, and Peru, called by the dyers Peachwood, apparently from the colour for which it is used, is shipped in logs sometimes as large as 18 in. diameter, and 6 ft. long. Both are similar to Brazil-wood in colour, and are generally too unsound for turning.

The trees yielding Nicaragua and Peach woods have not been yet ascer-tained, but havo been supposed to be species of *Catalpinia,* or of *Ilmnalojy-lon,* but they may be Tory distinct, as coloured woods belong to other genera. NUTMEG-WOOD. *See* Palm. OAK, *(Quer-cus).* Of this valuable timber there are two kinds common to England, and sev-eral others to the Continent and Ameri-ca. Oak of good quality is more durable than any other wood that attains the same size; its colour is a well-known brown. Oak is a most valuable wood for ship-building, carpentry, frames, and works requiring great strength or expo-sure to the weather; also for the staves of casks, spokes of wheels generally, and the naves of waggon-wheels, for tree-nails, and numerous small works. The red varieties are inferior, and are only employed for ornamental furniture. The English oak is one of the hardest of the species; it is considerably harder than the American, called white and red Canada oak, or than the wainscot oak from Memel, Dantzic, and Riga; the lat-ter, which are the more interspersed with the ornamental markings or flower, from the septa or medullary rays in the wood, are the least suitable as timber.

The wainscot oak of Norway is re-markably straight, and splits easily; so much so, that it is the practice of the country to bore a small hole in the top of the tree at the beginning of the win-ter, and to fill it with water, the expan-sion of which in freezing rends the tree from top to bottom.

Considerable quantities of oak are imported from Italy, Istria, and Styria, and they are considered to be of good growth and perhaps equal to the English in quality; they are used in our Govern-ment dockyards.

00 ALPHABETICAL AND DESCRIPTIVE OAK—*continued.*

The Live Oak is a fine tree, that is met with in the Southern States of North America; it is very different in appear-ance from the others, as the veins are small, and more evenly distributed throughout the wood: it is used in America, along with the North Ameri-can red cedar, for their finest ships; it is considered to be durable when dry, but not when exposed to wet.

"The sea air seems essential to its ex-istence, for it is rarely found in the forests upon the mainland, and never more than 15 or 20 miles from the shore." "The live oak is commonly 40 or 50 feet in height, and from 1 to 2 feet in diameter, but it is sometimes much larger."—*Mejihemon's Ciril Engineering of North AmeiHca,* p. 181.

There is also a fine evergreen oak in the Cordilleras of the Andes.

An Asiatio Oak, *Querent Amherstiana,* (No. 341, Dr. Wallich's collection,) from Martaban, appears to be a fine dense wood, and as dark as our black walnut.

The Afbican Oak is well adapted to the construction of merchant-vessels, but it is apt to splinter when struck by shot, it is therefore less used for ships of war. They are all softened by steaming, and are then much more easily cut or bent; the African bends less than the others, and is the darkest in colour, but it has not the silver grain nor the variegated appearance of the others, it is sometimes called Teak (which see).

Of the British Oak there are two distinct species according to modern botanists.

The *Quercns Hobai;* sometimes called *ieduneutato,* has acorns which are supported on long footstalks or peduncles; this timber is considered by some superior to that of the other species, *Q'. testrililorp,* but this probably depends on situation,

as the strength and toughness of this kind, as well as its durability, have been proved to be great. Dr. Liudley says its wood may be known by its medullary rays or silver graiu being so far apart that it cannot be rent, and this gives it quite a peculiar aspect.

Quemu Ilex, the evergreen or helm Oak, is common to the South of Europe; the wood is hard, heavy, and tough. *Q. Saber* is the cork troe. *Q. Cerrit,* called the

Turkey Oak, is common in the southeast of Europe; its timber is ornamental, being beautifully mottled, in consequence of the abundance of its silvery grain,

and is supposed to be often as good asany other; the Sardinian oak is apparently produced by it. The Wainscot Oak is supposed by some to be produced by *Q. Ctr-*

rit. Dr. Lindley considers it to bo a variety of *Q. tetsihftora,* grown fast in rich oak land. *0. hitpanka,* the Spanish oak, and *Q. auttriaea,* the Austrian oak, are found in the countries from which they are named; and *Q. JEgilopt* is the Valouia oak, abounding in Greece and Asia Minor, from which countries such large quantities of its acorns are imported into England. *Q. Crinita* is common in Asia Minor,

yields excellent timber, and is employed by the Turks in naval architecture.

The American Oaks are numerous, but the timber of *Querent abba,* or the white oak, comes nearest to the English Oak, and is largely exported to England as well as to the West Indies. *Q. vtrene,* the live oak, is confined to the southern of the United States, and is also found in Texas; it is said to yield the best oak in America, the timber being heavy, compact, and fine grained.

Q. tinctoria, dyers' or black oak, is best known from its inner bark being used as a yellow dye, under the name of Quercitron; its wood is strong but ooane. The other American oaks are inferior in the quality of their timber. Besides these there are Indian and Himalayan oaks: the timber of some of the latter is excellent in quality. The African Oak, or Teak as it is also called, is not a species of *Qnereu;* V. Teak, OLIVE-WOOD, principally imported from Leghorn, is the wood of the fruittree *(Olea eurojiea);* it is much like box, but softer, with darker grey-coloured veins. The roots have a very pretty knotted and curly character; they are much esteemed on the Continent for making embossed boxes, pressed into engraved metallic moulds. OLIVE-WOOD—*continued.*

There is another wood, apparently from South America, called Olive wood, but it does not agree in colour, either with the fruit or wood of the olive-tree, but is of a greenish orange, with broad stripes and marks of a darker brown tint; it is a handsome wood for turning, but not very hard.

Elttoderidron fflaucum is called *boit tColive,* bnt there is no proof that it yields the olive-wood alluded to, as the country from which this is imported is not distinctly known. OMANDER. See Coromandel. OHANGE-TREE. The orange, lemon, and lime trees, *(Citins,)* are evergreens that seldom exceed about 15 feet in height. The wood is only met with as an object of curiosity: it is of a yellow colour, but devoid of smell. See Apricot-trie.

The orange is *Citrus Aurantium,* the lemon, *C. Limonwu,* the lime, *C. Limetta,* and the citron, *C. Medita.* PALM-TREES. Two or three varieties only, of the four or five hundred which are said to exist, are imported into this country from the East and West Indies: they are known in England by the names palm, palmetto, palmyra, and nutmeg, leopard, and porcupine wood, &c., from their fancied resemblances, as when they are cut horizontally, they exhibit dots like the spice and when obliquely, the markings assimilate to the quills of the porcupine.

The trunks of the palms are not considered by physiological botanists to be true wood, they all grow from within, and are always soft and spongy in the centre, but are gradually harder towards the outside: they do not possess the medullary rays of the proper woods, but only the vertical fibres, which are held together by a much softer substance, like *pith* or cement, so that the horizontal section is always dotted, by which they may be readily distinguished from all true woods. The colours and hardness of the two parts differ very materially, and I am enabled, through the kindness of Sir James Sutherland and Colonel Sykes, to give the distinctive names of three to which I shall advert.

The *Areca Catechu,* or betle-nut palm, is remarkably perpendicular; it grows to the height of about 30 feet, and rarely exceeds 4 or 5 in. diameter; it bears a small tuft of leaves, and the fruit is in clusters like grapes. The betlenut is chewed by the Indians along with quicklime, and the leaf of the Piper Betle, in the manner of tobacco. The general colour of the wood is a light yellow brown; the fibres are large, hard, and only a few shades darker than the cementitious portions.

The *Cocos nucifera,* or cocoa-nut palm, flourishes the best in sandy spots near the sea-beach, and sometimes grows to 90 ft. in height and 3 ft. in diameter, but is generally less; it is rarely quite straight or perpendicular, and has broad pendent leaves from 12 to 14 ft. long, in the midst of which is a sort of cabbage, which, as well as the fruit, the cocoa-nut, is eaten: the husk of the nut supplies the material for coir-rope and matting. No part of this interesting tree is without its grateful service to tho Indian: the VOL I. H PALM-TREES— *continued,* leaves are used for making baskets, mats, and the covering of his dwelling; he also obtains from this tree oil, sugar, palm-wine, and arrack; and although the upper part of the trunk is soft and stringy, the lower supplies a useful wood, the fibres of which are of a chesnut brown, and several shades darker than the intermediate substance; the wood is employed for joists, troughs for water, and many purposes of general carpentry. The Asiatic Society has specimens marked, male, 1st, 2nd, 3rd, 4th sorts, and the same number of female varieties; no material distinction is observable between them.

The *Niepere* palm is much darker than either of the preceding kinds; the fibres are nearly black and quite straight, and the cement is of a dark brown, but in other varieties with these black fibres, the softer part is very light-coloured, and so friable that it may be picked out with the fingers. Colonel G. A. Lloyd informs me, that at the Isthmus of Darien they use the fibres of some of the palms as nails for joinery-work.

Palmyra-wood, or that of *Bortutua fiabelliformit,* says Mr. Laird, is largely imported into Madras and Pondicherry, from the Jaffna district at the northern part of Ceylon, for the construction of flat roofs, the joists of which consists of two slabs, the third or fourth part of the tree, bolted together by their flat sides so as to constitute elliptical rafters. They are covered first with flat tiles, and then with a white concrete called *Chunam,* consisting of shell lime, yolks of eggs, and *Jaggrcc,* (sugar,) beaten together with water in which the husks of cocoa-nuts have been steeped.

The prickly pole *(Cocos guianemk)* of Jamaica, &c., a palm growing 40 feet high, and of small diameter, is said to be very elastic, and fit for bows and rammers.—*Capt. Symonds.*

The palm woods are sparingly employed in England for cabinet and marquetry work, and sometimes for billiard cues, which are considered to stand remarkably well; they are also turned into snuff-boxes, &c. The smaller kinds are imported under the names of Partridge canes, (called also Chinese or fishing canes,) Penang canes from the island of that name.together with some other small palms which are used for walking-sticks, the roots serving to form the knobs or handles. The knobs of these sticks exhibit irregular dots, something like the scales of snakes; these arise from the small roots proceeding from the principal stem, which latter shows dotted fibres at each end of the stick, and streaks along the side of the same.

The *twisted* palm sticks, are the central stems or midribs of the leaves of the date palm; they are twisted when green, and stretched with heavy weights until they are thoroughly dry: they are imported from the Neapolitan coast, but are considered to be produced in Egypt.

The bamboos, which like the palms are endogeus, are used in India and China for almost every purpose in the arts; amongst others, in working iron and steel, as the bamboo is preferred as fuel in this art, the large pieces serve as the blowing cylinders, the small as the blast-pipe, and also, when combined with a cocoa-nut shell, constitutes the *hookah* of the artizan. In PALM-TREES— *continued.*

England the bamboos, and several of the solid canes, are used as walking-sticks and for umbrella and parasol sticks.

The shells of the cocoa-nut and coquilla-nut, and the kernels of the areca or betle-nut, and those of the corosos or ivory-nut, have likewise their uses in our workshops. *See* Supplement, pages 111 and 112, of this Catalogue.

PALISANDER, a name used on the Continent for rosewood.

There is considerable irregularity in the employment of this name; in the work of Bergeron a kind of stripped ebony is figured as *boil de Palixandrt,* in other French works this name is cousidered a synonym of *boit violet,* and stated as a wood brought *by* the Dutch from their South American colonies, and much esteemed.

PARTRIDGE-WOOD is the produce of the Brazils, and the West Indian

Islands; it is sent in large planks, or in round and square logs, called from their tints red, brown, and black, and also sweet partridge; the wood is close, heavy, and generally straight in the grain. The colours are variously mingled, and most frequently disposed in fine hair-streaks of two or three shades, which in some of the curly specimens cut plankways resemble the feathers of the bird; other varieties are called pheasant-wood. The partridge woods are very porous; cut horizontally the annual rings appear almost as two distinct layers, the one hard woody fibre, the other a much softer substance, thickly interspersed with pores: this circumstance gives rise to its peculiar figure, which often resembles that of the palm-tree woods.

Partridge-wood was formerly employed in the Brazils for ship-building, and is also known in our dockyards as Cabbage-wood: the red-coloured variety is called *Angelim* and *Cangelim* in the Brazils, and *Tava* in Cuba: a specimen in one of the collections at the Admiralty is marked " Bastard

Cabbage-wood," *Andira incrmis.*

It is now principally used for walking-sticks, umbrella and parasol sticks, and in cabinet work and turning; the ladies have patronised it also for fans.

The partridge-wood imported from the West Indies is yielded by *Ifeisteria eoccinea*

The wood of several trees is no doubt included under this name.

PEACH-WOOD. *See* Nicaragua-wood.

PEAR-TREE, *(Pynu communis,)* is a native of Europe. The wild trees are principally used, and they may be obtained from 7 to 14 inches diameter. The colour is a light brown, approaching that of pale mahogany or cedar, generally less red than the apple-tree.

It is one of the brown woods of the Tunbridge-turner, by whom it is much used; and it is esteemed a very good wood for carving, as it cuts with nearly equal facility in all directions of the grain, and many of the old works are cut in it. It is now much used for the engraved blocks for calico-printers, paper-stainers, and pastry cooks; it does not stand very well, unless it is exceedingly well-seasoned.

Some pieces of pear-tree much resemble lime-tree from being, in the language of the workmen, " without grain," but the pear-tree is harder and tougher, and has a few darker streaks: they are used however for similar purposes.

100 ALPHABETICAL AND DESCRIPTIVE

PERNAMBOUCA. *See* Brazil-wood. PERUVIAN-WOOD, a fine sound wood Bo called, is of the rose-wood character, and measures about 12 to 16 inches through; it is harder, closer, and lighter in colour than rosewood, with a straighter distribution of its dark red-brown and black veins; it has no scent. Its true name and locality are unknown. PIGEON-WOOD. Mr. Loddiges' specimen is of the colour of walnut-tree, with blackish cloudy marks; another from Jamaica, at the Society of Arts, is of a brown orange-colour; the latter is the more general tint of the woods thus named. *See* Zebra-wood. PINES and FIRS, *(Pinut,)* constitute a very numerous family of cone-bearing timber trees, that thrive the best in cold countries. The woods differ somewhat in colour, partly from the greater or less quantity of resinous matter or turpentine contained in their pores, which gives rise to their popular distinctions, red, yellow, and white firs or deals, and the red, yellow, and white spruce, or pitch pines, and larches. They are further distinguished by the countries in which they grow, or the ports from whence they are shipped, as, Norway, Baltic, Memel, Riga, Dantzic, and American timber; Swiss deal; &c. The general characters of the wood, and its innumerable uses besides those of ship and house carpentry, are too generally known to call for any description in this place; but those who may require it will find abundant information in Tredgold's Carpentry, pages 208 to 218. The Swiss deals, imported under the name *Belly-boards,* are used for the sounding-boards of musical instruments. The larch is particularly durable, from the quantity of turpentine it contains; it has of late been considerably employed in her Majesty's dockyards for naval architecture, as likewise the Hackmetack larch: larch is considered the best wood for the sleepers of railways; its bark is also used for tanning. "The American pitch-pine is likewise exceedingly durable, and is much used in the West Indies, &c., for flooring, as it is free from the attacks of the white ant." The white hemlock, from St. John's, New Brunswick, Halifax, contains very little turpentine, and is remarkably free from knots: it is sometimes imported from 2 to 3 feet square, and 60 to 70 feet long, and is suitable for piling, the staves of dry casks, &c.; it stands extremely well.

The Cowdie, Kaurie, or New Zealand Pine, or *Dammara atutralu,* is the most magnificent of the coniferous woods, although not a true pine. It is said to grow from 4 to 12 feet diameter; one that had been blown down by the wind was found by Brown to measure upwards of 170 feet. The Norfolk Island pine, *Araucaria excelta,* has enormous knots, which were noticed at page 87.

In Norway, when they desire to procure a hard timber with an overdose of tiirpentine, they ring the bark of the branches just before the return of the sap; the next year they ring the upper part of the stem; the third CATALOGUE OF THE WOODS.

101 PINES and FIRS—*continued.* year the central, and lastly, the lower part near the ground. By these means the sap or turpentine is progressively hindered from returning, and it very much increases the solidity and durability of the timber. The roots of some of the red deals Bo abound in turpentine, that the Scottish Highlanders, the natives of the West Indies, and of the Himalayas, use splinters of them as candles. The knots of deal, especially white deal, are particularly hard: they are altogether detached from the wood in the outer planks, and often fall out when exposed in thin boards.

The pines and firs being so numerous, and the timbers of many being known in, commerce by such a variety of names, it is difficult to ascertain the trees which yield them.

The *Pinut sylvestris,* however, called the *wild* pine, or *Scotch Jlr,* yiolds the red deal of Riga, called yellow deal in London, *Abies cxcctta,* or Norway spruce fir, yields white deal, *Abies pkca,* or silver fir, has whitish wood, much used for flooring; *Larix Eurofxa,* is the larch common on the Alpine districts of Germany, Switzerland, and Italy. Several other pines, as *P. Pinaster, Pineat Cembra, austriaca* and *pyrenaica,* are found in the south of Europe, but their timber is less kuown iu commerce.

The North American pines, *P. strobas,* or Weymouth pino, colled white pine in North America, and much used througheut the Northern States; *P. miiis,* or *lntta,* the yellow pino. is chiefly employed in the Northern and Middle States fnr heuse and ship-building; it is considered next in durability to *P unstralis.* Southern pine, called also *P. paiustris,* and yellow pine, pitch pine, and red pino in different districts: it is said to form fonr-fifths of the heuses in the Southern States, and to be preferred for naval architecture. Its timber is exported to tho West Indies, and to Liverpool, where it is called Georgia pitch-pino. *Pmuttcedal* fraukincenso pine, called white pine in Virginia; *P. rvjida,* Virginian or pitchpine; *P. banksUina,* Hudson's Bay or Labrador pine; *P. inapt,* Jersey or poor pine, and *P. resinota.* The American pitch pine or red pine, called Norway pine In Canada, and yellow pine-in Nova Scotia, and many others, yield deals of various qualities, more or less used in different districts.

The American spruce firs are the *Abies alba, nigra,* and *rnbra,* the white, black, and red spruce firs: the last is sometimes called Newfoundland red pino, and employed in ship-building; both it and the black pine are exported to England; *Abies canadensis,* hemlock spruce fir, and *A. basalmea.* Balm of Gilead fir, are also employed, altheugh lees valued for their timber, but the American Larch, *Larix americaaa,* is much esteemed. On the west coast of America somo magnificent pines have been discovered, as *P. Douglasiii* and *Lambertiana,* and others in Mexico. In the southern hemisphere the Cowdio pine, or New Zealand pitch tree, *Dammara aastralis,* considered so valuable for masts, belongs to the same genus as the *Dammar* tree, *D. Orienialis.* The Himalayas abound in true pines: a splendid species is the *Pinut Dtodara* already mentioned under Cedar, so also are *Pinas excelta, Khutrow longifolia,* with *Abies Webbiana, Pindrow,* and others.

PLANE-TREE, (the *Platanut ocddentalus,)* is a native of North America; it is abundant on the banks of the Mississippi and Ohio. This, perhaps one of the largest of the American trees, is sometimes 12 ft. in diameter; it is much used in that country for quays. The colour of the wood resembles beech, but it is softer. In Stephenson's Civil Engineering of North America, this is called Butterwood-tree, and he gives the dimensions of some, measured by Michaux, fully equal to the measure quoted-The American variety, which is that more commonly grown in England, is sometimes called water-beech and sycamore. Plane tree is used for musical instruments and other works requiring a clean light-coloured wood.

The *Platanus orientalist* called also lace-wood, is a native of the Levant, and other Eastern countries; it is smaller, softer, and more ornamental than the above; the beauty of its septa gives it the damasked appearance from which it is sometimes named. It is commonly used by the Persians for 102 ALPHABETICAL AND DESCRIPTIVE PLANE-TREE—*continued.* their doors, windows, and furniture, and is suitable to ornamental cabinet-work and various kinds of turnery. The first kind also has septa, but they are smaller.

Tho tnio lnco-wood tree is the *Daphne Lagdta.* PLUM-TREE, *(Prunus domettica* and *P. tpinosa,)* Europe, similar in general character to pear-tree, is used principally in turning. This is a handsome wood, and is frequently used in Tunbridge-works: in the endway of the grain it resembles cherry-tree, but the old trees are of a more reddish-brown, with darker marks of the same colour. It begins to rot in small holes more generally away from, rather than in the centre of the tree, and it is very wasteful on that account.

POON-WOOD, or Peon-wood, of Singapore, is of a light porous texture, and light greyish cedar colour; it is used in ship-building for planks, and makes excellent spars. The Calcutta poon is preferred. *Calophyllum inophpllum* is called Poona in the peninsula of India, and *C. angiuti' foliam,* Dr. Ruxburg says, is a native of Penaug and of countries eastward of tho Bay of Bengul, and that it yields the straight spars commonly called Poon, and which in these countries are used for the masts of ships.

PRINCES-WOOD, from Jamaica, is generally sent in logs like cocoa-wood, from 4 to 7 in. diameter, and *4* to 5 ft. long; it is a light veined wood, something like West India satin-wood, but of a browner cast; the sap-wood resembles dark birch-wood. It is principally used for turning.

The Princes-wood of Jamaica, called also Spanish elm, is *Cordia Geraicanihus,* but the above appears to be different.

POPLAR *(Popidut).* There are five species common to England, of which the Abele, or great white poplar, and the Lombardy poplar are the most used. The woods are soft, light, easy to work, suited to carving, common turnery and works not exposed to much wear; the woods of poplar-trees are sometimes used in temporary railway works, but not for the ordinary purposes of timber. It is considered to be very durable when kept dry, and it does not readily take fire. The bark of the white poplar is almost as light as cork, and is used by the fisherman to support his nets.

The wooden polishing wheels of the glass-grinder are made out of horizontal slices of the entire stem, about one inch thick, as from its softness it readily imbibes the polishing materials.

The wood of the *Abele,* or white poplar, is also commonly known as Aps; it is extensively used for toys and common turnery, and is frequently of a uniform reddish colour, like red deal, but with very small veins.

Populus alba is the white poplar, or Abele, *P. caneteens* the gray, or common white,

P. *tremala* is the aspen, and *P. pyramUlalu,* or *fastiyiaia,* the Lombardy poplar.

There are other species iu North America and the Himalayas.

PRIZE-WOOD. A large ill-defined wood, from the Brazils, apparently of the cocus-wood kind, but lighter, and generally of reddish colour. PURPLE-HEART is mentioned by Dr. Bancroft, *(tee* Greenheart;) it is perhaps the more proper name for the wood next described. CATALOGUE OF THE WOODS. 103 PURPLE-WOOD, or Amaranthus, from the Brazils, is imported in logs from 8 to 12 in. square and 8 to 10 ft. long, or in planks: its colour is dark grey when first cut, but it changes rapidly, and ultimately becomes a dark purple.

Varieties of King-wood are sometimes called purple and violet woods: these are variegated; but the true purple-wood is plain, and principally used for ramrods, and occasionally for buhl-work, marquetry, and turning. A few logs of purple-wood are often found in importations of King-wood; it is probable also that the purple-heart is thus named occasionally.

QUASSIA-WOOD. The quassia-tree is a beautiful tall tree, of North and South America and the West Indies. The wood is of a pale yellow, or light brown, and about as hard as beech; its taste is intensely bitter, but the smell is very agreeable; the wood, bark, and fruit are all medicinal.

"This wood is well known in the Isthmus of Darieu, and is invariably carried by all the natives as a 1 contra' against the bite of venomous snakes: itis chewed in small slices, and the juice is swallowed."—*Col. 0. A. Lloyd. Qnania amara* is a small tree. *Simaruba amara* is the Mountain damson of the West Indies, aud *Picrana trcelm,* the lofty Bitter-wood. All have a similarly-coloured wood, which is intensely bitter.

QUEEN-AVOOD, from the Brazils, a term applied occasionally to woods of the Qreenheart and Cocoa-wood character. QUINCE-TREE, *(Cydonia vulgarU.) See* Apbicot-tbee. RED GUM-WOOD. *See* Gum-wood. RED SAUNDERS, or RUBY-WOOD, an East Indian wood, the produce of *Pterocarput tantaliuut,* is principally shipped from Calcutta in logs from 2 to 10 in. diameter, generally without sap, and sometimes in roots and split pieces; it is very hard and heavy; it is very much used as a red dye-wood, and often for turning. The logs are often notched at both ends, or cut with a hole as for a rope, and much worn externally from being dragged along the ground; other woods, and also the ivory tusks, are sometimes perforated for the like purpose.

The wood of *Adenanthera pavania,* (see Coral wood,) is similar in nature, and sometimes confounded with the red sauuders.

ROSETTA-WOOD is a good-sized East Indian wood, imported in logs 9 to 14 in. diameter; it is handsomely veined; the general colour is a lively red-orange, (like the skin of the Malta orange,) with darker marks, which are sometimes nearly black; the wood is close, hard, and very beautiful when first cut, but soon gets darker. ROSE-WOOD is produced in the Brazils, the Canary Isles, the East Indies, and Africa. It is imported in very large slabs, or the halves of trees that average 18 inches wide. The best ia from Rio de Janeiro, the second quality from Bahia, and the commonest from the East Indies: the latter is called East India black-wood, although it happens to be the lightest and most red of the three; it is devoid of the powerful smell of the true rose-wood, which latter Dr. Lindley considers to be a species of Mimosa. The pores of the East India rose-wood appear to contain less or none of the resinous matter, in 101 ALPHABETICAL AND DESCRIPTIVE ROSE-WOOD— *continued..* which the odour like that of the flower *Acacia armata,* arises. Rose-wood contains so much gum and oil, that small splinters make excellent matches.

The colours of rose-wood are from light hazel to deep purple, or nearly black: the tints are sometimes abruptly contrasted, at other times striped or nearly uniform. The wood is very heavy; some specimens are close and fine in the grain, whereas others are as open as coarse mahogany, or rather are more abundant in veins: the black streaks are sometimes particularly hard, and very destructive to the tools.

Next to mahogany, it is the most abundant of the furniture woods; a large quantity is cut into veneers for upholstery and cabinet work, and solid pieces are used for the same purposes, and for a great variety of turned articles of ordinary consumption.

In the Brazils the ordinary rose-wood is called *Jacaranda Cabuna;* there is a sort which is much more free from resinous pores that is called *Cabana* only: and a third variety, *Jacaranda Tarn,* is of a pale red, with a few darker veins: it is close, hard, and very free from resinous veins, its colours more resemble those of tulip-wood. There are six, if not ten, varieties in Mr. 's collection at the Admiralty.

Mr. Edwards says that at the time when rose-wood was first imported there was on the scale of Custom-House duties, "Lignum Rhodium, per ton, £40," referring to the wood from which the "oil of Rhodium" was extracted, which at that time realised a very high price. The officers claimed the like duty on the furniture rose-wood; it was afterwards imported as Jacaranda, Palisander, and Palaxander-wood, by which names it is still called on the Continent. The duty was first reduced to six guineas, then in 1842 to one pound, and in 1845 the duty was entirely removed; the consumption has proportionally increased. It is now only known as rose-wood, some logs of which have produced as much as £150, when cut into'

Hose-wood is a term as gonorally applied as iron wood, and to as groat a variety of plants in different countries, sometimes from the colour and sometimes from the smell of tho woods. The rone-wood which is imported in such large quantities from Bahia and Rio Janeiro, called also Jacaranda, is so named according to Prince Maximilian, as quoted by Dr. Lindley, because when fresh it has a faint but agrotable smell of roses, and is produced by a *Mimota* in the forests of Brazil. Mr. G. Loddiges informs me it is the *Mimoia Jacaratula.*

The rose-wood, or candle-wood, of the West Indies, is *Amyrit baltamifera* according to Brown, and is also called Sweet-wood, while *Anij/ris montana* is tulled Tellow candle-wood, or rose-wood, and also yellow saunders. Other plants to which the name is also applied, are *Litaria guianentu* of Aublet, *Jtrythroxtflnm areolatum, Colliguaya odorifera,* Molina, &c.

The rose-wood of New South Wales is *Trichilia glandtUota* ; that of the East Indies, if tho same as what is there called Blackwood, is *Dalbergia lattfolia.*

The lignum rhedium of the ancients, from which the oil of the same name and having the odour of roses was prepared, has not yet been ascertained; it has beon supposed to be the *Genista canarieruis,* and by others, *Convolvulu t teopariut.* RUBY-WOOD. *See* Red Sausdeks. SALLOW, *(Salix caprea,)* is white, with a pale red cast, like red deal, but without the veins. The wood is soft and only used for very common works, such as children's toys: like willow, of which it is a variety, it is planed into chips, and made into bonnets and baskets; it splits well, *See* Willow. CATALOGUE OF THE WOODS.

105 8ANDAL-WOOD is the produce of *Santalum album,* a tree having somewhat the appearance of a large myrtle. The wood is extensively employed as a perfume in the funeral ceremonies of the Hindoos. The deeper the colour, which is of a yellow-brown, and the nearer the root, the better is the perfume. Malabar produces the finest sandal-wood; it is also found in Ceylon, ami the South Sea Islands. It is imported

in trimmed logs from 3 to 8 and rarely 14 in. diameter; the wood is in general softer than boxwood, and easy to cut. It is used for parts of cabinets, necklaces, ornaments, and fans. The bark of the sandal-wood gives a most beautiful red or light claret-coloured dye, but it fades almost immediately when used as a simple infusion; in the hands of the experienced dyer it might, it is supposed, he very useful.

There are woods described in the French works as red sandal-woods, and one specimen is so marked in Baker's collection; probably they are varieties of red saunders or sapan woods. *See* Calembeu.

The sandal-wood tree of the Malabar coast is the *Santalum album;* that of the South Sea Islands is considered to bo a distinct species, and has been named *Sanialum Frtycinetianum*; there is a' spurious sandal-wood in the Sandwich Isles, called by the natives *liaihio (Myoporum Ienuifolium).*

SAPAN-WOOD, or Buckum-wood, *(Casalpinia Sapan,)* is obtained from a species of the same genus that yields the Brazil-wood. It is a middle-sized tree, indigenous to Siam, Pegu, the coast of Coromandel, the Eastern Islands, 4c. It is imported in pieces like Brazil-wood, to which, for the purposes of dyeing, it is greatly inferior; it is generally too unsound to be useful for turning.

SATIN-WOOD. The best variety is the West Indian, imported from St. Domingo, in square logs and planks from 9 to 20 in. wide; the next in quality is the East Indian, shipped from Singapore and Bombay in round logs from 9 to 30 in. diameter; and the most inferior is from New Providence, in sticks from 34 to 10 in. square; the wood is close, not so hard as boxwood, but somewhat like it in colour, or rather more orange; some pieces are very beautifully mottled and curled. It was much in vogue a few years back for internal decoration and furniture; it is now principally used for brushes, and somewhat for turning; the finest kinds are cut into veneers, which are then expensive; the Nassau wood is generally used for brushes. Satinwood of handsome figure was formerly imported in large quantities from the island of Dominica. The wood has an agreeable scent, and is sometimes called yellow saunders. Bergeron mentions a *"bois saline" range."*

The satin-wood of Guiana is stated by Aublet to be yielded by his *Fmlia puianenrit,* which has both white and reddish-coloured wood, both satiny in appearance. The satin-wood of India and Ceylon is yielded by *Chloroxyltm Swirtenia.* SASSAFRAS-WOOD is a species of laurel, *(Sassafras officinalis;)* the root is used in medicine. The small wood is of a light-brown, the large is darker; both are plain, soft, and close. Sassafras-wood measures from *i* to 12 in. diameter; it is sometimes chosen for cabinet-work and turning, on account of its scent.

SAUL, or SSI, an East Indian timber-tree, the *Shorea robusta; (See* 877, Dr. Wallich's Catalogue): this wood is in very general use in India for beams, rafters, and various building purposes; Saul is close-grained and heavy, of a light brown colour, not so durable but stronger and tougher than teak, and is one of the best timber-trees of India. Captain Baker considers Saul to resist strains, howsoever applied, better than any other Indian timber; he says the Morung Saul is the best. The Sissoo appears to be the next in esteem, and then the teak, in respect to strength. *See* Baker's Papers. SAUNDERS. *See* Red Saunders. SERVICE-TREE. This is a kind of thorn, and bears the service-berry, which is eaten: it is very much like English sycamore in every character as regards the wood.

Bergeron describes the service-tree as a very hard, heavy, and useful wood, of a red-brown colour, and well adapted to the construction of all kinds of carpenters' tools. He says they will glue slips of the service-tree upon moulding planes, the bulk of which are of *oak,* on account of its hardness and endurance. He also speaks of a foreign service-tree, *(Cormier da laid,)* which is harder, but more grey in colour, and more veined: these appear to be totally different woods.

SISSOO, *(Dalbergia Sinoo,)* is one of the most valuable timber-trees of India, and with the Saul, is more extensively employed than any other in north-west India. The shipbuilders in Bengal select it for their crooked timbers and knees; it is remarkably strong; its colour is a light greyish brown, with darker coloured veins. "In structure it somewhat resembles the finer species of teak, but it is tougher and more elastic." There are two kinds used respectively in Bengal and Bombay; the latter is much darker in colour. The Indian black rose-wood, *(Dalbergia latifolia,)* is a superior species of Sissoo from the Malabar coast. *See* Baker's Papers. SNAKE-WOOD, Letter or Speckled wood, is used at Demerara, Surinam, and along the banks of the Orinoko, for the bows of the Indians. The colour of the wood is red hazel, with numerous black spots and marks, which have been tortured into the resemblance of letters, or of the scales of the reptile; when 6ne it is very beautiful, but it is scarce in England, and chiefly used for walking-sticks, which are expensive; the pieces, that are from 2 to 6 in. diameter, are said to be the produce of large trees, from three to four times those diameters, the remainder being sap.

Dr. Bancroft says, "*Bourra courra,* as it is called by the Indians, by the French *boii da lellre,* and by the Dutch *Letter hunt,* is the heart of a tree growing 30 feet in height with many branches," &c. *Canjica paiie,* No. 64 in Mr. Morney's collection of Brazilian woods, is somewhat like snake-wood, but less beautiful; it is much less red, and the marks are paler and larger. If not an accidental variety, the wood would be worth seeking.

"The above must not be confounded with the Snake-wood of the West CATALOGUE OF THE WOODS.

107 SNAKE-WOOD—*continued.*

Indies and South America, the *Cccropia,* of which there are three species all furnishing trees of straight and tall growth, and a wood of very light structure, presenting sometimes distinct and hollow cells. The *Balsat,* or floats, used by the Indians of South America for fishing, &c., are very commonly constructed of this wood."—*J. Myen.*

It is theught by some to be the *Tapura gaainentit* of Aublet.

SPECKLED WOOD. *See* Snake-wood. SPANISH CHESNUT. *See* Chesmjt. SPINDLE-TREE, *(Eaonymua europa,)* is a shrubby tree, with a yellow wood, similar to the English box-wood, but straighter and softer: it is turned into bobbins and common articles. Bergeron says the wood is used in France for inferior carpenter's rules, and that its charcoal, prepared in a gun-barrel or any closed vessel, is very suitable to the artist, as its mark may be readily effaced. SYCAMORE, the *Acer pteudo-plataniu,* common to Europe, is also called great maple, and in Scotland and the north of England, plane-tree; its mean size is 32 ft. high. Sycamore is a very clean wood, with a figure like the planetree, but much smaller, it is softer than beech, but rather disposed to brittleness. The colour of young sycamore is silky white, and of the old brownish white; the wood of middle age is intermediate in colour, and the strongest; some of the pieces are very handsomely mottled. It is used in furniture, pianofortes, and harps, and for the superior kinds of Tunbridge turnery: sycamore may be cut into very good screws, and it is used for presses, dairy utensils, &c. *See* Maple.

A variety of sycamore, which is called harewood, is richer in figure aud sometimes striped, but it is in other respects similar to the above, Some of the foreign kinds are very beautifully rippled or waved, almost as richly so as satinwcod; such pieces are selected for the backs of the handsomest violins, the sounding boards of which, and of most other instruments, are made of the Swiss deal, which is probably the produce of a Larch.

TEAK-WOOD is the produce of the *Teclona grandis,* a native of the mountainous parts of the Malabar coast, and of the Rajah in un dry Circars, as well as of Java, Ceylon, and the Moulmein and Tenasserim coasts.

It grows quickly, straight, and lofty; the wood is light and porous, and easily worked, but it is nevertheless strong and durable; it is soon seasoned, and being oily, does not injure iron, and shrinks but little in width. Its. colour is light brown, and it is esteemed most valuable timber in India for ship-building and house-carpentry; it has many localities. The Malabar teak grown on the western side of the Ghaut mountains is esteemed the best, and is always preferred at our Government dock-yards. Teak is considered a more brittle wood than the Saul or the Sissoo.

In 25 years the teak attains the size of two feet diameter, and is considered serviceable timber, but it requires 100 years to arrive at maturity. 103 ALPHABETICAL AND DESCRIPTIVE TEAK-WOOD *—contin wd.*

There is a variety, says Dr. Roxburgh, which grows on the banks of the Oodavery in the Deecan, of which the wood is beautifully veined, closer grained and heavier than the common teak-tree, and which is well adapted for furniture.

Some of the old trees have beautiful burrs, resembling the Amboyna, which are much esteemed. I have an excellent specimen of the burr of the teak-wood, through the kindness of Dr. Horsfield, of the East India House.

The woods in general do not very perceptibly alter in respect to length; Teak, says Colonel Lloyd, is a remarkable exception. He found the contraction in length in the beams of a large room he erected in the Mauritius, to be three quarters of an inch in 38 feet.

The teak-wood when fiesh has an agreeable odour, something like rosewood, and an oil is obtained from it. He adds, "The finest teak now produced comes from Moulmein and other parts of liurmah; some of this timber is unusually heavy and close-graiued, but in purchasing large quantities care must be taken that the wood has not been tapped for its oil, which is a frequent custom of the natives, and renders the wood less durable."

"At Moulmein, so much straight timber is taken and the crooked left, that thousands of pieces called ' shin logs,' and admirably adapted for shiptimbers, arc lcft. Tcak contains a large quantity of siliceous matter, which is very destructive to the tools."

African tTMk does not belong to the same *genua* as the Indian teak; by somo it U tbought to be a *Euphorbiaceout* ulstut, and by Mr. Don to be-a *Vilex.* TOON-WOOD has already been mentioned under the head of Cedar, as being similar to the so-called Havannah cedar, thc *Ccdrcla odorata.* Thc toon-trcc is *C. Toona;* its wood is of a reddish-brown colour, rather coarse-grained, but much used all over India for furniture and cabinet-work.

TULIP-WOOD is the growth of the Brazils. The wood is trimmed and cut like King-wood, but it is in general very unsound in the center; its colour is flesh red with dark red streaks; it is very handsome, but it fades. The wood, which is very wasteful and splintery, is used in turnery, Tunbridge ware manufactures, and brushes; it is often scarce. The specimen in W. Loddiges' collection from Rio Janeiro, (also called St. Sebastian,) bore the Portuguese name of *Sebattiao Aruda;* that in

Mr. Morney's, at the Admiralty, *Sebcutiao d'Arruda,* and Mr. 't *St. Sebaatine cCArouda,* evidently the same; that in my German collection,

Perolia arbor.—Lignum in modo marmorit varicgatum.

A wood, sometimes called French tulip-wood, from its estimation in that country, appears to resemble a variegated cedar: it is much straighter and softer in the grain than the above, the streaks arc well contrastcd, thc light bcing of an orange red; it appears to be a very excellent furniture and turnery-wood, but has no smell; it contains abundance of gum, and is considered to come from Madras, but which peninsula has no pines.

CATALOGUE OF THE WOODS. 109 VINHATICO. The Portuguese name for several yellow and yellow-brown woods. *See* Canaiiy-wood. VIOLET-WOOD. *See* Kino-wod. VINE-WOOD. *See* Apricot Tree. WALNUT. The Royal or Common Walnut, *(Jiiglans regia,)* is a native of Persia, and the North of China. Walnut was formerly much used in England before the introduction of mahogany. The heart wood is of a greyish brown, with black-brown pores, and often much veined with darker shades of the same colour; the sap-wood is greyish white. Some of the handsome veneers are now used for furniture, but the principal consumption is for gunstocks, the prices of which in the rough vary from a few pence to one and two guineas each, according to quality. An inferior kind of walnut is very much used in France for furniture, frames of machines, &c.; it is less brown than the fine sort.

The Black Virginian Walnut, *(Juglans nigra,)* is a native of America, and is found from Pennsylvania to Florida. It is a large tree, has a fine grain, is beautifully veined, and is the most valuable of the American kinds for furniture.

The White Walnut is the Hickory, which *see.* WILLOW. There are many varieties of the willow *(Salix).* It is perhaps the softest and lightest of our woods. Its colour is tolerably white, inclining to yellowish-grey; it is planed into chips for hat-boxes, baskets, and wove bonnets; it has been attempted to be used in the manufacture of paper. The small branches of willow are used for hoops for tubs, the large wood for cricket-bats. From the facility with which it is turned, it is in demand for boxes for druggists and perfumers, which are otherwise made of small birch-wood.

The wood of the willow is described by Mr. Loudon as soft, smooth, and light; the wood of the larger species, as *Salix alba* and *Kusulliana* is sawn into boards for flooring. The rod wood willow, S. *fragilit,* is said to produce timber superior to any other species; it is used for building light and swift-sailing vessels; 5. *Russelliana* being closely allied to *S. fragili t* is probably allied to it in properties. The wood of 5. *caprta* is heavier than that of any other species. Hats are manufactured in France from strips of the wood *H. alba.* YACCA WOOD, or Yacher, from Jamaica, is sent in short crooked pieces like roots, from 4 to 12 in. thick. The wood is pale brown, with streaks of hazel brown; it is principally used in this country for cabinet and marquetry work, and turning; some pieces are very handsome.

YELLOW WOOD. There is a fine East India wood thus called, it appears to be larger and straighter than box-wood, but not so close-grained. I should think it would be found to be a valuable wood for the arts: my specimen agrees almost perfectly with *Murraya,* No. 275 of Dr. Wallich's collection.

This is probably a *Nauclea.* The wood of *Nauelea cordifolia,* according to Dr. Roxburgh, is exceedingly beautiful in colour, like boxwood, but much lighter, and at the same time very close-grained. It Is used by the inhabitants of Northern India to make combs of.

1 110 ALPHABETICAL AND DESCRIPTIVE CATALOGUE. YEW. The yew-tree is common in Spain, Italy, and England, and is indigenous to

Nottinghamshire. The tree is not large, and the wood is of a pale yellow red colour, handsomely striped, and often dotted like Amboyna. It has been long famed for the construction of bows, and is still so employed, although the undivided sway it held in the days of Robin Hood has ceased. The

English species, *(Ta.nis baccata,)* is esteemed a hard, tough, and durable wood: it is a common saying amongst the inhabitants of the New Forest in

Hampshire, that a post of yew will outlive a post of iron; it would appear the yew-tree lives to a great age, as some of those in Norbury Park are said to have been recorded in Domesday Book. The yew-tree is used for making chairs, handles, archery-bows, and walking-sticks. Some of the older wood is of a darker colour, more resembling pale walnut-tree, and very beautifully marked; the finer pieces are reserved for cabinet-work, and it is a clean wood for turning. The Irish yew is preferred for bows.

The burrs of the yew-trees are exceedingly beautiful, and although larger in figure they sometimes almost equal the Kiabooca.

The American yew, *Taxut canadensis,* is supposed to be only a variety of *T. baccata* the Iliinalayau species aro closely allied to this and to *T. nneifera,* ZANTE, or Young Fustic, from the Mediterranean, is a species of sumach, *(RJws Cotinu.t).* It is small and of a golden yellow, with two thirds sap; it is only used for dyeing, and is quite distinct from the *Marat tinctoria,* or old fustic.

Speaking of this tree, Dr. Bancroft says: "A distinction was improperly created at least 130 years ago, (now 180,) calling that of the *Venice sumach Young* Fustic, (as being manifestly the wood of a small shrub,) and that of *Moms tinctoria* (which is always imported in the form of large logs or blocks,) *Old* Fustic."—*Bancrofts Phil, of Colours,* v. i. p. 413. The Zante is also called *Chloroxylon;* its modern Greek name is *Impporc* ZEBRA-WOOD is the produce of the Brazils, and Rio Janeiro; it is sent in logs and planks, as large as twenty-four inches. The colour is orange-brown, and dark.brown variously mixed, generally in straight stripes; it is suitable to cabinet-work and turnery, as it is very handsome. A wood from New South Wales bearing some resemblance to the above is sometimes called by the same name, as are also some other woods in which the stripes are of a distinct and decided character.

The zebra-wood is considered by upholsterers to be intermediate in general appearance between mahogany and rose-wood, so as to form a pleasing contrast with either of them. The Portuguese name for the zebra-wood appears from Mr. G. Loddiges' collection to be *Burapinima,* and from Mr.

's *Goncalo do para;* No. 63 of the last group, *Casco do tartarua,* is like zebra, but heavier, more handsome, and of a rich hazel-brown, with black wavy streaks. The pigeon-woods are usually lighter, and of more yellow browns.

Zebra-wood is also called Pigeon-wood by Browne; one kind of Pigeon-wood in Jamaica is *Qurttarda speciota* ; another kind, called also Zebra-wood, is described by Browne, but he was unable to make out the genus.

SUPPLEMENT. MATERIALS DERIVED FROM THE VEGETABLE KINGDOM. BETLE-NUTS, or Areca-nuts, are the fruit of the *Areca Catechu,* or *Favfel;* they have a thin brown rind, and in size are intermediate between walnuts and hazel nuts; their general substance is of a faint oily grey colour, thickly marked with curly streaks of dark brown or black. The Betle-nuts, although softer, resemble ivory, as regards the act of turning; they are made into necklaces, the tops of walking sticks, and other small objects. The substance of the betle-nut, together with quick lime, is chewed by the generality of the natives of India.

Fig. 24 is the section of the betle-nut, full-size, and at right angles to the stalk. Fig. 25 is the section through the line of the stalk, which shows the central cavity. Externally the marks constitute a tortuous running pattern, as seen in the turned knob, fig. 26.

COCOA-NUT SHELL. The general characters of this fruit, the produce of the palm *Cocoa tmcifera,* are too well known to need particular description: in India its thick fibrous husk is made into the *coir* rope, and in Europe into rope, matting, brushes, &c.; the substance of the shell is very brittle, and its structure is somewhat fibrous, but it admits of being turned in an agreeable manner. Those shells which are tolerably circular are used for the bodies of cups and vases, the feet and covers being made of wood or ivory. Common buttons are also made of the cocoa-nut shell, and are, considered better than those of horn, as they do not, like that material, absorb moisture, which causes them to swell and twist. COQOILLA NUTS are produced in the Brazils by *Attalca funifera,* according to Martius, or the *Cocoa lap 'td-ca* of Gaertner; the latter title is highly descriptive. The coquilla nut is represented in section, half size, in fig. 27: the shell is nearly solid, with the exception of the two separate cavities represented, each containing a hard, flattened, greasy kernel, generally of a disagreeable flavour: the cells occasionally enclose a grub or chrysalis similar to that figured, which consumes the fruit. The passages leading into the 112 SUPPLEMENT TO THE CATALOGUE OF WOODS. COQUILLA NUTS—*continued.* chambers are lined with filaments or bristles, and this end of the shell terminates exteriorly in a covering of these bristles, which conceal the passages; this end is consequently almost useless, but the opposite ia entirely solid, and terminates in the pointed attachment of the stalk. Sometimes the shell contains three kernels, less frequently but one only, and I have heard of one coquilla nut that was entirely solid.

The substance of the shell is brittle, hard, close, and of a hazel brown, sometimes marked and dotted, but generally uniform. Under the action of sharp turning tools it is very agreeable to turn, more so than the cocoanut shell; it may be eccentric turned, cut into excellent screws, and admits of an admirable polish and of being lackered. On the whole it is a very useful material, and suitable for a great variety of small ornamental works, both turned and filed; coquilla nuts are extensively manufactured into the knobs of umbrellas and parasols, small toys, &c.

COROSOS, or IVORY-NUTS, are produced by *Phytehphas macrocarpa,* growing in central America and Columbia, (Humboldt.) They are described as seeds with *osseous albumen;* the tree is a genus allied to the *Pandanece,* or Screw Pines, and also to the Palms. The nuts are of irregular shapes, from one to two inches diameter, and when enclosed in their thin husks, they resemble small potatoes covered with light brown earth; the coat of the nut itself is of a darker brown, with a few loose filaments folded upon it. The internal substance of the ivory-nut resembles white wax rather than ivory; it has, when dried, a faint and somewhat transparent tint, between yellow and blue, but when opened it is often almost grey from the quantity of moisture it contains, and in losing which it contracts considerably. Each nut has a hole, which leads into a small, central angular cavity; this, joined to the irregularity of the external form, limits the purposes to which they are applied— principally the knobs of walking-sticks, and a few other small works. Fig. 28 is the section of the ivory-nut at right-angles to the stalk, and half size; and fig. 29 is the section through the stalk itself, which proceeds from *s.* SUPPLEMENT TO THE CATALOGUE OF WOODS.

113

"THE NEW WOOD." This curious name was applied by the late Mr. Marshall, Upholsterer of Soho Square, to some very beautiful wood in his possession, apparently of several distinct kinds, the proper names for which are unknown, although he submitted specimens of them to various institutions in England, Scotland, France, &c.

It is in pieces eight or ten feet long, and about twelve inches diameter, which are externally of a dark purple brown, entirely divested of bark and sap. The wood is very fine in the grain, hard and heavy, and although oily, free from scent; splinters of it blister the workmen's hands very quickly.-These woods, (of which Mr. Marshall favoured me with specimens,) display all the characters and colours of the most handsome kinda with which we are acquainted, without being atrictly like any of them; they well deserve the inspection of the curious, or those who may think their description overrated.

MEMOIR ON THE PRESERVATION OF WOODS. A paper bearing this title was lately read before the French Academy of Sciences, by its author, Dr. Boucherie, and I propose, as an appropriate sequel to the foregoing pages upon the woods, to attempt to convey a general notion of the numerous experiments referred to.

He contrasts the increasing consumption and the rapid decay of timber, with its slow rate of production, which make it necessary to economise its employment. He adverts to the many projects for its preservation, enumerated by our countryman Mr. John Knowlea, *(tee* Note, p. 22,) and the methods subsequently propoaed, to many of which he objects from their uaelessness; to others from the slow and superficial manner in which timbers part with their contained fluids, or absorb new ones by simple immersion, (circumstances long since proved by Duhamel;) and to all from their *expense,* which ia of course the ultimate test of general application.

Dr. Boucherie argues that all the changes in woods are attributable to the soluble parts they contain, which either give rise to fermentation or decay, or serve as food for the worms that so rapidly penetrate even the hardest woods. As the results of analysis he Hays, that sound timbers contain from three to seven per cent. of soluble matters, and the decayed and worm-eaten rarely two, commonly less than one, per cent.; he therefore concludes that "since the soluble matters of the wood were the causes of the changes it undergoes, it is necessary to its preservation, either to abstract the soluble parts in any way, or to render them insoluble by introducing substances which should render them infermentable or inalimentary;:' which he considers may be done by many of the metallic salts and earthy chlorides.

Dr. Boucherie shows, by parallel experiments upon "vegetable matters very susceptible of decomposition, as flour, the pulps of carrot and beet-root, the melon, Jtc, (which only differ from wood, of which they possess the origin and constitution, by the greater proportion of soluble matter which they contain,)" that in the natural sfcitea they rapidly alter, but are Vol. r. i 114 SUPPLEMENT TO THE CATALOGUE OP WOODS.

MEMOIR ON THE PRESERVATION OF WOODS—*continued.* preserved by the pyrolignite of iron, (pyrolignite *brat* de fer,) a cheaper material than the corrosive sublimate commonly used, and one very desirable in several respects. He presumed that by immersing the end of a tree

immediately after it was felled into a liquid, the vital energies not having ceased, the tree would then absorb such fluid through all its pores, by a process which he calls aspiration; and in this fortunate surmise he was entirely successful. This led step by step to numerous practical results,

which their inventor enumerates as follows, and describes in separate chapters.

1st. ' For protecting the woods from the dry or wet rot."

2nd. "For augmenting their hardness."

3rd. "For preserving and developing their flexibility and their elasticity." 4th. "For rendering impossible the changes of form (*jeu)* they undergo, and the splits *(ditjonctioni)* which take place when they are brought into use, or are submittedto atmospheric changes." 5th. "For greatly reducing their inflammability and oombustibility." 6tb. "For giving them various and lasting colours and odours."

I shall endeavour to convey a general notion of the methods in the same order. 1. Durability. He took a poplar tree, measuring 28 *mkres* in height and 40 *centimetret* diameter, simply divided from its root with its branches and leaves undisturbed, and immersed it erect to the depth of 20 *centimetret* in a vessel containing pyrolignite of iron; in six days it was entirely impregnated even to the leaves, and had absorbed the large quantity of three *hectolitret* (p. 132). This method required powerful lifting apparatus, and a support for the tree to lean against, and was therefore objectionable.

He repeatedly operated upon trees lying on the ground, by attaching to their bases waterproof bags containing the liquid: the experiments were varied in many ways; sometimes portions of the branches were lopped off, but the crown or tuft was always left upon the principal stem; at other times the aspiration was effected by boring detached holes near the earth supplied with different fluids, which give rise to all kinds of diversities in the result; and other trees were pierced entirely through, and a horizontal cut extending to within an inch or so of each side was made with a thick saw, leaving only sufficient wood for the support of the trees.

For fear of losing the trees upon which he had the opportunity of experimenting, the process was not deferred beyond 24, 86, or 48 hours after they were felled, as the vigour of the absorption was found to abate rapidly after the first day, and that at about the tenth day it was scarcely perceptible: it was also found the aspiration entirely failed in *dead* wood, whether occurring at the heart of old trees, or at parts of others from any accidental interruption of the flow of the sap during the growth: and also that resinous trees absorbed the fluids less rapidly than others. Observations were also made of the quantities of the liquids taken up; SUPPLEMENT TO THE CATALOGUE OP WOODS.

115 MEMOIR ON THE PRESERVATION OF WOODS—*continued.* these fluids, when o£ a neutral kind, as the chloride of soda, often equalled in bulk that of the wood itself, without causing any addition to its weight; the acid and alkaline fluids were less abundantly absorbed, apparently from contracting the vessels by their astringent action. It is stated that the pyrolignite of iron effected the preservation of the substance when equal to less than a fiftieth of the weight of the green wood. These points are all separately treated in the original paper. 2. The hardness of the wood was considered by various workmen to be more than doubled by the aotion of the pyrolignite. 3. The flexibility, (due to a oertain presence of moisture,) was increased in a remarkable manner by the chloride of lime and other deliquescent salts, the degree of elasticity depending upon their greater or less concentration. As a cheap substitute for the above, the stagnant water of sait marshes was adopted, with a fifth of the pyrolignite, for the greater certainty of preservation. Pieces of prepared deal, 3 *millimilra* thick and 60 *centimitrei* long, were capable of being twisted and bent in all directions, as into screws, also into three circular coils; the wood immediately regained its figure when released; this condition lasted eighteen months, that is, until the time his paper was read. 4. The warping and splitting, principally due to the continual effect of the atmosphere in abstracting and restoring the moisture, was stayed by impregnating the wood with a weak infusion of the chloride, so as always to retain it to a certain degree moist; one-fifth of pyrolignite was also added in this case. The seasoning of the wood was also considered to be expedited by the process, and which was not found to interfere with the ordinary use of oil-paint, 4o. Large boards of the prepared wood, some of which were painted on one or both sides, and similar boards of unprepared wood, were compared; at

the end of twelve months, the former were perfect as to form, the latter were warped and twisted as usual. 5. The inflammability and combustibility of the woods were also prevented by the earthy chlorides, which fuse on their surfaces by the application of heat, and render them difficult of ignition. Two similar cabins were built of prepared and of ordinary wood respectively, and similar fires were lighted in each; the latter was entirely burned, the other was barely blackened. 6. In respect to colours infused by the aspiratory process, the vegetable colours were found to answer less perfectly than the mineral, and the latter succeeded best when the colour was introduced at two processes, so tha the chemical change, (that of ordinary dyeing,) occurred in the pores of the wood itself. Odorous matters, could only be infused in weak alcoholic solutions, or essential oils: they were considered to be equally durable with those supplied by the hand of nature. Resins, similarly introduced, were found to increase amazingly the inflammability of the woods, and to render them impervious to water. 116 SUPPLEMENT TO THE CATALOGUE OP WOODS. MEMOIR ON THE PRESERVATION OF WOODS—*continued.*

On the whole the method is considered to promise the means of working almost any desired change in the constitution and properties of woods, when the fluids are presented to them before the vitality of the tree has ceased. It is true we have as yet only two years' trial of these experiments, but they have been scientifically deduced, and their inventor is still engaged in prosecuting them. It is to be hoped, and also expected, that these interesting and flattering promises of success will be realised, and even extended, when tried by that most severe of all tests, time.

4 Siuoe I collected tbe above particulars from the number for June, 1840, of *Ltt Annalet dt Chimie it de Physique,* pp. 113 to 157, I have been favoured by J. E. Puddock, Esq., with a printed copy of the English translation of the original paper, preceded by the report of Messrs. De Mirbel, Arago, Poncelet, Audonin, Gambey, Bouading.auIt, and Dumas, on the part of *VAcadtmie dee Sciences,* confirming the value of the invention. In France, Dr. Boucherie hai relinquished his *brevet,* and thrown the process open to the public in consideration of a national reward; and immense preparations have been there mode for the employment of the preservative process for the French navy. In England Dr. Boucherie and Company have obtained two patents, and Mr. Puddock, their agent, has specimens of pine, plane-tree. *Sic.,* variously prepared and coloured, with the pyrolignite of iron, the prussiate of iron, tbe prussiste of copper, and various other metallic salts, *&c.*

See Appendix, Noto B. page 459; and also Appendix, Vol. II. Note H, page 9S8.

SECOND DIVISION OR PART. » CHAPTER VII.

MATERIALS FROM THE ANIMAL KINGDOM.

SECT. I. PORCELANOUS AND NACREOUS SHELLS, BONES, ETC.

The bard solid substances derived from the Animal Kingdom, are parts of the external or internal skeletons, as shells and bones; or of the instruments of sustenance and defence, as horns, hoofs, nails, claws, and teeth: these, together with the various coverings of animals, whether hair, feathers or scales, are alike composed of animal and earthy matters, almost exclusively albumen, gelatine, and lime, combined in various proportions, and with a structure more or less interspersed with animal fibre. Many of these are either formed by the deposition of successive annual layers, or they are altogether yearly renewed.

A brief consideration of the chemical difference between their component parts, and of their respective proportions, in such as are used in the arts, will show the reasons for their various characters, and different treatment with tools.

Albumen, the principal ingredient of these animal substauces The author begs to premise, that having but little personal experience in the subject-matters of this present chapter, he has not hesitated to draw largely from the *Manuel du, Tourneur* and also from two valuable papers "On Horn and Tortoiseshell," and "On Bone, and its uses in the Arts," by Arthur Aikin, Esq., read before the Society of Arts in 1832, 1838, and 1839, and published in their Transactions, Vol. HI., Part 2, pp. 334—379: in which, in addition to the information on the points here to be discussed, are contained many interesting particulars, on the physiology, and on the historical and present uses of these substances. An article on Tortoiseshell, in Gill's Technological Repository, 1827, Vol. I., p. 332, (derived from the Franklin Journal and *tEncyclopedic M&hodique,)* has likewise been consulted: the several extracts will be respectively noticed.

The author has been further enabled by the kindness of various practical friends, to advance other examples and particulars, and to procure specimens of several of the materials in their different stages of manufacture. In accordance with the prescribed plan of the work, he has dwelt more at length on those parta which the amateur mat practise with comparatively few apparatus.

and which exists in the purest form in the white of eggs, is hardened by a degree of heat less than the boiling temperature of water, and is insoluble in the same. Gelatine, of which jelly and glue are different examples, is softened by heat, and rendered fluid by the addition of water; both are easily cut and scraped, in all their various stages from soft to hard, and during this change they contract very materially, but without entirely losing their elasticity.

The earthy matters of the animal solids, principally the phosphate and carbonate of lime, are widely different from the foregoing, and also from the substances of the woods and metals. They are inelastic, and often crystalline, and therefore incapable of being cut into shreds or shavings: as when they are divided, they become smaller fragments or particles which are always angular: they are comparatively uninfluenced by water or small changes of temperature, and are incapable of contraction.

When the earthy and crystalline

structures prevail, the animal substances are harsh, incapable of absorbing moisture, or of alteration of size or form; when the animal and fibrous characters prevail, they are easily cut, and they absorb moisture, soften, and swell.

In some of the shells, the quantity of animal matter is so small and the lime is in so hard and compact a form, that they are very brittle, partially translucent, generally they have smooth surfaces, and are incapable of being cut with a knife or tools; such shells are called *porcelanous,* from their resemblance to porcelain; they include most of the univalve shells, such as the whelks, limpets, and cowries. Most of these can only be worked upon after the manner of the lapidary, with emery and other gritty matters harder than themselves, by which means The numbers attached to the following substances show, in a rough manner, the rate per cent. of animal matters respectively contained; the remainder, principally carbonate and phosphate of lime, being neglected.

Enamel of teeth, the hardest of the class, contains from 2 to 3 J per cent. of animal matter, (Berzelius). Porcelanous shells are nearly similar. Nacreous shells, 24 per cent. (Hatchett). Ivory, 24 per cent. (Ure); 25 per cent. (Merat Guillot). Bone, 83 per cent. (Berzelius). Horn, is coagulated albumen and lime, with per cent. of phosphate of lime, (Ure). Tortoiseshell is nearly the same aa horn. The horn of the buck and hart are intermediate between bone and ordinary horn (Ure).

they are cut and polished, as will he explained in speaking of that art; by analysis, porcelanous shells are considered closely to agree with the enamel of the teeth.

The nacreous shells, thus named from *nacre,* the French for mother-of-pearl, are most commonly known in the shells of the pearl-bearing oyster of the Indian Seas *(Ostrcea margaritifera),* but they include the generality of the bivalve shells, as the various oysters, muscles, &c.; within they are smooth and iridescent; without they have a rough coat or epidermis.

These kinds contain a larger proportion of animal matter, which is considered to be arranged in alternate layers with the carbonate of lime; and as these shells also are impenetrable to water, they neither shrink nor swell. The pearl shells are less frangible and hard than the porcelanous shells, and they admit of being sawn, scraped, and filed, with ordinary tools; but they are harsh, scratchy, and disagreeable under the operation.

The beautiful iridescent appearance of the pearl shells is attributed to their laminated structure, which disposes their surfaces in minute furrows, that decompose and reflect the light; and owing to this lamellar structure, they also admit of being split into leaves, for the handles of knives, counters and the purposes of inlaying. As the pieces are very apt to follow, and even to exceed the curvature of the surface, splitting is not much resorted to, but the different parts of the shell are selected to suit the several purposes as nearly as possible; and the excess of thickness is removed upon the grindstone in preference to risking the loss of both parts in the attempt to split them.

The usual course in preparing the rough pearl shell for the arts, is to cut out the square and angular pieces with the ordi» nary brass-back saw, and the circular pieces, such as those for buttons, with the annular or crown saw, fixed upon a lathe mandrel. The sides of the pieces are then ground flat upon a wet grindstone, the edge of which is turned with several grooves, as the ridges are considered to cut more quickly than the entire surface, from becoming less clogged with the particles ground off. The pieces are finished upon the flat side of the stone; and are then ready for inlaying, engraving and polishing, according to the purposes for which they are intended. Cylindrical pieces are cut out of the thick part of the shell, near the joint or hinge, and rounded upon the grindstone, ready for the lathe, in which they may be turned with the ordinary tools used for ivory and the hard woods.

In the bones of animals, the earthy and animal matters arc more nearly balanced; they are therefore less brittle than the shells, but prior to being used they require the oil with which they are largely impregnated to be extracted by boiling them in water, and bleaching them in the sun or otherwise. This process of boiling, in place of softening, robs them of part of their gelatine, and therefore of part of their elasticity and contractibility likewise; they become more brittle, and having a fibrous structure, they break in splinters.

The forms of the bones are altogether unfavourable to their extensive or ornamental employment; most of them are very thin and curved, contain large cellular cavities for marrow, aud are interspersed with vessels that are visible after they are worked up into brushes, spoons, and articles of common turnery. The buttock and shin bones of the ox and calf, are almost the only kinds used. To whiten the finished works, they are soaked in turpentine for a day, boiled in water for about an hour, and then polished with whiting and water.

The following are considered by an experienced dealer to be the respective qualities of the pearl shells. The Chinese, from Manilla, are the best; they are fine, large, and very brilliant, with yellow edges. Singapore, fine large shells, deadwhite. Bombay, a common article. Valparaiso, also common, with jet-black edges. South Sea pearl shells, common, with white edges.—*See* note on page 42.

The very beautiful dark green pearl shells, are known as ear-shells or sea-ears; they are unlike the others in form, being more concave, and with small holes around the margin, and are the coverings of the *Haliolu,* found in the Californian, South African, and East Indian Seas. Cameos are cut in the conch-shell, *Strombm Gigat,* of the southern coast of America, and the West Indian Islands.

The Rev. Essex H. Bond informs me that he has seen the Chinese work the largest of known shells, the *Chama Gigat* of Linnaeus, the *Tridacna Gigat* of Lamarck, into snuff-bottles, tops of walking-sticks, bangles (a kind of bracelet,) and similar articles, some of which he possesses. The shell is a bivalve and not nacreous, generally white, sometimes pale blue; it may be beautifully polished, and is less readily scratched than mother-of-pearl; its localities are the Indian Seas, New Holland, and the Med Sea, but the largest are obtained from Sumatra, one pair from whence, described in Sir Joseph Banks' MSS. Library, is said to weigh, the one valve 285, the other 222 pounds, but the more usual weight is about 100 pounds each valve. Mr. Bond considers the useful portions of the shell, already prepared, might be obtained from China.

HORNS OP THE STAG AND OX. 121

Bone is far less disagreeable under the tools than the pearl shell, but it is nevertheless hard, harsh and chalky; the screws cut on bone are imperfect and soon injured. It is harder, often whiter, but much less pleasant to work than ivory, which beautiful material will be treated of separately in the next chapter.

SECT. II.—HORN.

The horns of animals have next to be considered, for which, says Mr. Aikin,

"In the English language we have only one word to express two quite different substances; namely, the branched bony horns of the stag genus, and the simple laminated horns of the ox genus, and other kindred genera."

"The bony horns are called in the French *bois*, from their likeness to the branch of a tree: they are annually renewed."

"The other sort of horn to which the French appropriate the term *come,* and which is the subject of our present inquiry, is found on the ox, the antelope, the goat, and sheep kinds."

These two kinds will be considered separately.

The stag-horn closely resembles the ordinary solid bones, both in its chemical characters, and also in structure, as it is spongy and cellular in its central parts. The horn is sawn into pieces, filed to the required shapes, and used without any further preparation, the natural rough exterior of the horn being left in the original state; its appearance is neat and ornamental, and from. its uneven surface is very suitable for the handles of knives, and other instruments requiring to be held with a firm grasp.

"When short pieces of stag-horn are used entire, as for the handles of table-knives, the hollow cellular part is concealed by the addition of the metal cap, and those parts of the white internal substance, which are necessarily exposed, are browned with a hot iron, or the flame of a blowpipe, so as nearly to match the other parts. The deer horns are imported from Ceylon and Bombay, and the finest from Germany. See Appendix, Note M, p. 957, Vol. ii., on straightening stag-horn and buck-horn.

The horns of the ox tribe are deposited in annual layers upon the bony cores that project from the foreheads of the animals; Tnuu. Soc. of Arts, Vol. LIL, Part 2, p. 831.

122 OPENING 01' THE HOKN OP THE OX.

whence it results, that the general form of the horn, (neglecting its curvature,) is conical, the portion beyond the core is solid, and the other extremity tapers off so as to terminate at the base in a single plate, or extremely thin edge.

Horn consists almost entirely of animal matter, chiefly membranous, namely coagulated albumen with a little gelatine, and an inconsiderable portion of phosphate of lime: had the horns much more earth they would be brittle like bones, had they much more gelatine they would be soluble like jelly or glue; as they are constituted, the quantity of gelatine is only sufficient to allow them to be considerably softened by a degree of heat not exceeding that of melted lead, after which they may be cut open with knives or shears, flattened into plates, divided into leaves, and struck between dies like metal. Their gelatine serves as a natural solder, so that neighbouring surfaces, when perfectly free from greasy matter, may be permanently joined together by moisture, heat and pressure, the union becomes perfect, but horn being a cheap material the process of joining it is seldom practised.

Our own supply of the horns of the ox and cow is insufficient for the numerous uses to which this substance is applied, and they are largely imported from Buenos Ayres and the Cape of Good Hope, also those of the bison and buffalo from the East Indies; the latter are sometimes very beautiful, and reserved for superior purposes. The straight conical horn of the rhinoceros is also occasionally used; it is solid, and formed as of a group of hairs cemented together; the transverse section of the upper part of the horn exhibits small dots. The horns of the chamois and antelope, and those of some other animals, arc generally looked upon as natural curiosities, and are only polished exteriorly, without any strictly manufacturing process being applied to them.

The first step in operating upon horn is the separation of the bony core, which is effected by macerating the horns in water for about a month, when, from the putrefaction of the inter mediate membrane, the core may be readily detached; this is not thrown away, but burnt to constitute the bone earth used for the cupels for assaying gold and silver.

The solid portion or tip of the horn is usually sawn off, and the remainder, if not cut into short lengths, is softened by immersion GENERAL TREATMENT DRINKING-HORNS.

123 for half an hour iu boiling water; it is then held in the flame of a coal or wood fire, until it acquires nearly the heat of melted lead, when it becomes exceedingly soft, after which it is slit up the side with a strong pointed knife, and opened out by means of two pairs of pincers applied to the edges of the slit; and lastly, the *"fiats"* are inserted between iron plates previously heated and greased, which are squeezed tight in a kind of horizontal frame or press by means of wedges; wooden boards may be used.

For general purposes, as for combs, the pressure should be moderate, otherwise, in the language of the workman, it *breaks the grain,* or divides the laminae, and causes the points of the teeth to split; but great pressure is purposely used in the manufacture of the leaves for lanterns, which are afterwards completely separated with a round-pointed knife, scraped and polished. The heat and pressure when applied to the light coloured horn render it also transparent.

An improved mode of " opening horn" was invented by Mr. J. James, by which the risk of its being scorched or frizzled over the open fire is entirely removed; he employs a solid block of iron with a conical hole, and an iron conical plug: these are heated over a stove to the temperature of melted lead, and the horn, after having been divided lengthways with a saw or knife, is inserted in the hole, the plug is gradually driven in with a mallet, and in the space of about a minute the horn is softened and ready for being opened in the usual manner, f

In making drinking-horns, and some few other turned works,, the material is cut to the appropriate length, brought to the circular form, and allowed to cool in the mould; the process is similar to that just described, although the old methods of the open fire and wooden cones are commonly used. The horn is then fixed in the lathe by its larger end, and turned on its inner and outer surfaces, and the groove, or *chime,* for the bottom, is cut with an appropriate tool. A thin plate, previously cut out of a flat piece of horn with a crown saw, is dropped into the horn, and forced into the groove, after the horn has been sufficiently heated before the fire to allow the necessary expansion; in cooling the contraction fixes the bottom water-tight.

The shavings thus removed are used for the Chinese sensitive leaves, which curl up when placed on the hand, from the temporary evaporation of their moisture, t Trans. Soo. of Arts, Vol. XLV., p. 165. *t* Vol. LIL, p. 841. 124 THE EXPANDING SNAKE.

As an illustration of the peculiar properties of horn, and a mode of its employment in the lathe, may be mentioned the expanding snake: this toy is well known to consist of a conical piece of horn, the one end of which is carved to represent the head, and the remainder is cut iuto a single spiral shred, so as to admit of great expansion in imitation of the body of the reptile. I find the elastic portion of the one before me to measure when compressed, barely one inch and a quarter in length, and that it expands to upwards 6f three feet and a half, or thirty-five times: no mean proof of the elasticity of the material.

In making this trifle, the material is first turned to a conical form, after which a hole of about one-eighth or one-sixth of an inch diameter, is pierced from the tail almost through the head; the horn is then soaked for about two days in cold water to soften it, and the spiral incision or screw is made at one single cut, by means of a tool extending from the center to the circumference; the cutter is not required to be very thin, as the shaving will bend away to make room for the same. One of the three following modes of proceeding is recommended in the *Manuel du Tourneur.*

First, by the employment of a sliding rest, adapted to cutting screws, by which the tool is *traversed,* or guided mechanically along the horn during the rotation of the mandrel of the lathe; and to prevent the fracture of the toy during its construction, a stick of wood, with a button on the end of it, is put up the aperture, to receive and support the spiral as it is produced.

Another method is by the employment of a lathe with a *traversing* or screw-cutting mandrel, upon which latter the horn WORKS MOULDED IN HORN. is fixed, the tool being kept *stationary* in the slide rest. Both methods require expensive apparatus, the principles of which 125 will be explained in the chapter on screw-cutting tools in the second volume.

The third plan is extremely simple, and appears on inspection to have been the one pursued in this instance; it is ascribed to the German toy-makers. The horn is prepared as before, but the lathe and slide-rest give way to the ordinary carpenter's brace, which carries the piece of horn as in fig. 30. A small tool is fixed in the vice or bench; it consists of a piece of wood, to which is screwed a hardened steel plate about one-twelfth of an inch thick, it has a hole equal to the diameter of that in the horn, for the passage of the supporting wire; the plate is divided radially; the one edge is sharpened very keenly, and bent so much in advance of the other, that their *difference* of level or agreement, shall be equal to the intended *thickness* of the continuous shaving of the body of the snake, and therefore the projecting edge assimilates to the mouth of a plane; the last processes in every case being to carve the head and to attach a little piece for the end of the tail.

It is necessary the coils of the snake should be of a conical form, or dished, as in fig. 31, instead of heing quite flat, as it increases the strength of the toy; this is accomplished by making the cutting edge of the tool *oblique* to the axis of the snake. Fig. 32 shows the tool for the lathe. The several details are too simple to require further explanation.f

The handles for knives, razors, and other works moulded in horn, are thus made: the horn is first cut into appropriate pieces with the saw, and when heated these are pared with a knife or spokeshave, to the general form and size required; in this state horn works as easily as a piece of deal; after having been pared the pieces are pressed into moulds.

An idea of the moulds will be conveyed by imagining two dies, or pieces of metal, parallel on their outer surfaces, and with a cavity sunk entirely in the one, or partly in each, according to circumstances: the cavities made either straight, curved, twisted, rounded, bevilled, or engraved with any particular device, according to the pattern of the work to be produced.

The pressure is applied to the dies, by enclosing them in a kind of clamp, made like a very strong pair of nutcrackers, but with a The actual apparatus described in the *Manuel* is slightly different, but less convenient than the latter, which is far more likely to be now employed.

t Manuel du Tourneur, 1816. Vol. II, pp. 117—123. 126 DYEING HORN: TORTOISESHELL. powerful screw at the end opposite to the joint; the mould, dies, and horn, are dipped into boiling water for a few minutes, and then screwed as fast as possible immediately on removal from the same, and in about twenty minutes the work is readyfor finishing; some handles are made of two pieces joined together "Horn is easily dyed by boiling it in infusions of various coloured ingredients, as we see in the horn lanterns made in China. In Europe it is chiefly coloured of a rich red-brown, to imitate tortoiseshell, for combs and inlaid work. The usual mode of effecting this is to mix together pearl-ash, quicklime, and litharge, with a sufficient quantity of water and a little pounded dragon's blood, and boil them together for half an hour. The compound is then to be applied hot on the parts that are required to be coloured, and is to remain on the surface till the colour has struck; on those parts where a deeper tinge is required, the composition is to be applied a second time. This process is nearly the same as that employed for giving a brown or black colour to white hair; and depends on the combination of the sulphur, (which is an essential ingredient in albumen,) with the lead dissolved in the alkali, and thus introduced into the substance of the horn." f The horn which is naturally black is less brittle than that which is so stained. SECT. III. TORTOISESHELL.

Tortoiseshell comes next under consideration. "The animal which produces this beautiful substance, is a marine tortoise, called the *Testudo imbricata,* or hawk's-bill turtle. Its Latin name is derived from the mode in which the scales on its back are placed, over-lapping one another like the tiles on the roof of A patent has been taken out by the Messrs Deakina for making knife-handles, door-knobs, large rings, &c., of horn, by an ingenious combination of moulding and dovetailing; the material from its greasy nature being leas adapted to the cementing or soldering process than tortoiseshell.

"On referring to French authorities," says Mr. Aikin, "I find it stated that horn, steeped for a week in a liquor, the active ingredient of which is caustic fixed alkali, becomes so soft that it may be easily moulded into any required shape. Horn shavings subjected to the name process become semi-gelatinous, and may be pressed in a mould in the form of snuff-boxes and other articles. Horn, however, so treated becomes hard and very brittle, probably in consequence of its laminated structure being obliterated by the joint action of the alkali and strong pressure." —Trans, of the Society of Arts, Vol. LIL, page 340. t Idem, 341. DESCRIPTION OP TOUTOISE-SHELL. 127 a house." In this circumstance it seems to differ from almost all others of its genus."

The usual size of the full-grown animal is about a yard long and three quarters of a yard wide; its covering consists of thirteen principal plates, five down the center of the back, and four on each side, and in a tortoise of the above size, the largest, or main-plates, weigh about nine ounces, and measure about thirteen by eight inches, and one quarter of an inch thick at the central parts; but they are thinned away at the edges where they overlap, owing to the deposition of the substance of the shell in annual layers, each extending beyond the previous one. Very rarely, the shells are three-eighths thick and proportionately heavy. Others are very thin and appear to consist of only one single layer; this is supposed to occur when the animal loses a plate by accident, or that it is stripped and thrown back again into the sea whilst alive; such shells are usually very light coloured and are called "yellow belly." There are also twenty-five small pieces of shell which envelope the edge of the animal, but these can only be applied to very small purposes.

Some of the tortoiseshell is of very dark brown tints running into black,and interspersed with light gold-coloured dashes and marks; these are considered the best; others are lighter, even to pale red-browns, yellow, and white; the last are not valued, the yellow are used for covering the works of musical snuff-boxes,! and the light red and brown shells are manufactured into ladies' combs, for exportation to Spain, where they obtain double the price of those made of the darker coloured tortoise-shell. The shell of the turtle is also used, but it has not the transparent character of the foregoing; the colours are lighter, less beautifully marked, and it is little valued.J Trans. Soo. of Arts, Vol. LIL, p. 843.

+ Some musical boxes are covered with horn or the belly shells of the turtle. + A tortoiseshell merchant of considerable experience informs me that he considers the following the qualities of the different varieties. Manilla, fine and large; Singapore, nearly as good; West Indian, large and heavy, but red; Honduras, better coloured, darker, but with large dark red spots; Calcutta, dark, heavy, and of bad colour; Bombay shell, the worst and very scabby, from the attachment of barnacles, limpets, &c. There is also the Loggerhead tortoiseshell, which is almost useless, and the Yellow Belly, which plates are very thin and yellow.

Of the Turtle shell, that from Colombia is best, fine and dark; Jamaica, light coloured red and very inferior; East Indian, of middling quality and but seldom met with. The small *fisha* are called *chicken,* and their markings are diminutive. 128 GENERAL TREATMENT OP TORTOISESHELL,

The treatment of tortoiseshell is essentially the same as that of horn, but on account of its very much greater expense, it is economised so far as possible. Before the shells are worked they are often dipped in boiling water to *temper* them; three or four minutes commonly suffice, but they require a longer period when they are either thicker or more brittle than usual: excess of boiling spoils the colours of the shells, renders them darker, and covers the outside with an opaque white film. Others, flatten and temper the shells with hot irons, such as are used by laundresses; the shell is continually dipped in cold water to prevent its being scorched; but as a general rule the less tortoiseshell is subjected to heat, or to being pulled about, the better, as from its apparent " *want of grain* " or fibre, it becomes in consequeuce very brittle.

Many of the works in tortoiseshell are made, partly by cutting them out of the shell, and partly by joining or adhesion, called by the French *souder.* Thus in the Manuel du Tourneur, the artist is directed to form the ring of tortoiseshell for the rebate of a box, by cutting out a long narrow slip of the shell; the ends are then to be filed with a clean rough file to thin feather edges, to the extent of three quarters of an inch of their length, the one on the upper and.the other on the lower surface, to constitute the lap or joint; the slip is dipped into boiling water, and when softened it is bent into an oval form with the intended joint on the flat side, the ends are held in firm and accurate contact with the finger and thumb, and the piece is dipped into cold water to make it retain the form.

Fig. 33.1

I

A pair of tongs is required, such as those in fig. 33, with flat ends measuring about one inch wide and three or four long, and that spring open when left to themselves, but fit perfectly close and even, when compressed; these are made warm. In the Manuel du Tourneur, 1816. Vol. I., p. 454.

AND SOLDKUING OR JOINING THB SAME.

129 mean time the ends of the ring are sprung asunder sideways, to bring the *scarfs* or parts to be joined to the iuner and outer surfaces respectively, that they may be retouched with the file to remove any small portion of grease which may have been accidentally picked up, and the joiut is restored to its proper position. A piece of clean linen is then soaked in clean water, squeezed dry with the fingers, folded in ten to twenty thicknesses, to about the size of one and a half inch wide, and three or four long; the ends are now folded together, placed on each side the joint, the whole is inserted between the tongs, and fixed moderately tight in the jaws of an ordinary vice. The softening, and consequent adhesion of the shell, will be known by the flexibility of the ring when the loose part is wriggled about with the fingers; the work is either allowed to cool in the vice, or after a time is dipped into cold water.

The success of the process will depend on three different circumstances; the parts to be joined must be entirely free from grease and dirt, on which account the surfaces should not be touched after being filed; the temperature of the tongs should, be just sufficient to colour writing-paper, of a pale orange tint; and moisture or vapour must be present, apparently to liquefy the gelatine of the tortoiseshell at the surfaces of union.

The ring when cold, is pressed with the fingers into the circular form, or even into an oval in the opposite direction, which would cause the ends of the joints to start, if the soldering were imperfectly performed; should this happen, the application of the moistened rag and heated tongs must be repeated until the result is perfect; the ring is made circular by warming it in boiling water, and gently forcing it on a wooden cone of small angle.

Another mode, the invention of an amateur, is also described; the strip of shell is chamfered off at the ends and bent round a piece of wood, a compress of linen in six or eight folds is put upon the joint, and the whole is tightly bound round with string, and immersed in boiling water for ten minutes. The contraction of the string, and the expansion of the wood, from being wetted, supply the needful pressure, and the process is said to be quite successful, f

Moulding and soldering tortoiseshell are also performed under water in various other ways; for example in attaching the back 130 SOLDERING, FLEXIBILITY, COMBS, EYE-GLASSES.

of a large comb to that piece which is formed into the teeth; the two parts are filed to correspond; they are surrounded by pieces of linen and inserted between metal moulds, connected at their extremities by metal screws and nuts; the interval between the halves of the mould being occasionally curved to the sweep required in the comb. Sometimes also the outer faces of the mould are curved to the particular form of those combs in which the back is curled round, so as to form an angle with the teeth; the joint when properly done cannot be detected either by the want of transparency or polish at the part.

Considerable ingenuity is shown in turning to economical account the flexibility of tortoiseshell in its heated state; for example, the teeth of the larger descriptions of combs are *parted,* or cut one out of the other with a thin frame-saw, when the shell, equal in size to two combs with their teeth interlaced as in fig. 34, is bent like an arch in the direction of the length of the teeth as in fig. 35. The shell is then flattened, the points are separated with a narrow chisel or *pricker,* and the *two* combs are finished whilst flat, with coarse single-cut files, and triangular scrapers; and lastly they are warmed, and bent on the knee over a wooden mould, by means of a strap passed round the foot, in the manner a shoemaker fixes the shoe last. Smaller combs of horn and tortoiseshell are parted whilst flat, by an ingenious machine with two chisel-formed cutters placed obliquely, so that every cut produces one tooth, the repetition of which completes the form ation of the comb. Ivory and boxwood combs cannot be thus parted; they are cut in the old way, one tooth at a time, by various contrivances of double saws, as will be explained.

In making the frames for eye-glasses and spectacles, the apertures for the glasses were formerly cut out to the circular See Mr. Rogers' Comb-cuttiDg Machine, Trans. Soc. Arts, Vol. 49, part 2, pp. 150—8: since remodelled and improved by Mr. Kelly.

TOOLS FOR MOULDING TORTOISESHELL BOXES. 131 form, with a tool something like a carpenter's center-bit, or with a crown saw in the lathe; the disks were in either case preserved, to be used for inlaying in the tops of boxes, and the outside of the frame was then shaped with saws and files. This required a piece of tortoiseshell of the entire size of the front of the spectacles, but a piece of a third that width is made to answer for inferior spectacles, as the *eyes* are *strained,* or *pulled.* A long narrow piece of the material is cut" out, and two slits are made in it with a thin saw; the shell is then warmed, the apertures are pulled open, and fashioned upon a taper triblet of the appropriate shape: figs, 36, 37, and 38, explain this method: the groove for the edge of the glass is cut with a small circular cutter, or sharp-edged saw, about three-eighths or half an inch diameter, and the glass is sprung in when the frame is expanded by heat.

Tortoiseshell is also manufactured into

boxes and a variety of moulded works, but the process calls for extensive preparations, and is not often followed by the amateur.

The construction of tortoiseshell boxes requires a copper, with a fire-place beneath the same; a trough with cold water; and the all-important parts, a press and moulds. The former may be compared to the ordinary coining press, or to a strong rectangular frame, usually of wrought iron, with a screw in the center of the upper cross-piece; the base of the press is fitted into a square recess in the center of a bench, fixed so firmly to the floor or wall, as to resist the efforts of two or three men at the end of a lever five or six feet long, whose entire force is sometimes required in tightening the mould. For the convenience of transferring the heavy press, from the hole in the bench to the hot or cold water, a crane, the center of which is equally distant from the three, is added to the establishment.

The mould required for a round box consists of a thick wrougbt-iron ring *a a,* fig. 39, turned interiorly to the diameter of the box, it stands loosely upon a plate *b*; it is accurately fitted with several pieces commonly of brass, as *c* the bottom die, *d* the This arrangement, described from an English workshop, (Mr. Vanham's,) is in some respects more convenient than the apparatus described in the Manuel du Tourneur. In the Encyclopedia Metropolitana, part Mechanics, articlc 428, plate xxxdv., is described a press for horn and tortoiseshell, by Holtzapffel and Co., in which the press, crane, and boiler are combined; the cast-iron pedestal contains water and a fire-place beneath, and the press is lowered into the water by racks and pinions; as the screw of the press has a multiplying power, the force of one individual suffices, and the machine does not require to be fixed to the flow of the 132 VARIOUS PROCESSES FOR top die for a plain box, e a plain block for flat plates, and / a die engraved with any particular device to be impressed upon the work.

In the *Manuel du Tourneur* the methods of making four different kinds of boxes are minutely given. Thus in the *"Bottes a feuilles,"* the best kind, in which the cover and bottom part are each made out of a single leaf of shell, the circular pieces are to be cut out of the shell as much larger than the size of the box, as the vertical height in addition to the diameter; so that a box of three inches diameter and one inch deep, would require pieces of four and five inches respectively for the cover and bottom.

The round plate of shell is first placed centrally over the edge of the ring, as in fig. 39; it is slightly squeezed with the small round edged block g, and the entire press is then lowered into the boiling water; in one quarter, or half an hour, it is transferred to the bench, and *g,* is pressed entirely ilown, which bends the shell into the shape of a saucer, as at fig. 40, without cutting or injuring the tortoiseshell, after which the press is cooled in the water-trough. The same processes are repeated with the die *d,* which has a rebate turned away to the thickness of the shell, and perfects the angle of the box to the section fig. 41, ready Fig. 39.
for completion in the lathe; it is however safer to perform each of these processes at twice, with two boilings.

When the shell is insufficiently thick, two pieces are joined together, and should they from the nature of the shell be of irregular thickness, the thick and thin parts respectively are placed in contact; for such cases the dies *c e,* of a larger mould arc used. The piece *d,* is adapted to boxes of various depths, or to the tops or bottoms respectively, by slipping loose rings upon it to contract the length of its smaller part.
MOULDING BOXES IN TORTOISESHELL. 133
When the box is required to have a device, an engraved die/, is substituted in place of *c.* The same tools are also used for horn boxes, and for the embossed wooden boxes, but the latter process is mostly performed in the dry way, the warmth being supplied by heated plates, put above and below the two parts of the mould which are then compressed, and the whole is allowed to cool in the air.

The *Manuel* describes the construction of inferior boxes, *"Tabatieres de morceaux,"* in which small pieces of tortoiseshell with bevelled edges are carefully fitted together with the file and arranged along the bottom and up the sides of the mould; or else they are first pressed into a flat plate, and made into a box as a separate process, but the joints, from the manner in which they are made, can scarcely ever escape observation.

The " *Boites de tres-petits morceaux"* are made of still smaller fragments, which are often cemented on a thin leaf of shell to ensure their better union; and lastly, the " *Boites de drogues"* are made of the fine dust and filings, which are passed through a sieve, and treated in other respects much in the same manner as the foregoing, but these boxes are quite opaque and brittle; a thin hoop of good tortoiseshell is sometimes inserted in the mould, to form the rebate of the box, which alone is then transparent; at other times, the shavings are mixed with mineral colouring matters, to imitate granite, lapis-lazuli, and other stones, arts that are scarcely at all practised in this country.

After the lapse of ten days or a fortnight, it sometimes happens the box shows a tendency to recover its primary form, that of a flat plate, and from being cylindrical on the edge, it becomes in a slight degree conical and larger without. After being again returned to the mould, boiled and pressed, its figure is in general permanent.

This disposition is turned to useful account in restoring the fitting of a box that may have become loose, as by dipping the lower part, or the rebate, into warm water, it will expand and fill out the lid, but it requires care that it be not overdone.

The tortoiseshell boxes usually made in England are those which are veneered upou a body or fabric of wood, for which The reader will find details of tho luethoda of making all these kinds of boxes in the Manuel du Tourneur, 1816, vol. ii. pp. 460—477. And in Gill's Tech. Repos., 1827, p. 336, the apparatus of a wholesale manufacturer of the *boiies dc drntjnu,* since established iu tho Brazil-!, is minutely expUhie 1.

134 VENEERED AND INLAID BOXES. purpose the plates are scraped and filed to an uniform thickness, and glued on much the same as veneers of wood; generally fine glue is the only cement used, but various compositions are resorted to by different manufacturers. To improve the appearance of the shell, and to conceal the glue and wood beneath, the back of the veneer is rubbed with a mixture of lamp-black, vermilion, green, chrome, or white, in fish glue; the colours are applied over the entire surface, or partially, to modify the effect, and thus prepared the veneers are glued upon the boxes.

In tortoiseshell works inlaid with mother-of-pearl and gold or silver plates or wire, the substances to be inlaid are first prepared, and for pearl-shell a paper is pasted on a thin piece of pearl; the pattern is drawn thereupon, and the small pieces are cut out with a fine buhl saw; gold and silver plates are sometimes also thus sawn out.

A plain mould similar to fig. 39, but rectangular, and with plain dies, as *c* and e, is used; a few shavings of tortoiseshell are first placed on the piece *c,* to make a bed or cushion, then a piece of paper to prevent them from adhering to the thin leaf of tortoiseshell, which is next inserted in the mould. The small pieces of pearl-shell, &c. to constitute the pattern, are then carefully arranged in their intended positions, and the top plate *e* is very carefully lowered into the mould above the pieces, so that it may not misplace any of them. The mould is then slid into the press, slightly squeezed, and plunged into the copper for an hour, carried to the bench and screwed moderately tight; the work is now examined to see that nothing is misplaced, it is returned to the cauldron for a time, and the final squeeze is given by the entire force of three men, after which, whilst still under pressure, the whole is plunged into the cold water. The tablet is then fit to be smoothed and glued on the wooden box.

It will be readily conceived, that the force required depends upon the dimensions of the work: pieces of three or four inches square require all the appliances described; whereas the little *shields,* or *escutcheons,* as they are called, upon razors and knives, may be pressed in with much slighter apparatus, such in fact as were previously described as being used in moulding them. In cutlery, a different method is generally resorted to, which applies equally well to ivory and pearl-shell, substances which cannot be submitted to the softening and moulding processes employed for horn and tortoiseshell.

INLAID CUTLER WORKS. WHALEBONE. 135

The cutlery works which are dotted all over with little studs of gold or silver, are drilled from thin pattern plates of brass or steel iu which the series of holes have been carefully made; the drill or *"passer"* has an enlargement or stop, which, by encountering the surface of the pattern-plate, prevents the point of the drill from penetrating beyond the assigned depth into the handle; the holes in the ivory or pearl-shell are then filled with silver or gold wire, which is either filed and polished off level with the general surface, or allowed to project as little studs.

For shields and escutcheons they use pattern plates or templets of hardened steel, pierced with the exact form of the shield. The cutting tool somewhat resembles an ordinary breastdrill eight or ten inches long, and like it, is used with the breastplate and drill-bow; but the extremity of the tool is cleft, or made in two branches, which left to themselves spring open to the extent of an inch or more; each half of the tool has a shoulder or stop, which bears upon the surface of the steel guard-plate, as in the drill, and a rectangular cutting part that protrudes through the shield-plate as far as the required depth of the recess, and is sharpened both at the end and side, or at the ends only.

When the elastic tool, or *"spring passer"* has been compressed, so as to enter the guard-plate, it is put in motion, and flounders about in all directions, so far as it can expand, and routs or cuts out the shallow recess; the escutcheons are punched out, fixed by two rivets, and smoothed off; these processes are very expeditious, and produce accurate copies of the respective patternplates employed. The tortoiseshell when unnecessarily thick for a single scale for a penknife, is sawn to serve for two; and the colours are brightened up by placing a piece of Dutch leaf beneath the same; they are finally polished on the various wheels used by the cutler, as will be explained in the pages of the next volume, devoted to polishing. Tortoiseshell has been manufactured into hollow walking-sticks, and in ISil Her Majesty was graciously pleased to present to the British Museum a tortoiseshell bonnet, made in Navigator's Island.

SECT. IV.—WHALEBONE.

This material is scarcely at all used by the amateur, and the few remarks that are offered are principally extracted from Mr. Aikin's paper.

136 WHALEBONE AND SOME OF ITS USES.

"Whalebone, as I have already stated, (says Mr. Aikin,) may be considered as a kind of horn; which latter substance it resembles perfectly, both in its chemical and principal physical properties; and is particularly interesting as forming the transition from horn to hair."

"It is the substitute for teeth in the Greenland whale, and in the black southern whale; but is not found in any of the cetaceous animals that have teeth.

"From the roof of the mouth hang down on each side the tongue about three hundred plates of whalebone, all the blades on one side being parallel to each other and at right angles to the jaw-bone." "The average length of the middle blades is about nine feet, but they have occasionally occurred of the length of fourteen or fifteen feet."

"The general colour of whalebone is a dusky greyish black, intermixed with thin stripes or layers of a pale colour, which arc often almost white—very rarely the entire flake is milk-white."

"The preparation of the whalebone for use is very simple. It is boiled in water for several hours, by which it becomes soft enough to be cut up while hot, in lengths of different dimensions according to the use to which it is to be applied." "Whalebone that has been boiled, and has become cold again, is harder and of a deeper colour than at first; but the jet-black whalebone has been dyed, and by the usual processes it takes very bright and durable colours."

Whalebone is now principally used for the stretchers for umbrellas, and as a substitute for bristles in common brushes; it is also plaited into whips, and solid pieces of mixed shades are twisted for walking-sticks; but it does not admit of being soldered or joined together like tortoiseshel). Whalebone also furnishes a very neat and durable covering for pocket telescopes. Narrow pieces of the material are grooved or made into ribs, by drawing them like wire through a corresponding aperture in a steel plate, after which they are wound round the tube, and "*tucked under*" the rings at the extremities. Broad flat strips of the parti-coloured whalebone, (the light portions of which absorb the green dye,) are also used: these are secured by narrow black bands which overlap the two edges, and other bands are wound around the ends also.

Trans, of the Society of Arts, Vol. LI I., pp. 347—9. CHAPTER VIII. MATERIALS FROM THE ANIMAL KINGDOM, CONCLUDED. SECT. I. DESCRIPTION OF THE VARIOUS KINDS OF IVORY.

Ivory, the tusk or weapon of defence of the male elephant, and of which each animal has two, is placed by the chemists intermediately between bone and horn, and its mechanical characters corroborate the position. It is generally considered that the male elephant alone possesses tusks, commercially known as elephants' teeth, but this appears questionable, as by many the female is reported to have tusks likewise, but of smaller size, and some consider the latter produce the small solid tusks called "ball ivory," used for making billiard balls.

Ivory has less gelatine than bone; but as it leaves the animal in a state fit for use, without the necessity of removing any of its component parts for its purification, its elasticity and strength are not impaired by such abstraction. Ivory is not therefore so brittle as bone, neither does it splinter so much when broken, but its greater ultimate share of animal matter leaves it more sensible to change of form and size.

The shape of the tusk is highly favourable to its use; as it is in general solid for above half its length, and of circular or elliptical section; it is entirely free from the vessels or pores often met with in bone, and although distinctly fibrous, it cannot be torn up in filaments like horn, nor divided into thin flexible leaves as for miniatures, otherwise than by the saw.

Its substance appears very dense, and without visible pores, as if beautifully cemented by oil or wax; and notwithstanding that it possesses so large a share of lime, it admits of being worked with exquisite smoothness, and is altogether devoid of the harsh meagre character of bone. It is in all respects the most suitable material for ornamental turning, as it is capable of receiving the most delicate lines and cutting, and the most slender proportions.

Set foot note, page 118. 138 IVORY OF THE ELEPHANT, HIPPOPOTAMUS,

The general supply of ivory is obtained from the two present varieties of the animal, the Asiatic and the African: they are considered by physiologists to be distinct species, and to be unlike the extinct animal from which the Russians are said to obtain their supply of this substance; which, although described as *fossil* ivory, does not appear to have undergone the conversion commonly implied by the first part of the name, but to be as suitable to ordinary use as the ivory recently procured from the living species. An extract from the interesting account of "The Elephant of the Lena," is subjoined as a note.

The hippopotamus, or river-horse, supplies the ivory used by the dentist, which is imported from the East Indies and Africa; the animal, in addition to twenty grinders, has twelve front teeth, the whole of which agree in substance with ivory, but not in their size or arrangement. The six in the upper jaw are small and placed perpendicularly; in the lower jaw of the hippopotamus, the two in the center are long, horizontal, and straight; the two next are similar but shorter; but the two external *semi-circular* teeth are those so highly prized by the dentists on account of their superior size, and are those usually referred to when the "*sea-horse*" or hippopotamus tooth is spoken of, although the animal is in reality a quadruped inhabiting rivers and marshy places.

The circular hippopotamus teeth are covered exteriorly with a "The Mammoth, or Elephant's bones and tusks, are found throughout Russia, and more particularly in Eastern Siberia and the Arctic marshes. The tusks are found in great quantities, and aro collected for the sake of profit, being sold to the turners iu the place of the living ivory of Africa, and the warmer parts of Asia, to which it is not at all inferior."

"Almost the whole of the ivory-turner's work, made in Uusaia, is from the Siberian fossil ivory; and sometimes the tusks, having hitherto always been found iu abundance, are exported from thence, being less in price than the recent. Although for a long series of years, very many thousands have been annually obtained, yet they are still collected every year in great numbers on the banks of the larger rivers of the Russian empire, and more particularly those of farther Siberia."—The Naturalist's Library, 1836. Mammalia, Vol. v. p. 133.

The Mammoth teeth are but rarely exposed for sale in this country; I only learn of two; the one weighed 186 lb. , was 10 feet long, of fine quality, and, exoept tho point which was cracked, was cut into keys for pianofortes: the other also was largo, but very much cracked and useless; of the latter I havo a specimen: the substance of the ivory between the cracks appears quite of the ordinary oharacter, alth6ugh the interstices are filled with a dry powder resembling chalk. Both teeth were *solid unto within six inches of the root.* AND SOME OTHER ANIMALS.

139 thick coat of enamel, which entirely resists steel tools, and will even strike fire with that metal; it is usually re-

moved upon the grindstone in order to arrive at the beautiful ivory within, which, owing to the peculiarity of its section, is better adapted to the construction of artificial teeth than the purposes of turning; the other teeth are tolerably round, and fit for the lathe.

The ivory of the hippopotamus is much harder than that of the elephant, and upwards of double the value; in colour it is of a purer white, with a slight blue cast, and is almost free from grain. The parts rejected by the dentists are used for small carved and turned works.

In texture it seems almost intermediate between the proper ivory and the pearl-shell; as when it is turned very thin, it has a slightly curdled, mottled or damasked appearance, which is very beautiful; the general substance is quite transparent, but apparently interspersed with groups of opaque fibres, like some of the minerals of the chatoyant kind.

The teeth of the walrus, sometimes called the sea-cow, which hang perpendicularly from the upper jaw, are also used by the dentists; the outer part, or the true ivory, nearly resembles the above, but the oval center has more the character of coarse boue; it is brown, and appears quite distinct. The long straight tusks of the sea-unicorn or narwal, which are spirally twisted, also yield ivory; but they are generally preserved as curiosities. These two kinds are principally obtained from the Hudson's Bay Company.

The masticating teeth of some of the large animals are occasionally used as ivory; those of the spermaceti whale are of a flattened oval section, and resemble ivory in substance; but they are dark-coloured towards the center, and surrounded by an oval band of white ivory, like that of the aquatic varieties generally; they are not much used.

The grinders of the elephant are occasionally worked; but their triple structure of plates of the hard enamel, of softer ivory, and of still softer cement, which do not unite in a perfect The enamel of the hippopotamus is much thicker, but similar to that of the generality of masticating teeth, which ia found upon analysis to agree very nearly with the hard porcelanous shells. The enamel is sometimes scaled off by driving a thin chisel between it and the ivory; and I learn, the flame of the blowpipe is likewise frequently used for the purpose of separating it. Several of the other teeth have enamel, but the semicircular tooth by far the most abundantly.

140 ANTIQUE STATUES IN IVOltY, ETC. manner, renders them uneven in texture. Owing to the hardness of the plates of enamel, the grinders are generally worked by the tools of the lapidary; they are but little used, and when divided into thin plates are disposed to separate, from change of atmosphere, the union of their respective parts being somewhat imperfect. They are made into small ornaments, knife-handles and boxes, which are occasionally imported. The tusk of the elephant is, however, of far more importance than all these other kinds of ivory, and appears to have been extensively used by the Greeks and Romans. Amongst the former, Phidias was famous for his statues, thrones, and other works of embellishment, made in ivory combined with gold, an art described as the *Toreutic*. In reference to the construction of ivory statues, *Monsieur Quatremiere de Quincy,* in his great work on ancient sculpture,f advances some curious speculations of their having been formed upon centers or cores of wood, covered with plates of ivory; and also that the ancients were enabled to procure larger elephants' teeth, or possessed the means of softening and flattening out those of ordinary size, from which to obtain the pieces presumed to have been thus employed.

These questionable suppositions, particularly the last, scarcely seem called for, as solid blocks of ivory of the sizes commonly met with, would appear to be sufficient for the construction of colossal figures, in the mode ingeniously demonstrated by *M. de Quincy* in his plates 26 to 31. It is much to be regretted that none of these statues have descended to our times.

One of the constituent parts of ivory being animal matter, we should naturally expect it to be less durable than the inorganic materials, in which numerous fine specimens of ancient art still exist in great comparative.perfection. J Ivory appears not to suffer very rapid decay, in the lengthened deposition in the frozeu These grinders contain in themselves the principal characters of the entire group of teeth and tusks, and when used by the animal their unequal degree of hardness continually preserves the ridges of the enamel, that refuse to wear away so readily as the softer parts, and thus converts the surfaces of the teeth into furrows like mill stones for bruising graminivorous food.

t Le Jupiter Olympien, ou l'Art de la Sculpture Antique. Paris, 1515.

J At the present day, Sir. Benjamin Cheverton copies various works sculptured in marble, and other materials, upon a reduced scale in ivory, alabaster, and even marble, by means of mechanism perfected by himself. His miniature busts possess a degree of faithfulness and perfection that leaves nothing to be desired.

THB SUPPLY OP IVOUY. 141 earth of Siberia, nor when immersed in water; but various specimens in the British Museum, apparently less favourably situated, and in contact with the air, exhibit the effect of time, the ivory being decomposed and divided into flakes and pieces which exhibit its lamellated structure in a very satisfactory manner. In reference to the supply of the recent ivory, from Asia and Africa, Mr. M'Culloch says,—

"The duty of one pound per hundred weight f on elephants' teeth, produced, in 1830, 3721/., showing that the quantity consumed amounted to as many hundred weights, or to 416,752 lb. The average weight of an elephant's tusk may be taken at about 60 lb.; so that 69-16 tusks must have been required to furnish this supply of ivory,—a fact which supposes the destruction of 3473 male elephants! But the destruction is really much greater; and would probably amount to, at least, between 5,000 and 6,000 elephants. If, to the quantity of ivory required for Great Britain, we add that required for other countries of Europe, America, and Asia, the slaughter of this noble animal will appear immense; and it may well excite surprise, that the breed has not been more diminished." %

It is probable Mr. M'Culloch's estimate of 60 lb., as the average weight of teeth, is far too high, and from the observation of merchants well qualified to judge, it appears that 15 or 16 lb. would be nearer the average; if so, even the above numbers would be quadrupled.§ North Gallery, Boom II., Wall case 2.

+ The duty on Ivory was reduced one shilling per cwt., and then removed in 1845. *t* M'Culloch's Dictionary of Commerce, 1832, p. 691.

§ Africa is considered to produce ivory in much greater abundance than Asia, and generally of far better quality. The finest transparent ivory is principally collected along the Western coast of Africa, within ten degrees North and South of the equator. On this coast the ivory is considered to become more and more inferior in quality, and more broken, (apparently from hostile encounters,) with the increase of northern latitude: that from Mogador being perhaps the worst.

The best white ivory is, for the most part, the produce of the eastern coast of Africa generally, and until recently was imported almost exclusively from Bombay; of late years it has been partially collected along the coast, and on the Island of Madagascar; very inferior ivory is however sometimes received from these localities, and only a small quantity is now obtained from the Cape of Qood Hope.

The Asiatic teeth shipped from Calcutta, Madras, and part of those from Bombay, are the produce of India generally; they are called Asiatic, East Indian, Siam, 142 DIFFERENCES IN SIZE AND

Elephants' teeth differ considerably in their size, weight, and appearance. The outsides of the African teeth run through all the transparent tints of light and deep orange, hazel, and brown, and some are almost black. Those from Asia are similar, although generally lighter, and frequently of a kind of opaque fawn, or stone-colour; they have seldom the transparent character of the African teeth; and they commonly abound in cracks of inconsiderable depth from which the others are comparatively free.

Some teeth are as long as from eight to ten feet, and as heavy as 150, rarely 180 lb. each tooth; sometimes they are only as many inches long, and about one inch in diameter, and of the weight of five or six ounces, and even lighter; tbe teeth less than from 10 to 14 lb. are called " scrivelloes." In section, the tusks are rarely quite circular, sometimes nearly elliptical, seldom exceeding the proportions of four to five, but commonly less exact than either of these forms; figs. 43, 44, and 45, are accurately reduced from sections of teeth; the largest tusk the author has met with measured eight and a quarter inches the longest, and seven inches the shortest diameter of the irregular oval.

The curvature of the teeth is sometimes as much as the halfcircle, as in fig. 46, and occasionally even so little or less than

Singapore, and Ceylon teeth, &c. The last two are described as ivory of a fine grain with a "pearly blush" appearance, and seldom large.

It appears that many of the better teeth, and the superior parts of others, are selected by the natives of India for their own consumption, and for exportation to China, as numerous hollow pieces, and other portions of teeth, (the ends of which are generally covered with wax to protect them from the air,) are imported along with the entire teeth; and this selection seems confirmed, as the ivory obtained from Madagascar direct contains generally a larger proportion of superior teeth.

The whole of the ivory shipped from Bombay is scored or *scriered* with characters, apparently Arabic, and is named Bombay ivory, from the commercial habit of describing imports from the ship's papers, which quote the port whence the vessel cleared. Many persons however considered this ivory to be exclusively Afritan, although it seems rather improbable that the whole of the ivory from its own immediate vicinity should be excluded from this mart. The commercial regulations regarding the East Indian and East African ivory are alike; those for the sale of the West African ivory arc different. The foregoing remarks, (say the Messrs. Fauntleroy, from whom principally they were obtained,) should however be received with some latitude, as frequent exceptions in the teeth themselves, and differences of opinion amongst those habitually engaged in the commerce exist; although, considered generally, persons in the constant habit of inspecting ivory can distinguish pretty correctly not only the continent, but also the territory from which any tooth was obtained.

The author has an elephant's tusk which is very hollow and weighs *2* ounces. EXTERNAL FORM *OV* ELEPHANTS TUSKS. 143 the sixth, as in fig. 42; they are sometimes finely tapered off, especially in the African teeth; at other times their ends are very much worn away, in rare instances, to the extent of a third of their apparent length, and Figs. 42. generally more so on the one side of the center than the other.

Other teeth end very ahruptly, as if they had been broken and repointed before they left the head of the animal, in which they are generally inserted for about one-fourth their length.

The teeth are hollow about half-way up, and a speck, sometimes called the nerve, but in reality the apex of the successive hollows, is always visible throughout the length of the tusk to its extreme end, the tooth being formed by layers deposited on a vascular pulp after the manner of teeth generally.

The inner and outer surfaces of-the teeth are in general tolerably parallel, and exteriorly they are curved in the one direction only, so as to lie nearly flat on the ground; but occasionally they arc much curved in both directions, as represented in figs. 47 and 48; the smaller is beautifully formed, and resembles in 144 DIFFERENCES IN THE QUALITY OF IVORY.

shape a handsome bullock's horn the other is furrowed throughout its length, and appears the result of disease or injury.

The choice of ivory in the teeth is admitted by the most experienced to be a very uncertain matter; of course for the purposes of turning, a solid coue would be the most economical figure, but as that form is not to be met with, we must be satisfied with the nearest approach that we can find to it, and select the tooth as nearly straight, solid, and round as possible, provided the other prognostics are equally favourable.

The rind should appear smooth and free from cracks, and if the heart should be visible at the tip, the more central it is the better; by the close inspection of the tip, from which the bark is always more or less worn away, it may be in general learned whether the tooth is coarse or fine in the grain, transparent or opaque, but the colour of the exterior coat prevents a satisfactory judgment as to the tint or complexion of the ivory within.

After the most careful scrutiny on the outside of the tooth, however, the first cut is *always* one of a little anxious expectation, as the prognostics are far from certain, and before proceeding to describe the preparation of ivory, I will say a few words of its internal appearance when exposed by the saw.

The African ivory, when in the most perfect condition, should appear when recently cut, of a mellow, warm, transparent tint, almost as if soaked in oil, and with very little appearance of grain or fibre; it is then called transparent or green ivory, from association with green timber; the oil dries up considerably by exposure, and leaves the material of a delicate, and generally permanent tint, a few shades darker than writing-paper.

The Asiatic ivory is of a more opaque dead-white character, apparently from containing less oil, and on being opened it more resembles the ultimate character of the African, but it is the more disposed of the two to become discoloured or yellow. The African ivory is generally closer in texture, harder under the tools, and polishes better than the Asiatic, and its compactness These are drawn from two teeth in my possession: the first measures round the curve 41 inches, and in a straight line 29 inches; its weight is 8J pounds; the other measures respectively 894 and 30 inches, and weighs 274 pounds: these may be considered as extreme cases. Others are altogether as straight: I have one that is 49 inches long, and only rises *i* inches from the straight line, and others approach the right line still more nearly.

BULLETS, ETC., FOUND IN ELEPHANTS' TEETH. 145 also prevents it from so readily absorbing oil, or the colouring matter of stains when intentionally applied.

The rind is sometimes no more than about one-tenth of an inch in thickness, and nearly of the colour of the inner ivory, but occasionally it is of double that thickness, dark-coloured, and partially stains the outer layers. As we do not find all specimens of the most perfect kind, we must be prepared to expect others, especially amongst the larger teeth, in which the grain is more apparent, but it generally dies away towards the center of the tootb, the outside being the coarser; the regularity of the grain sometimes gives it the appearance of the engine-turning on a watch-case.

In some teeth the central part will appear of the transparent character, the outer more nearly white; and the transparent teeth often exhibit, at the solid parts, white opaque patches, which are frequently of a long oval form. Amongst the white ivory, the teeth are often found to be marked in rings alternately light and dark coloured, these are called *"ringy* or *cloudy."*

In those teeth in which there appears to be a deficiency of the animal oil, the intervals between the fibres occasionally assume the chalky character of bone, and are disposed to crumble under the tools unless they are very sharp; in this they resemble the softer parts of woods when worked with blunt tools; sometimes the ivory is not only coarse but dark or brown, and the two defects not unfrequently go together.

The cracks occasionally penetrate further than they appear to do when viewed from the outside, and more rarely a very considerable portion of the tooth is injured by a musket-ball, although the gold and silver bullets, said to be used by the Eastern potentates, are exceedingly scarce, or else transmuted into iron, of which metal they are commonly found, and less frequently of lead.f The ball generally lacerates the part very much, and a new deposit of bony matter is made that fills up all the interstices, encrusts the hollow, and leaves, a dotted mottled mass I have retained the commonly received terms, *African* and *Asiatie* ivory, although the greater part of both kinds appears to come from Africa. Perhaps a more practical distinction would be for "African" ivory, *trantparenl ivory;* and for " Asiatic " ivory, *opaque ivory.* See foot note, p. 141.

t I have only heard of two golden bullets thus found, the one Mr. Fauntleroy says was cut through by a comb-maker in dividing a tooth, and was in the possession of his uncle: the other was found by Mr. Brain, its value was seventeen shillings.

Vol. L' 146 PREPARATION OF IVORY extending many inches each way from the ball, and which completely spoils that part for any ornamental purpose.

SECT. II.—PREPARATION OP IVORY, ETC.

On account of the great value of ivory, it requires considerable judgment to be employed in its preparation, from three conditions observable in the form of the tusk; first, its being curved in the direction of its length; secondly, hollow for about half that extent, and gradually taper from the solid state to a thin feather edge at the root; and thirdly, elliptical or irregular in section. These three peculiarities give rise to as many separate considerations in cutting up the tooth with the requisite economy, as the only waste should be that arising from the passage of tlic thin blade of the saw: even the outside strips of the rind, called spills, are employed for the handles of penknives, and many other little objects; the scraps are burned in retorts for the manufacture of ivory black, employed for making ink for copperplate printers, and other uses; and the clean sawdust and shavings are sometimes used for making jelly.

The methods of dividing the tooth either into rectangular pieces, or those of circular figure required for turning, are alike in their early stages until the lathe is resorted to: I propose, therefore, to begin with the former. The ivory saw, fig. 49, is stretched in a steel frame to keep it very tense; the blade generally measures from fifteen to thirty inches long, from one and a half to three inches wide, and about the fortieth of an inch thick; the teeth are rather coarse, namely, about five or six to the inch, and they are sloped a little forward, that is between the angle of the common hand-saw tooth and the cross-cut saw. The instrument should be very sharp and but slightly set; it requires to be guided very correctly in entering, and with no more The substance of the ivory is not in all cases thus injured by the balls, and Mr. Combe (Philos. Trans. 1801, p. 165,) explains in a very satisfactory manner, how a bullet may enter the tusk of an elephant and become embedded in the ivory without any opening for its admission being perceptible. This he elucidates on the supposition of the ball entering at the root, descending into the hollow, and being covered up by the growth of the layers, which are successively deposited upon the central vascular pulp, in the formation of the tooth according to the process commonly observed in similar cases. He cites an instance of a spear head seven inches long, having been thus embedded in the tusk of an elephant, from entering at the thin part near the skull of the animal.

FOR VARIOUS RECTANGULAR WORKS

147 pressure than the weight of its own frame, and it is commonly lubricated with a little lard, tallow, or other solid fat.

The cutler generally begins at the hollow, and having fixed that extremity parallel with the vice, with the curvature upwards, he saws off that piece which is too thin for his purpose, and then two or three parallel pieces to the leugths of some particular works, for which the thickness of the tooth at that part is the most suitable; he will then saw off one very wedge-form piece, and afterwards two or three more parallel blocks.

In setting out the length of every section, he is guided by the gradually increasing thickness of the tooth; having before him the patterns or gages of his various works, he will in all cases employ the hollow for the thickest work it will make. As the tooth approaches the solid form, the consideration upon this score gradually ceases, and then the blocks are cut off to any required measure, with only a general reference to the distribution of the *heel,* or the excess arising from the curved nature of the tooth, the cuts being in general directed, as nearly as may be, to the imaginary center of curvature. The greater waste occurs in cutting up very long pieces, owing to the difference between the straight line and the curve of the tooth, on which account the blocks are rarely cut more than five or six inches long, unless for some specific object.

In subdividing those blocks which are entirely solid, no great difficulty is experienced: in those which are nearly solid, as in fig. 50, the first step is to cut a central slice just thick enough to avoid the hollow, unless the pieces *a, b,* are required to have A very convenient method of marking out the tooth into equal lengths, is by the employment of a parallel strip of thin sheet copper, as *vide* as the *length* of the pieces; it is bent on the inner side of the tooth half round the same, and a line is scored on each side; when several pairs of marks are thus made, they show angular spaces betwixt them, and the saw is sent exactly midway through the angles. The blocks then truly represent so many key-stones of an arch.

Fig. 50.

some particular size; *c, d,* would serve for leaves for miniatures or veneering, and the remainder would be cut up of any required sizes, as sketched beyond *c, d.* For square pieces of similar size, the block is cut into parallel slabs; for bevelled pieces, as the taper handles of knives and razors, the slabs are cut out wedgeform, the thick end of one against the thin end of the next, as at *e, f:* these slabs are afterwards divided with parallel or inclined cuts, either with the frame saw, fig. 49, or the circular saw, as will be explained in the second volume.

In flat works, such as razor and knife handles, the broad surfaces, if cut radially, would show the edges of the rings or layers of the ivory; but cut parallel with the curve, or as the tangent, the grain is much less observable, and the ivory appears finer. In the keys for piano-fortes this is particularly attended to, the finest broad keys are always cut upon the flat side of the oval, as at /, those upon the long diameter are cut into the narrow pieces called tails, (used between the black keys,) and the intermediate parts, are cut obliquely as at *g,* this causes much waste.

For such pieces as have large hollows more management is necessary, as the thickness and curvatures of the material have to be jointly considered. When the hollows are thin, they are cut into squares or handles as large as the substance will allow; but on account of the circular section of the tooth, some of the pieces, if not all, must necessarily be angular or wedge-form; as regards pieces for the lathe this is of little consequence.

In all cases the entire division of the block or ring should be determined upon, and carefully marked in pencil upon the end of the piece, before the saw is used.

When a tusk is cut up exclusively for turnery-work, the first cut is more generally made where the hollow terminates, which spot is ascertained by thrusting a small cane or a wire up the tooth, and every cut is directed as nearly as possible at right angles to FOB, THE PURPOSES OP TURNING.

149 the curve of the tusk, or to the center of the circle as before described. Unless the tooth is very far from circular, it is usual to prepare the principal quantity into cylinders or rings, as large as they will respectively hold, and the diagrams, figs. 51, 52, and 53, are intended to explain the best mode of centering the pieces, or placing them in the lathe.

If, as in preparing an ordinary block of wood, a circle were made at each end, and the work were chucked from the centers

Figs. 51. 52 (i 4 *d ' e /* of the circles, as at o, tig. 51, the largest cylinder that could be obtained would be that represented by the four sides of the dotted rectangle within that figure. Very much less waste would result from placing the centers so much nearer to the convex side, as to obtain the cylinder represented in fig. 52, by allowing the waste to be equally divided between the points *a, c,* and *e.*

It is however more economical, to cut those teeth which are much curved into the shortest blocks, and the figure 53, which represents the proportions more commonly adopted, shows the small comparative degree of waste that would occur in a piece of half the length of the others, when centered in the most judicious manner.

The first process in prepariug to rough-turn the block is to fix it slenderly in the lathe between the prong chuck and the point of the popit-head, and its position is progressively altered by trifling blows upon either end, until when it revolves slowly, and the common rest or support for the tool is applied against the most prominent points *a, e,* and *c,* respectively, the vacancies or spaces opposite to each, at *d, b,* and /, shall he tolerably equal, so that, in fact, about a similar quantity may have to be turned away from the parts *a, e,* and *c,* for the production of the cylinder, represented by the dotted lines within the figure.

The centers having been thus found, they should be made a little deeper with a small drill; and then the one end of the block being fixed upon the prong chuck, the opposite extremity, 150 METHODS OF CHUCKING TO AVOID WASTE.
supported by the center, is turned for a short distance slightly conical, ready for fixing in a plain boxwood chuck, or a brass chuck lined with wood, to complete the rough preparation, unless indeed it is entirely performed upon the prong chuck.

"With the decrease in length, less attention is requisite in the centering on account of the interference of the curvature of the tooth, and the pieces may be at once rasped to the circular form, and then chucked either in a hollow chuck/ or else by cement or glue, against a plain flat surface; the full particulars of which processes will be found in the fourth volume in the chapters devoted to the various methods of chucking, which should have been rendered familiar to the amateur on less costly materials, before he largely employs ivory, at any rate in the unprepared state.

When the blocks of ivory are long and much curved, a thin wedge-form plate may be sometimes sawn from the end, hi preference to turning the whole into shavings; the end is turned cylindrically for a short distance, just avoiding to encroach on the lower angle of the block, and, as soon as practicable, a parting tool is used for cutting a radial notch for the admission of the saw, which may be then employed in removing a thin taper slice. The process is at any rate scarcely attended with more trouble than turning the material into shavings, and thin pieces are retained for a future purpose, such in fact as those represented beyond the dotted lines at the ends of the figures.

The hollow pieces of ivory are treated much in the same manner as those which are solid, and into which latter condition they are sometimes temporarily changed, by rasping a piece of common wood such as beech to fit into the hollow, driving it in pretty securely, but so as not to endanger splitting the ivory; the work is then centered as recently explained, the chuck and center being in this case received in the wood.

With the hollowpieceSthc process of turning must be repeated, on their inner surfaces, for which purpose a side cutting tool with a long handle for a secure grasp should be used; the tool should be held very firmly so as to withstand the jerking intermittent nature of the cut, until the irregularities are reduced.

For this purpose the sliding-rest is very desirable, as the tool is then held perfectly fast without effort on the part of the individual, and if the chucking be correctly done, the greatest possible economy of the material is attained; the hand tools succeed very well on the outer surface, as the rest or support upon which they are then placed is so close to every point of the exterior surface, that they may be held securely with less effort, although the sliding-rest is nevertheless desirable there also.

When the ivory hollow is thin, and far from circular, the material would be turned entirely into shavings, in attempting to produce a circular ring; the circular dotted lines, in figs. 43 and 44, page 143, are intended to explain this. Fig. 43 might be turned into an oval ring; but it is more usual to cut such irregular hollows into small square and round pieces, as explained.

When thin rings or short tubes are required, they are frequently cut one out of the other in the lathe, in preference to wasting the material in shavings; this is done with the parting tool as in fig. 54; an incision being made of uniform diameter from each end, and coutinued parallel with the axis, until the two cuts meet in the center: very short pieces may be thus divided from the one end only. When the rings are large and thin, it is desirable to plug them at one or both ends, with a thin piece of *dry* wood, turned as a plug to fill the diameter, and prevent the ivory from becoming oval in the course of drying.

Fig. 55 explains the mode of preparing such an object as a snuff-box out of a solid block; that is, with the ordinary parting tool entered from the front, and the inside parting tool entered from within; the incisions of which meet and remove a series of rings. The clotted lines represent the paths of the respective tools, the shaded parts the ring obtained, and the black lines the tools themselves. An aperture must necessarily be made in the center, of a diameter equal to the extreme width of the tool; but after the removal of the first or central ring, a tool of con Solid pieces of ivory about 4 to 6 inches long, and 1J to 2 inches diameter, cut out in this manner, are often imported from India.
152 SEASONING AND SHRINKING OP IVOKY. siderably larger size may be used to extract a much wider ring; and a little tallow or oil applied to the parting tools will, in a great measure, prevent the shavings of the ivory from sticking to them and impeding their progress.
Before quitting the subject of the preparation of ivory for the lathe, let me advise those amateurs, who may be desirous to produce either one large spec-

imen of ivory work, or several pieces forming a set, as of draftsmen or chess-men, to endeavour if possible to make the whole of the work from one tooth; as although the colour of ivory may be considered as yellowish white, and therefore, like writing-paper, pretty much alike, such is not the case, and it is often extremely difficult or almost impossible, to match pieces from two different teeth, so that the colour, transparency, and fibre, shall exactly agree.

Ivory requires a similar drying, or seasoning, to that recommended for wood; as when the pieces cut out of the tooth are too suddenly exposed to hot dry air, they crack and warp nearly after the same manner as wood, and the risk is the greater the larger the pieces; and on this account ornaments turned out of ivory or wood, especially those composed of many parts, should not be placed upon those chimney-pieces which, from their size, are so close to the fire as to become heated thereby in any sensible manner.

Notwithstanding the difference between the component parts of wood and ivory, and that the latter does not absorb water in any material degree, it is subject to all the changes of size and figure experienced by the woods, and in one respect it exceeds them, as ivory alters in *length* as well as *width,* whereas from the former change wood is comparatively free.

The change however is very much less in the direction of the length than the width; this is particularly experienced in billiardballs, which soon exhibit a difference in the two diameters, if the air of the apartment in which they are used, differ materially from that in which the ivory had been previously kept. The balls are usually roughly turned to the sphere, for some months *See* foot note, p 47. The Tithe Commissioners there referred to, refuse to sanction ivory drawing scales at all, as although "they may appear to be correct at the time of observation, they find them subject to a considerable variation from atmospheric influence. *See* their Papers. IVORY CANNOT BE BLEACHED OR SOFTENED. 153 before they are used, to allow the material to become thoroughly dry before being turned truly spherical; and in some of the clubs they even take the precaution of keeping the rough balls in their own billiard-room for a period, to expose them to the identical atmosphere in which they will be used.

Ivory agrees likewise with wood, in shrinking unequally upon the radius and tangent when cut out of quarterings, as explained by the fig. 14, p. 49: on this account, billiard-balls are always made out of teeth scarcely larger than themselves, which sized teeth are called "ball teeth," and procure an advanced price.

It may be asked what means there are of bleaching ivory which has become discoloured; the author regrets to add that he is unacquainted with any of value. It is recommended in various popular works to scrub the ivory with Trent sand and water, and similar gritty materials; but these would only produce a sensible effect, by the removal of the external surface of the material, which would be fatal to objects delicately carved by hand or with revolving cutting instruments applied to the lathe.

Perhaps it may be truly advanced that ivory suffers the least change of colour when it is exposed to the *light,* and closely covered with a glass shade. It assumes its most nearly white condition when the oil with which it is naturally combined is recently evaporated; and it is the custom iu some thin works, such as the keys of piano-fortes, to hasten this period, by placing them for a few hours in an oven heated in a very moderate degree, although the more immediate object is to cause the pieces to shrink before they are glued upon the wooden bodies of the keys. Some persons boil the transparent ivory in pearl-ash and water to whiten it; this appears to act by the superficial extraction of the oily matter as in bone, although it is very much better not to resort to the practice, which is principally employed to render that ivory which is partly opaque and partly transparent, of more nearly uniform appearance.

It is imagined by some that ivory may be softened so as to admit of being moulded like horn or tortoiseshell, its different analysis contradicts this expectation; thick pieces suffer no change in boiling water, thin pieces become a little more flexible, and thin shavings give off their jelly, which substance is 154 IVORY VENEERS. DIAMOND CEMENT. occasionally prepared from them. Truly the caustic alkali will act upon ivory as well as upon most animal substances, yet it only does so by decomposing it; ivory when exposed to the alkalies, first becomes unctuous or saponaceous on its outer surface then soft, if in thin plates, and it may be ultimately dissolved provided the alkali be concentrated; but it does not in any such case resume its first condition.

Ivory is not in all cases used in solid pieces, to which the foregoing remarks principally apply; but is frequently cut into thin leaves and glued upon fabrics of wood, for the manufacture of small ornamental boxes, and works of various kinds, after the manner of the veneers of wood, or the plates of tortoiseshell; it is also used in buhl works, combined with ebony. Such thin plates are usually cut out of the solid block, parallel with the axis of the tooth, as at *c, d,* in fig. 50, page 14-8, with a fine feather-edge veneer saw; but the mode introduced in Russia for cutting veneers spirally from a cylindrical block of wood, with a knife of equal length, (as if the veneer were uncoiled like a piece of silk or cloth from a roller,) has been latterly applied to the preparation of ivory into similar veneers, converting the cylinder of ivory into one ribbon, probably by the action of a reciprocating saw.

The ivory thus divided must be necessarily very thin to be sufficiently pliant; and as ivory is admitted to be more transparent than writing-paper of equal thickness, this introduction promises to be of more use to the artist in water-colours than for veneering, as such thin leaves from their transparency are apt to show the colours of the wood or glue: on this account the ivory for veneering should not be extremely thin, and the woods and glue should be selected of very light colours.

The modes pursued in these veneered

works are analogous to those already described in the fifth chapter in reference to the woods; it is therefore only necessary to add a few words on the white-fish glue, or " Diamond cement," as it is sometimes called, Monsieur H. Pape, of Paris, piano-forte manufacturer, has taken out patents for this method of cutting ivory spirally into sheets. A specimen, 17 inches by 33 inches, and about one-thirtieth of an inch thick, glued upon a board, may be seen at the Polytechnic Exhibition in Regent Street, and M. Pape advertises to supply sheets as large as 30 by 150 inches. He has veneered a pianoforte entirely with ivory. Similar veneers of ivory are now also cut in England.

FACTITIOUS IVORY.—EGGS OF BIRDS. 155 which is very often used for ivory-work, both in attaching ivory to ivory, and ivory to wood.

This cement is made of isinglass, (which is prepared from the sound or swimming-bladder of the sturgeon,) dissolved in diluted spirits of wine, or more usually in common gin. The two are mixed in a bottle loosely corked, and gently simmered in a vessel containing boiling water, in about an hour the isinglass will be dissolved and ready for use; when cold, it should appear as an opaque, milk-white, hard jelly, it is remelted by immersion in warm water, but the cork should be at the time loosened, and it may be necessary, after a time, to add a little spirit to replace that lost by evaporation. Isinglass dissolved in water alone, soon decomposes. See Appendix, note N, p. 957, vol. ii.

Factitious ivory, and tortoiseshell, have been prepared in France in thin plates or veneers. I have procured some pieces about a foot square, of the so-called ivory; it looks exactly like a dull opaque cement, the tortoiseshell like a piece of stained horn; neither of them at all approach the beauty of the true substances. Plaster casts may, by Mr. Franchi's process, be made very closely to resemble ivory, the method is described in the catalogue with which Vol. III. commences.

Having adverted in the present and last chapter to many animal substances suitable to the mechanical arts, obtained from various inhabitants of the land and water, let me in conclusion mention some that arc obtained from the feathered tribes, namely the eggs of birds, which although of limited application in the arts of embellishment, have in all ages served as models or standards of beautiful form.

They maybe made to answer in a very perfect manner for the bodies of vases, the feet and upper parts of which are turned out of wood or ivory; for this purpose the egg-shells have been commonly used in their entire state; a hole having been made at the top and bottom for the extraction of their contents, and the attachment of the remaining parts. I have now the pleasure to bring before the reader a method of cutting the shells of the eggs of our various domestic fowls, and other birds, for the formation of vases with *detached* covers.

Mr. G. D. Kittoe, the inventor of this ingenious, skilful, and elegant application of the lathe, has at my request kindly 156 MODE OF DIVIDING EGG-SHELLS, FOR furnished the sketches of the figures and the annexed description.

"In the accompanying drawing is represented the nose of a lathe with an egg chucked ready for cutting. Fig. 56, is the section of a chuck for holding the eggs to prepare them for the chuck represented in fig. 57.

"Fig. 56, is what is generally termed a spring chuck, and is made by rolling stout paper thoroughly moistened with glue, upon a metal or hard-wood cylinder, the surface of which has been greased to prevent the paper adhering to it, and upon which it must remain until perfectly dry; when it may be removed, and cut or turned in the lathe as occasion may require." Figs. 56. *a* 57.

"This sort of chuck is very light, easily made, and well adapted for the brittle material it is intended to hold. Before fixing the egg in it, the inner surface should be rubbed with some adhesive substance, (common diachylon answers exceedingly well;) when this is done, the egg should be carefully placed in the chuck, the lathe being slowly kept in motion by one hand, whilst with the other the operator must adjust its position, until he observes that it runs perfectly true: then with a sharp-pointed tool he must mark the center, and drill a hole sufficiently large for the wire in the chuck, fig. 57, to pass freely through."

"When this is done, the egg must be reversed, and the same operation repeated on the opposite end; its contents must then be removed by blowing carefully through it; it is now ready for cutting, for which purpose it must be fixed in the chuck shown in fig. 57, which is made as follows:"

"A, is a chuck of box or other hard wood, having a recess turned in it at *a, b,* into which is fitted a piece of cork, as a soft substance for the egg to rest against. B, is a small cup of wood, with a piece of cork fitted into it, serving the same purpose as VASES WITH DETACHED COVERS.

157 that in A. A piece of brass, *d,* is to be firmly screwed into the chuck A, and into that a steel wire, screwed on the outer end; on this a small brass nut, *c,* is fitted to work freely in a recess in the piece B; when the egg is threaded on the wire through the holes which have been previously made in it, this nut is to be gradually tightened up until it presses the cup, B, against the egg, sufficiently to hold it steady and firm enough to resist the action of a finely-pointed graver used to cut it."

Mr. Kittoe adds that the tool requires to be held very lightly, as a little unduc violence would crush the shell; neither should the latter be pinched unnecessarily tight in the chuck, as otherwise when the point of the tool divides the shell, the two parts might spring together and be destroyed by the pressure.

It requires some delicacy of hand to attach the rings to the edges of the shell to constitute the fitting; the foot and top ornaments are fixed by very fine ivory screws, the heads of which are inserted within the shell.

In the next chapter I have to proceed to the materials obtained from the Mineral Kingdom.

Mr. Kittoe has done me the favour to give me a very elegant example of this fragile manufacture, one rather different from those to which his mind is professionally directed in his study as an engineer, in the works of Messrs. Haudslay, Son and Field. He informs me, the shells, when soiled with the fingers, are the most safely cleaned with a solution of citric acid, and that even the eggs of the sparrow may be successfully treated, as above. THIRD DIVISION OR PART. CHAPTER IX. MATERIALS FROM THE MINERAL KINGDOM.

The materials from the mineral kingdom may be divided so far as regards these pages into two groups; the earthy, and the metallic.

The earthy materials, the subject of this chapter, when employed in the mechanical and useful arts, are generally used iu their natural states.

The metallic minerals, consist in general of metallic oxides, combined with a larger quantity of some base, such as silex, clay, or sulphur, which are the most common mineralizers; the cohesion of the mass has in general to be overcome by heat, which destroys the affinity of the component parts, and allows of the separation of the metals in various ways. Of these processes the author will have scarcely anything to say; but the metals themselves, when so obtained, will be treated of at some length in the succeeding chapters.

The earthy and crystalline mineral substances are less frequently worked by the amateur, than the metallic, and therefore they will be noticed rather briefly, aud in the order of their hardness, as derived from the following table. The instructions on a variety of processes of grinding and polishing these earthy substances, will be found in the chapters of Vol. III., to be exclusively devoted to that subject under its numerous modifications, as applied to the materials of the three kingdoms generally; both as the exclusive means for the production of form, and also for the embellishment of surface, or polishing.

TABLE OF HARDNESS, 4c. 1. Talc. 2. Compact Gypsum 3. Calcareous spar. 4. Fluor spar. 5. Apatite.

Lead, Steatite or Soapstone, Meerschaum.. 23 ITin, Ivory, Potstone, Figure-stone, Cannel-coal,) Jet. &c 90 Gold, Silver, and Copper, when pure; soft Brass,

(Serpentine, Marbles, Oriental Alabaster, Sc. 71

Platinum, Gun Metal 53

Soft-Iron 43 6. Felspar.. 1 Soft-steel, Porphyry, Glass 52

„ „.. (Hardened-Steel, Quartz, Flint, Agate, Granite, '. bllex„ I j Sandstono, Sand 26 8. Topaz..! Hardest Steel 6 9. Sapphire.. i Ruby and Corundum...... 1 10. Diamond.. Cuts all substances 1 THEIR TREATMENT DEPENDENT ON STRUCTURE.

159

The annexed table exhibits the relative degrees of hardness of the several substances in the estimation of the mineralogist; thus talc may be scratched by gypsum, gypsum may be scratched by calcareous spar, the last by fluor spar, and so on throughout; in the next column are named some of the minerals, metals and other substances of similar degrees of hardness; and the last column contains the number of minerals, which in respect to hardness are ranked under each of the ten grades.

In the several practices of working these numerous substances, *structure* must also be taken into account, or the mode in which their separate particles are combined; thus hardened-steel, quartz, granite, and sandstone, are each included under the number 7. The particles of the steel, however, are much more firmly united than those of the glassy crystalline quartz, which is far more brittle; and still more so than the aggregations of crystals in the granites; the last may be wrought by sharppointed picks and chisels of hard steel, which crush and detach, rather than cut the crystals; and although sandstone consists almost entirely of particles of silex cemented with silex, still as the grains of the sandstone are but loosely held together, it may be turned with considerable facility with the tools used for turning marble, and which is the every day practice in turning the grindstone. Whereas granite which contains from half to threefifths felspar, a substance softer than silex, and porphyry which consists of crystals of felspar embedded in a base of felspar, cannot be turned with steel tools at all.

Several mineral substances are formed by the successive deposition of their component parts in uniform layers, as in mica and slate; or in alternate depositions, as in the Yorkshire flags or sandstones. Mica may in consequence be split into leaves even so thin as the one 50,000th of an inch; it is used by the optician in mounting objects for the microscope, and is often misnamed talc; slate may be split into very thin leaves of considerable size; and those sandstones which result from the recomposition of granite, are most readily split through the layers of minute scales of mica, which being lighter than the other ingredients, are deposited in separate layers.

Many hard substances, as the agates, carnelians, and flints, show neither the crystalline nor lamellar structure, and break with a fracture termed the conchoidal, of which the broken 160 ST-BTJCTUKES, CEMENTS, CLAY.

flakes of glass, flint, and pitch, may be taken as familiar examples. Hard crystalline gems, on the other hand, are formed of laminae, arranged in various directions, and may be readily split by the hammer and chisel through their natural cleavages or joints; but in most of the earthy minerals, grinding is resorted to for obtaining the ultimate and defined shapes, the consideration of which methods are for the present deferred. Should none of these processes be resorted to by my readers, they will at any rate serve to explain the broad features of the respective influence of the mineral materials, (amongst others) upon tools; which is undoubtedly an important link in our subject, and one full of general interest and variety, from the diversity of the methods which are pursued in such of the useful and ornamental arts as require these natural mineral substances, that include both the softest and hardest solids with which we are acquainted.

The hard mineral substances are mostly attached to the lathe, by resinous cements, as driving them into hollow chucks, like pieces of wood or metal, would endanger their being broken, from their crystalline nature. The soft cements consist of about half a pound of resin, one ounce of wax, and any fine powder, often the fine dust from the stones that are turned; pounded brick-dust and coarse flour are used, and pitch also enters into the composition of other kinds. Shell-lac either alone or mixed with half its weight of finely-powdered pumice-stone is sometimes employed; and fine sealing-wax, which is principally shell-lac, is used as well as many other kinds of cement. The stone is in general warmed to the melting point of the cement; but sometimes the latter is melted by friction alone.

CLAY (1).

This material is only worked in the soft and plastic state. In pottery, it is attached to the potter's wheel or horizontal lathe, by its own adhesiveness alone, and is turned by the hands and blunt wooden tools; it is also pressed into moulds of metal and plaster of Paris, some of which are mounted on the lathe when the objects are smoothed within and moulded without. Lathes Flint, when first raised, may be split with remarkable precision, with the *imperfectly flat conchmdal* fracture, as may be well observed in gun-flints, the perfection of the keen edges thus produced, exceeds that of any which may be obtained upon similar substances by art; when the water is completely dried out of the flint it breaks with the ordinary *conehoidal* fracture.

with vibrating mandrels, or possessing the movement of the roseengine, are likewise employed for the production of some works in pottery and china. The artists who model in clay, use blunt instruments, mostly of wood, which are rounded at the ends; and all artizans cut or divide this material with a stretched pack-thread, or a metal wire. The clay for superior pottery works and modelling, often called pipe-clay, is decomposed felspar, it is mostly obtained from Cornwall and Devon; and the importance of the Stourbridge, and some other refractory clays, in the construction of crucibles and fire-bricks for a variety of purposes in the arts, must not be overlooked. See Appendix, Note O, Page 957, Vol. II. MEERSCHAUM (1). AMBER (2).

These are principally used for smoking-pipes. Previously to being turned, the *meerschaum* is soaked in water, it is then worked with ordinary tools, and is described "to cut like a turnip;" after having been dried in a warm room, it is polished with a few of its own shavings, and rubbed with white wax, which penetrates its surface. Sometimes the pipes are dipped into a vessel containing melted wax.

Amber is principally found on the shores of the Baltic, also at Cromer in Norfolk, &c.: the most esteemed is the opaque kind, resembling the colour of a lemon, and sometimes called fat amber: the transparent pieces are very brittle and vitreous. The German pipe-makers by whom it is principally used, employ thin scraping tools, and they burn a small lamp, or a little pan of charcoal beneath the amber, to warm it slightly whilst The meerschaum pipes are made of a kind of fuller's earth, called *Ktff-lil,* (literally foam-earth,) formerly dug in pits in the Crimea, but now in Anatolia. The *Keffkil* is pressed into moulds upon the spot, dried in the sun, and baked in an oven; the pipes are then boiled in milk, and polished with a soft leather, after which they are carried to Constantinople. They are then bought up by German merchants, who transport them to Pest in Hungary, where, as yet large and rude, they are soaked in water for twenty-four hours, and then turned in a lathe: the sound ones are for the most part sent to Vienna, where they are finished and often expensively mounted in silver.

They afterwards find their way to all the German fairs, and obtain various prices, from two to forty pounds sterling, those being the most esteemed which, from having been long smoked, become stained of a deep yellow, or tortoiseshellcolour, from the oil of the tobacco. The *Krff-kil* is used in the public baths at Constantinople for cleansing the hair of the women. *Sec* Dr. Clarke's Travels in Russia, Tartary, and Turkey, e., vol. ii. 4th cd. p. 282.

VOL. I. M 162 JET, AND CANNEL COAL.

it runs in the lathe, and prevent it from chipping out; they also succeed in bending it, by means of heat. Amber is made into necklaces and ornaments, worn principally in Turkey and India. JET, CANNEL COAL, ETC. (2—2i).

Jet is found at Whitby, Scarborough, and Yarmouth, and is also imported from Turkey, but it is not generally met in large pieces. It may be turned with most of the tools for the soft and hard woods, and worked with saws and files, all used in the ordinary way. Jet, until polished, appears of a brown colour, and is manufactured by the lapidary into a variety of ornaments, such as necklaces, earrings and crosses.

Cannel coal is principally obtained in England from Yorkshire, Shropshire, Derbyshire, and Cumberland; it is also found in parts of Scotland and North "Wales. It occurs in seams, generally about three inches but occasionally one foot thick, amongst ordinary coal; sometimes, as at the Angel Bank Collier', near Ludlow, it constitutes the entire bed. Compared with jet it is much more brittle, also heavier, and harder; it is less brown when worked, less brilliant, but more durable when polished; neither of them are at all influenced by acids or moisture, although they temporarily expand by heat.

Cannel coal may be thought to be a dirty and brittle material, but this is only partially true; it is far better suited to the lathe than might be expected, although a peculiar treatment is called for in the entire management, which commences with the selection of pieces free from flaws, of a compact grain, and of a clean conchoidal rather than of a flaky structure.

As regards the artificer, cannel coal may be considered to he made up of horizontal layers, and to have a grain something like that of wood. The horizontal surface may be readily distinguished from the vertical, either by its splitting off in flakes, or by its appearing as if varnished at those parts contiguous to the ordinary coal, and which are chipped away as useless; sometimes hard fragments or crystals, apparently of iron ores, interrupt its uniformity of surface, and flaws which show when broken the varnished appearance, occasionally diminish its strength.

The material is cut out with a saw; that for ivory, fig. 49, p. 147, is most proper, but the hand-saw will answer; the pieces are then roughed to the shape with a chopper, or a parallel blade of steel, about eight inches long, one and a half wide, and one quarter thick, sharpened very keenly with two bevils at the one end, and used with the hand alone. For making a snuff-box, whether plain, screwed, or eccentric turned, the plankway, or the surface parallel with the seam, is most suitable; it is also proper for vases, the caps and bases of columns, &c. Cylindrical pieces, as for the *shafts* of columns, should be cut from either *edge* of the slab, as the laminae then run-lengthways, and the objects are much stronger; these latter pieces when prepared should be driven into conical chucks with a *mallet,* as the blow of a hammer near the edge would shiver off a flake lengthways. Cylindrical pieces cut the plankway should be chucked with a *hammer,* that the blow may be exactly central, otherwise the cylindrical piece would perhaps be broken in two transversely. Cement is also very much used for chucking the work.

All the tools for cannel coal are ground with two bevils exactly Kke the chisel for soft wood turning, but they are held *horizontally;* a small gouge, from one quarter to three-eighths of an inch wide, also slightly bevilled off from within, is used for roughing out, or rather bringing the work as near as possible to the shape, to save the finishing tools: these should be ground with *thin* and *very sharp* edges, otherwise they burnish instead of scrape the work. The ordinary tools for ivory and hardwood, if employed, must be held downwards at an angle of about twenty degrees; these tools are sometimes used with a wire edge turned up in the manner of a joiner's scraper.

The plankway surfaces turn the most freely, and with shavings much like those of wood, the edges yield small chips, and at last a fine dust, but which does not stick to the hands in the manner of common coal. Flat objects, such as inkstands, are worked with the joiners' ordinary tools and planes; but with these likewise, it is also better the edge should be slightly bevilled on the flat side of the iron. The edges of cannel coal are harder, and polish better than the flat surfaces.

Cylindrical pieces thus prepared, say three inches long, and three-eighths of an inch diameter, are so strong they cannot be broken between the fingers; similar pieces have been long since used for the construction of flutes; and iu the British Museum may be seen a snuff-box of cannel coal, said to have beeu turned in the reign of Charles I., and also two busts of Henry VIII., and hia daughter Lady Mary, carved in tho same material. ALABASTER. ALABASTER (2 21).

This is a sulphate of lime, or compact gypsum, which occurs in various places; in England the finest is found near Derby, where the pure white is employed for the purposes of sculpture, but the finest white alabaster is from Italy: the variegated kinds are turned into vases, pillars, and other ornamental works. Before the Reformation the alabasters were used for statuary, and many curious monuments in this material may be seen in the churches of the midland counties of England.

There are but few kinds of tools employed in turning alabaster, namely, points for roughing out, flat chisels for smoothing, and one or two common firmer chisels, ground convex and concave for curved lines. The point tools used in Derbyshire are square, and described under marble; the Italians prefer a triangular point, as an old triangular file driven into a handle and ground off obtusely at the end. The carved parts are done by hand with small gouges, chisels, and scorpers of various forms and sizes, drills, files, and saws, are also employed; and the surface, unless polished, is finished with fish-skin and Dutch rush.

The fibrous gypsum, called from its brilliaut appearance satiustone, is much softer; it is turned into necklaces and small ornaments, by a sharp flat chisel held obliquely, a square point would split off the fibres. All the above kinds of alabaster or gypsum produce, when calcined, the well-known plaster of Paris, a substance used for cementing together such of the vases as are made of detached parts; plaster of Paris also renders other The Italian alabaster, when first raised, is semi-transparent like spermaceti; it is wrought in this state; the works are generally rendered of a more opaque whito by placing them in a vessel upon little fragments of the stone, so that they may be entirely surrounded by the cold water, which is then poured in and very slowly raised to nearly the boiling temperature: this should occupy two hours. The vessel is then allowed to cool to 70 or 80 Fahr., the object is taken out, closely wrapped in cloth, and allowed to remain until dry. The alabaster at first appears little altered, but it gradually assumes the opaque white; for the first six months it is considered to remain softer than at first, but to become ultimately somewhat harder from the treatment.

Alabaster readily absorbs grease and dirt of any kiud, but it is cleaned by the Italians very dexterously; some use weak alkaliue and acid solutions. Soap aud water are not to be recommended, as the unpolished parts absorb the oil of the soap.

and far more important services in a variety of the useful and ornamental arts.

Oriental alabaster is a very different substance from the above; it is a stalagmitic carbonate of lime, compact, or fibrous; generally white, but of all colours from white to brown, and sometimes veined with coloured zones; it is of the same hardness as marble, is used for similar purposes, and wrought by the same means.

SLATE (2 J—3).

The common blue and red slates consist of clay and silex in about equal parts; the largest slate quarries, perhaps in the world, are at Bangor in Wales. The blocks when quarried are split into sheets, sometimes exceeding eight feet by four, by means of long, wide, and thin chisels, applied on the edge, parallel with the laminae, and struck with a mallet or hammer. The sheets are sawn into rectangular pieces and slabs, by ordinary circular saws with teeth, moved rather slowly; and these are afterwards plaued for billiard-tables, &c., in machines nearly resembling the engineer's planing machines for metal, but with tools applied at about an angle of thirty degrees with the perpendicular, f

Slate is also turned in the lathe with the heel or hook tools used for iron, and also with ordinary tools, used with or without the slide rest, which are however rapidly blunted when applied superficially: it is much tougher at the ends or edges of the laminae than at the flat sides. Slate has been recently worked into chimney-pieces, and a variety of objects for internal decoration, which are ornamented by a patent process,J in the manner of *papier mdchS* and china; imitations of marbles and granite are thus made at about one-third the prices of marble. Some of the In Sir John Sonne's museum there is a magnificent sarcophagus of stalagmite, purchased of Belzoui for one thousand guineas; and there are also fine specimens in the Egyptian Gallery in the British Museum.

t The process of sawing slate appears rather crushing than cutting, or a trial of strength between the tool and the slate, as the latter is carried up to the saw by machinery, and cannot recede from the instrument; the saw is sharpened about four times a day, and is worn out in about two months. The planing tools for common slabs are six inches wide, and when made of the best cast-steel and properly tempered, they last a day and a half without being sharpened; the jambs for chimney-pieces anil other mouldings, not exceeding about six inches wide, are planed with figured tools of the full width. *X* Invented by Messrs. Magnus and Co., Pimlico Slate Work. 166 SERPENTINES, MARBLES. substances known to mineralogists as slates are exceedingly hard, and vary from the hardness 21, to that of flint or 7. Many varieties, including the Turkey oilstones, are used for sharpening tools; and this family also includes the touchstones formerly used in assaying gold. SERPENTINE, POTSTONE, STEATITE (31).

These are natural compounds of magnesia and silica, they are generally worked immediately on being raised, being then much softer, but with the evaporation of their moisture they assume the general hardness of marble. The serpentine and steatite are found abundantly in Cornwall; serpentine is often called green marble, and by the Italians *Verde de prato,* it is much used, but some of the serpentines will not polish well.

Potstonc is an inferior variety of serpentine; in Germany it is abundantly turned into various domestic articles in common use, whence its name.

Steatite is called soapstone from its smooth unctuous feel, and when first raised may be scratched with the fingernail, but it becomes nearly as hard as the others. A variety of steatite is carved by the Chinese into images employed as household gods, and is named figurestone; until lately, it was supposed to be a preparation of *rice.* Steatite enters into the composition of porcelain.

MARBLES, LIMESTONES (31).

The term marble is applied by the mason to any of the materials that he employs, which admit of being polished; but the mineralogist designates thereby the compact carbonates of lime variously coloured. The principal kinds worked in the ornamental arts, are the white or statuary marble from Italy, a variety of coloured marbles, principally from Devonshire and Derbyshire, The moulds for Mr. Henning's exquisite models of the bas-reliefs of the frieze of the Parthenon, aud other antique marbles, are sunk in *intaglio* in the blue slate. In reference to the choice of this material Mr. Henning says: "Giving my son, Samuel, a lesson in arithmetic, I found him idling and drawing faces upon his slate. I handed him one of my gravers, desiring him to engrave one of his little sketches: I had previously done about eighteen pieces of the frieze in ivory in *relief* onetwentieth the full size. " The thought was immediately followed, and the lost portions of the originals were restored with great happiness and skill by engraving them in intaglio in slate, although for his works to bo appreciated, the original casts from the slate should be seen, and not their pirated copies obtained from plaster moulds, which it is to be regretted deprive the artist of his merited reward.

and the black bituminous marble from Derbyshire, Wales, and various parts of Ireland.

The marbles are turned with a bar of the best cast steel, about two feet long, and five-eighths of an incb square, drawn down at each end to a taper point, about two inches long, and tempered to a straw-colour; this point is rubbed on two opposite sides on a sandstone, and held to the marble at an angle of twenty or thirty degrees; the tool soon gets dull, and must be again rubbed on the sandstone, (a bit of Yorkshire flag,) to sharpen it, water should drop on the marble to prevent the tool from becoming heated and losing its temper. The point will keep getting broader by constant grinding, till it forms a kind of chisel an eighth of an inch wide, after which it will require drawing out again. For cutting in the mouldings, a more delicate point is used, and these are the only tools employed; a flat tool will not turn marble at all.

The black marble is subjected to several processes of embellishment referred to at foot.f The Society of Arts has a good collection of marbles. *See* also the note on page 172, respecting the Museum of Economic Geology.

t Black marble is etched with various figures, imitative of Etruscan vases, and Egyptian hieroglyphics: this is done after the work is polished. All the parts to be preserved are pencilled over with a varnish, in which colouring matter is ground that the artist may see his progress, mastic varnish, with vermilion or flake white will answer: the uncovered surface of the marble is dissolved with any diluted mineral acid, and the muriatic is perhaps the best, the acid is washed off with water, and the varnish is removed with turpentine, after which the etching appears of a dull grey, upon a bright black-ground.

Another method which does not destroy the polish, and in which a bronzywhite is given, is considered to be the removal of the colouring matter alone. Its effect is far more beautiful, but the method is not divulged by the few artists who practise it.

Some works in black marble are engraved with an ordinary graver, such as would be used for copper or steel plates, and the graver is also employed conjointly with the etching and discolouring processes above referred to.

By scratching the marble with a hard steel point it produces a kind of cream-colour, and by slighter scratching a fainter shade, by these means pretty landscapes are produced; but these delicate touches soon lose their colour, by acquiring a partial polish, if wiped for the removal of the dust, such specimens should be therefore kept under glass.

Black marble is also commonly selected for inlaid works, known by the name of *PUtra Dura;* the device, whether flowers, a running pattern, or any other is drawn xipon the marble, and cut in very carefully with a chisel and mallet; thin pieces of the various-coloured marbles and stones of nearly the same hardness, aro fitted accurately into the spaces and cemented with shell-lac; after which the whole is polished off together. When the stones thus inlaid differ materially from the 163

LIMESTONES, YLUOIl SPAK.

Many of the limestones, although chemically like the marbles, are less compact, and therefore do not readily admit of being polished; of these may be noticed the Bath stone and other oolites, which are aggregations of egg-shaped particles, like the roes of fish; when first raised they may be cut very readily with an ordinary toothed saw, and turned with great freedom. The Maltese stone, of which many beautiful turned and carved works were recently sold in London, belongs likewise to this group; it is very compact, and nearly as soft as chalk, from which in fact it scarcely differs in any respect except in its delicate brown cream-colour; the natives of the island of Malta display considerable taste in the objects turned and carved in this limestone.

I'LUOR Spar (4). Is a natural combination of lime and fluoric acid, and the workable variety is peculiar to Derbyshire, where the art of turning it is carried to great perfection. The most costly varieties are the deep blue and purple, found only at Castleton in that county. Fluor being an aggregation of crystals, all having a fourfold cleavage, is very difficult to turn, as the laminae are easily split; few even of the best workmen can turn it into very thin hollow articles; the following is the process.

The stone is first roughed out with a point and mallet. Then heated till it will readily frizzle yellow resin which is applied all over it; this penetrates about one-eighth of an inch and holds the crystals together. It is next rough-turned, and a little hollowed; it is again heated and resined, and turned still more into form, then it is bound round with a thin wire, and again resined, and so on till it is sufficiently thin to show the colours, it is then resined for the last time, and polished in the same manner as marble; but the process is more difficult, and ultimately but very little resin remains in the surface of the work. The only tool used is the steel point.

The blue colour of fluor is often so intense that the works marble in hardness, the soft wear away too readily, and the hard too slowly, Bo as to prevent a plane uniform surface from being produced; some judgment is therefore required in their selection.

This is easily seen in the cubical crystals of common fluor spar, of which any of the angles can be very readily split off with the penknife, leaving a triangular facet; the eight angles give *four* pairs of parallel cleavages, and when sufficiently pursued they convert the cube of fluor into the octohedron, its primary form.

FREESTONES, PORPHYRY, GRANITE, ETC.

169 cannot be wrought thin enough to show it: when this is the case the stone is very gradually heated in an oven until it becomes nearly red-hot, when the blue changes to an amethystine colour. Great care is required, for if suffered to remain too long the colour would entirely disappear. The white and lighter kinds of fluor are not worth one-tenth of the value of the blue, but are wrought in the same manner for commoner works.

FREESTONES, SANDSTONES (6—7).

In the introduction to this chapter, some remarks were given intended to explain the various *structural* differences between these substances, and those noticed in the following group.

Freestone is a term commonly applied by the mason to such of the sandstones used for building purposes as work *freely* under the tools; namely, the stone-saw, a smooth iron blade fed with sand and water, and the ordinary picks and chisels, which are too familiar to require more than to be named. The freestones are frequently turned into balustrades, pedestals and vases, in the modes next adverted to.

Sandstones, from their relatively slight cohesion, may be turned with the point tool used for marble, although in the workshop, the grindstone is commonly turned with an old file drawn down for some two or three inches to about one-eighth of an inch square, and held downwards upon the rest at the angle of 20 or 30 degrees; it is rolled over and over, which continually produces a new point; the stone is then smoothed with a flat piece of iron or steel, or rubbed with a broken lump of another grindstone.

PORPHYRIES, ELVANS, GRANITES (6—7).

The division and preparation of the softest of the former materials, namely clay, can be accomplished by the hands alone; in others, as alabaster and slate, with the ordinary toothed saws; and for those of a harder nature, the stone-saw fed with sand and water, is an economical mode of dividing them with great exactness and little waste, from their original forms to those in which they are ultimately required, and which is greatly facilitated by the structure of such as occur in stratified beds; but the use of the stone-saw may be considered to cease with the sandstones.

Different and far more troublesome methods of working arc necessary with those materials now to be considered, that arc much harder, and in which the existence of stratification is considered but rarely and imperfectly to exist; namely, in the compact and cemented porphyries, principally from Egypt and Sweden; the crystalline granites, abundant in Devonshire, Cornwall, and Aberdeenshire, and the elvans of Cornwall, which latter, and some other varieties, appear to merge from the porphyries to the granites, are used for similar purposes, worked by the same means, and ask for an intermediate position.

"In converting the rude masses of granite to their intended forms, the line of the proposed division is first marked, and holes from two to three inches deep, and four to six inches asunder, are bored upon this line, by means of an iron rod terminating at each end in chisel-formed edges of hardened steel, with a bulb in the middle to add weight; this tool, called *& jumper,* is made Tho harder crystalline rocks last referred to, rarely occur in stratified beds, although a remarkable case has of late years been found to exist in the Foggintor Quarry, worked by the Haytor Granite Company, on Dartmoor, the subjoined particulars of which have been kindly communicated to me by William Johuson, Esq., who read a paper before the British Association at Plymouth, 1841, on the Railways and Machinery connected with this interesting work. *See* the Report in the *Athcnaum,* p. 654.

"The distinct beds of the Foggintor Quarry follow pretty nearly the natural slope of the hill, and are thicker and more nearly horizontal the deeper they are situated. The beds are intersected by heads or natural joints, which are nearly perpendicular fissures, (sometimes solid seams,) of unknown depth, that separate the horizontal floor of the quarry into irregular figures, and the continuity of the beds is often broken at the different headings. The quarry, in some parts 100 feet high, is worked in benches, or like a huge irregular flight of stops.

"In detaching the masses of granite rock from their natural beds, the points of least resistance are first determined by an experienced eye, and holes are sunk at those points, vertically or inclined as circumstances may require: the diameters of these holes vary according to the mass and the amount of resistance; and their depths according to the thickness of the blocks to be detached.

"The holes are made with an iron rod, terminating at foot in a chisel-formed edge of hardened steel; the tool is held by one man, who changes its position at every blow received from sledge hammers worked by other men who stand around. When the holes are thus made sufficiently deep, they are charged with gunpowder, in order to effect a separation of the mass by blasting, the ordinary process of tamping confines the powder, and the fuse communicates the blast. The art of the quarryman consists in placing the blast, (or *Aot,)* where the smallest amount of powder will remove the largest mass of rock with the least breakage, simply dislodging or turning it over ready for converting. Three and even five thousand cubic feet have been removed by one discharge." to fall on one spot, it rebounds, and is partially twisted round to present the edge continually in a different angular position J in this manner a very expert workman will bore about a hundred holes in a day. Every one of the holes is then filled with two half-round pieces of iron *caNed feathers,* with an iron pointed wedge between them; the wedges are progressively and equally driven until the stone splits, and the fissure will be in general moderately flat, even should the mass be four or six feet thick, although in such cases the holes are sometimes continued round the ends also."

"The *scouters,* the next class of men, employ the jumpers' feathers and wedges for removing any large projections, by boring holes sideways, and thus casting off large flakes; the *spalders* employ heavy axe-formed or mwc£/e-hammers, for spalling or scaling off smaller flakes; and the *scabblers* use heavy pointed picks, and complete the conversion, so far as it is effected at the quarry, ready for the masons employed in erecting the buildings for which the blocks are used, who complete their formation on the spot." All these materials are likewise used in the ornamental arts.

Porphyry is worked in the lathe with remarkable perfection, and many excellent specimens from Sweden, of vases, slabSj pestles and mortars, and bearings intended for the gudgeons of heavy machinery may be seen in London. I learn that these objects are first worked as nearly as possible to the required forms with the pick, are then mounted in lathes driven by water power, and finished by grinding them with other lumps of porphyry, supplied with emery and water; the machinery is kept going day and night, and the gangs of men relieve one another at certain intervals.

Granite is incapable of being turned in the lathe: it is Various works in porphyry are contained in the Polytechnic Exhibition, Regent Street. Amongst them is a table five feet ten inches diameter, tastefully inlaid with a great number of pieces of various colours; the pedestal is fluted with great exactness, apparently by mechanism: it is said the construction of the table occupied five men during seven years. The sum asked for it is 5002.

The Emperor of Russia had four splendid vases of porphyry five or six feet high, two of which he gave to the Duke of Devonshire; these are at Chatsworth House. This manufacture has in a very few years converted the barren valley in which the porphyry is found, to a scene busy with the hum of industry.

The Saloon of Egyptian Antiquities, British Museum, contains various sculptures in porphyry, the work of almost inconceivable labour and perseverance.

therefore treated like porphyry, that is, shaped with heavy picks, and finally with smaller points used with a hammer; it is afterwards ground with circular or reciprocating motion, according to the figure, by means of iron plates fed with sharp sand, next with emery, progressively finer and finer, upon wooden rubbers, the endways of the grain; and lastly, the polish is perfected with felt rubbers and crocus. The process is tedious and difficult from the unequal hardness of the particles; in this respect granite is inferior to porphyry.

Of late years numerous vases and other circular and ornamental objects have been admirably executed in polished granites and elvans, which occur of various colours and degrees of hardness; when decomposed they are friable, and furnish the china stones extensively used as one of the materials for porcelain and china, and also for making very refractory crucibles.

AGATE, JASPER, CHALCEDONY, CAItNELIAN, ETC. (7).

Are all composed of silex nearly pure; they break in general with a conchoidal fracture, and to divide them into plates it is necessary to resort to the lapidary process. They may be slit with emery, but it is far more economical to employ diamond powder, as the time then required is only one third of that called for when emery Four beautifully polished granite columns may be seen in the King's Library, British Museum; and in the Museum of Economic Geology, there are also many ornamental polished works in granites, elvans, marbles, serpentines, &c., and likewise six inch cubes of a great variety of the same, and of numerous other building stones, collected under the "Commission appointed to visit the Quarries, and to inquire into the qualities of the stone to be used in building the New Houses of Parliament," The various materials are most elaborately detailed in the Parliamentary Keportof the Commission, together with every information respecting cost, supply, and transport; the dates of various buildings in which the stones have been used, and remarks on their present states of preservation: followed by the chemical and physical examination of the stones, and other points which concern the civil engineer, architect, and builder, to whom the report would appear to be almost invaluable. Ultimately four varieties of magnesian limestone were selected for the erection of the national edifice.

The Museum of Economic Geology also contains a most interesting series of specimens of all the metallic ores found in Great Britain and Ireland; and in many instances there are associated with them geological maps and models of the mines, working models of the machinery for mines, and also specimens explanatory of the various processes of manufacture from the ore to the marketable metal. For an infant museum it is most rich, interesting and instructive.

The Institution of Civil Engineers also has a collection of granites.

TOPAZ, SAPPHIRE, RUBY. 173 is used; these stones are always ground with emery, and polished with rotten-stone, as will be explained in the third volume. Agate is used as the bearing planes for the knife edges of delicate balances, for pestles and mortars, burnishers for gilders, and bookbinders, and also for some other purposes in the mechanical arts; the whole of the stones in this group are largely employed for the purposes of jewellery, the handles of knives, snuff-boxes, and a variety of ornaments.

TOPAZ (S). SAPPHIRE, RUBY (9).

These may be split with *plane* surfaces through their natural cleavages, and which method is continually employed; otherwise, they can be only slit with the diamond powder. The first and similar stones may be smoothed with emery, but emery being in hardness only equal to 9, produces but little effect upon topaz, upon sapphire aud ruby it is almost inert, and on diamond quite so; the sapphire and ruby and also diamonds are therefore always polished with diamond powder.

On account of the peculiar interest attached to the mechanical applications of the hard gems, it is proposed to depart a little from the subject and order of these pages, to advert to some few of their uses, which may not be generally understood. The sapphire, the ruby and also the diamond, are commonly used for the construction of certain parts of the best time-pieces and watches, such as the pivot-holes, pallets, and other parts of the escapements.

Fig. 58 represents upon a true, although very enlarged scale, the jewelled pivot-hole for the *verge,* or the axis of the balance of a marine chronometer, *a* is the hardened steel pivot, which is turned with a fine cylindrical neck, and made convex at the end; the jewelling consists mostly of two stones, the one, commonly sapphire or ruby, is turned to the form of the black figure *b,* that is, convex above and concave beneath, of two different 174 APPLICATIONS OF THE RUBY.

sweeps, to thin it away at the part where it is to be pierced with the hole, and which is made a little smaller in the middle to lessen the surface bearing.

The other, which is called the " *top-stone"* or " *end-stone,"* is generally a ruby, in the form of a plano-convex lens, or else it is a diamond cut into facets; the flat side of this touches the end of the pivot.

Each stone is burnished into a brass or steel ring, like some of the lenses of telescopes, and the two stones, (separated a slight distance for the retention of oil by capillary attraction,) are inlaid in a counter-sunk recess, *c, d,* in the side plate, or other part of the watch, and retained therein by two side screws as represented, although unimportant variations are made by different artists in the shapes and proportions of the parts.

The delicacy of these jewelled holes will be imagined, when it is added that in the axis above referred to, the side plate *e, e,* is only one-tenth of an inch thick; the rings from c, to *d,* one-sixth, and the pivot the one two-hundredth part of an inch diameter; in pocket-watches, and more particularly the flat Geneva watches, these measures, especially the first two, are amazingly reduced, although the same number of parts are nevertheless employed in every hole that is jewelled.

The wire for making the pendulum springs for chronometers, is sometimes drawn through a pair of flat rubies with rounded edges, as represented in fig. 59; the stones are cemented into the ends of metal slides having screw adjustments, not represented. Sometimes two pairs of rubies are placed one before the other, to constitute a rectangular hole of variable dimensions, for equalizing the wire both in width and thickness.

Rubies and other gems, are drilled with holes conical from both sides, as in fig. 60, for drawing the slender silver gilt, and silver wires used in the manufacture of gold and silver lace; the wires are afterwards flattened, wound spirally upon silk, and then woven into the lace. Ruby holes are also employed for rounding the leads of ever-pointed pencils; but for this use they are chamfered from the one side only, and the lead is pushed through from the small side, the ruby is then used as a cutting tool; whereas the hole in the draw-plate is slightly rounded upon the ridge, and acts more as a burnisher or compressor; the action of the wire, which is pulled through in the direction of the arrow, tends to draw the stone more firmly into its seat. The finest holes of all, are made by barely allowing the point of the drill to penetrate into the apex of the conical hole, previously formed on the opposite side of the ruby.

All these applications are adopted on account of the very great hardness of the stone, but they could scarcely exist were there not one substance still harder than the ruby to serve for the tools by which these several forms are wrought, and the brief consideration of which will now be proceeded with.

Diamond (10).

The diamond is the hardest substance in nature, and in common with some other crystalline bodies, it is harder at the natural angles and edges, and also at the natural coat or skin of the stone than within, or in its general substance. Its peculiar hardness is probably altogether due to its highly crystalline form, as by analysis the diamond, charcoal, and plumbago, are found to be nearly identical: the first is absolutely pure carbon, the others are nearly so.

The principal use of the diamond is for jewellery, its preparation for which will be touched upon in the slightest possible manner; but from its peculiar hardness the diamond fulfils some more really important although less *brilliant* services as tools, without which several curious and highly valuable processes must be altogether abandoned, and others accomplished in an inferior although more costly manner by other means.

The diamond is prepared for the purposes of jewellery by three Its analogy to these and other combustible bodies was inferred by Sir Isaac Newton, owing to the highly refractive power of the diamond; this has been subsequently proved; although as a combustible body it submits alone to the action of the oxy-hydrogen blowpipe, before which it is dissipated in carbonic acid gas.

The high refractive power of the diamond has led to its employment for the construction of small lenses for single microscopes, because compared with glass lenses of the same focal length, the diamond requires much less curvature of surface, it therefore admits of the employment of a larger pencil of rays, and gives more light. The ruby, sapphire, and garnet, have the same properties in inferior degrees, but the heterogeneous structure of these jewels has caused their employment as lenses to be abandoned. See article Microscope, in the Penny Cyclopedia; also Pritchard's works on the Microscope.

The numbers upon the Scale of Chemical Equivalents for Diamond and Quartz have been very recently pronounced to be alike; this has opened a wide field for speculation.

170 WORKING THE DIAMOND. THE GLAZIERS DIAMOND. distinct processes, namely splitting, cutting and polishing, which will be adverted to in a very few lines. In order to split off the portions not required, the stone is fixed in a ball of cement, about as large as a walnut, the line of division is sawn a little way with a pointed diamond fixed in another ball of cement, and the stone is afterwards split with the blunted edge of a razor struck with a hammer; the small fragments removed, when they are too small for jewellery, are called *Diamond bort.*

In cutting diamonds, two stones are operated upon at once; they are cemented in the ends of two sticks, which are supported on the edges of a box three or four inches wide, rested against two pins as fulcra, and forcibly rubbed against each other; hy which means they abrade each other in nearly flat planes and remove a fine dust called *diamond powder,* which falls through the fine holes in the bottom of the box, and is there collected.

The diamonds are lastly polished upon an iron lap or *skive,* charged with diamond pow der, the stone being guided mechanically; it is fixed by soft solder in a copper cup, or *dop,* attached by a stout copper-wire to the end of the *pincers,* a flat board terminating at the other extremity in two feet, which rest upon a fixed support, the whole forming a long and very shallow triangular stool, loaded at the end near the stone. In the last two processes the stone is re-adjusted for producing every separate facet. We will now proceed to the applications of the diamond as tools.

The invaluable instrument, the glazier's diamond,although employed for a considerable period, was for the first time investigated scientifically by Dr. "Wollastou in 1816,f who pronounced its operation to depend upon a peculiarity of crystallization in the diamond, the facets of which are frequently *round* instead of flat, and therefore the edges are *circular* instead of straight. The rounded edge first indents the glass, and then slightly separates its particles, forming a shallow fissure, with a split The reader will find an excellent account of splitting, cutting, and polishing, diamonds, read by the late Mr. Edmund Turrell before the Royal Institution, and published by him iu Gill's Technological Repository, 182", p. 1, &c. The subject is also noticed in Mawe's Treatise on Diamonds, 2nd Edition, p. 67.

t *See* Transactions of the Royal Society, 1816, p. 2G5. Transcribed by Turrell with additions, at pp. 66—8, of Gill's Tech. Repos., 1827. THE GLAZIER'S DIAMOND. 177 ting rather than a cutting action, none of the material being removed. 62. 63. The primitive form of the diamond is that of a regular oetohedron which is represented in fig. 61; it is like two square pyramids joined base to base; the four sides of the pyramids meet at the angle of 90, their bases at the angle of 109" or thereabouts. Many of the diamonds merge from the form of the oetohedron, into that of the sphere, or a very long egg, in which cases although a disposition to the development of the six points, each formed by the meeting of four surfaces, exists, they are curiously twisted and contorted. The Count de Bournon has published upwards of one hundred forms of crystallisation of the diamond, but the irregular ct'tohedrons with round facets are those proper for glaziers' diamonds.

The extreme point of any diamond may be employed to *scratch* glass with a broad white streak, and detach its particles in a powder, but such glass will break with difficulty, (if at all,) through such a scratch; whereas the almost invisible fiasure, made when the rounded edge is slid over the glass with but slight pressure and almost without causing any sound, is that which produces the effective *cut;* and the cut or split thus commenced will be readily extended through the entire thickness of the glass, when the extremities of the sheet are bent with the fingers or appropriate nippers.

If we could obtain a diamond in the form of a circular button, the edges of which were turned to the angle of 90 or 100 degrees, it would be the perfection of the instrument, as there would be then no point to interfere with its action, and any part of its edge might be used. But as the *natural diamond,* unaltered by the artisan, is always employed, it must be so applied upon the glass, that one of its curved edges bears upon the intended line of division of the glass, and with the extreme point just out of contact, this in so small an object, necessarily confines the position within very narrow limits.

The patent swivel diamond ensures the one condition, by placing the edge of the stone upon the line of the cut, and a few trials at different elevations, generally from 70 to 80 degrees, will soon give the other position. At the commencement a slight force is applied, until the stone appears to bite or hang to the glass; it is then drawn steadily along, with but little pressure, and the good cut will be scarcely either seen or heard.

To show that the diamond possesses nothing in itself that should adapt it to cutting glass, beyond its peculiar form and hardness, Dr. Wollaston succeeded with great labour in giving the same form to the ruby, the spinel ruby, topaz, and rock crystal, with all of which likewise he effected the cutting of glass, but they wer of course far less economical than the natural angle of the diamond itself, which requires no such tedious preparation, and lasts very much longer.

It must not be supposed however, the diamond endures for ever, the ordinary painter and glazier may use one diamond throughout bis lifetime, by having it TOL. I."

The following figures represent, say two or three times magnified, the forms of diamonds that would be most proper for various tools, hut it will be remembered they are only *selected* as near to the respective shapes as they can be found, either amongst imperfect diamonds, or from fragments split off good stones in the first stage of their manufacture for jewellery; these 66. 67. 68. 69. 70. 71. 72. 73. 74.
pieces are known as *diamond bort.* The diamonds are mostly fixed in brass wires, by first drilling a shallow hole for the insertion of the stone, which is embedded slightly below its largest part, and the metal is pinched around it. Shell-lac is also used for cementing them in, and spelter or tin solders may be fused around them with the blowpipe, but pinching them in annealed brass is preferred.

When diamond tools larger than those made of crystals or thin splinters are required, diamond powder is applied upon metal plates and tools of various forms, which serve as vehicles, and into which the particles of diamond powder are embedded, either by slight blows of the hammer, or by simple pressure.

In the construction of the jewelled holes represented in fig. 58, and in similar works, the rubies and sapphires, although sometimes split, are more commonly *flit* with a plate of iron three or four inches diameter, mounted on a lathe, and charged on the edge with diamond powder and oil. When sliced, they are ground parallel one at a time on a flat plate of copper (generally a penny-piece), mounted on the lathe, and into the turned face of which small fragments of diamond have been hammered, this is called a roughing mill; a similar plate with finely washed diamond powder is used for polishing them.

The rubies are afterwards cemented with shell-lac, on the end re-set to expose other angles; but in some glass works, where enormous quantities of this useful material are cut up, the consumption of diamonds amounts to one and two dozens or upwards *every week,* as the sides, from being convex, become rapidly oncave, and the principle is lost, of a small brass chuck, turned cylindrical on their edges, and bevilled for burnishing into the metal rings. They are also turned concave and convex on their respective faces, the turning tool being a fragment or splinter of diamond, fixed in a brass wire; fig. 64 represents the flat view, and 65 the edge view of such a tool, but of the form more usually selected for turning hardened steel, namely, an egg-shaped diamond split in two, the circular end being used with the flat surface upwards; the watch jeweller uses any splinter having an angular corner.f

In drilling the rubies they are chucked by their edges, and a splinter of diamond, also mounted in a wire is used; should the drill be too conical, the back part is turned away with a diamond tool to reduce it to the shape of fig. 66, and from the crystalline nature of the stone, some facets or angles always exist to cause the drill to cut. The holes in the rubies are commonly drilled out at two processes, or from each side, and are afterwards polished with a conical steel wire fed with diamond powder.

In producing either very small, or very deep holes, a fine steel wire, fig. 67, is used, with diamond powder applied upon the end of the same, the limit of fineness being the diameter to which the steel wire can be reduced.

In drilling larger holes in china and glass, triangular fragments of diamond are fixed in the cleft extremity of a steel wire, as in figs. 68 and 69, either with or without shell-lac. Another common practice of the glass and china menders, is to select a tolerably square stone, and mount it as in fig. 70 in the end of a taper tin tube, which wears away against the side of the hole so as to become very thin, and by the pressure, to embrace the stone by the portions intermediate between its angles.

A process employed by the late Sir John Barton, see foot note, p. 42. t The *convex* surfaces of the rubies are polished with *concave* grinders of the same sweeps; the first of copper, the next glass, and the last pewter, with three sizes of diamond powder, which is obtained principally from Holland, from the men who *cut* diamonds for jewellery, an art which is more extensively followed in that country than elsewhere. The watch jewellers wash this powder in oil, after the same manner that will be hereafter explained in regard to emery. *X* The stone is, from time to time, released by the wearing away of the metal, but these workmen are dexterous in remounting it; and that the process is neither difficult nor tedious to those accustomed to it, is proved by the trifling sum charged for repairing articles, even when many of the so-called rivets or rather staples are cemented in; they employ the upright drill with a cross staff.

A similar diamond drill mounted in brass, was used by Mr. Ellis, with the

For larger holes, metal tubes such as fig. 71, fed with diamond powder, are used; they grind out an annular recess, and remove a solid core; copper and other tools fed with emery or sand may be thus used for glass, marble, and various other substances. The same mode has been adopted for cutting out stone waterpipes from within one another by the aid of steam machinery.

Fig. 72 represents the conical diamond used by engravers for the purpose of etching, either by hand, or with the various machines for ruling etching grounds; for ruling medals, and other works. Conical diamonds are turned in a lathe by a fragment of another diamond, the outside skin or an angle being used, but the tool suffers almost as much abrasion as the conical point, from their nearly eqnal hardness; therefore the process is expensive, although when properly managed entirely successful.

To conclude the notice of the diamond tools, figs. 73 and 74 show the side and end views of a splinter suitable for cutting fine lines and divisions upon mathematical instruments. The similitude between this and the glazier's diamond will be remarked, but in the present case the splinter is selected with a fine acute edge, as the natural angle would be too obtuse for the purpose.

Mr. lloss, the inventor of the ingenious Dividing Engine rewarded by the Society of Arts in 1831, informs me that with a diamond point of this kind, presented to him by Mr. Turrell, he was enabled to graduate ten circles upon platinum, each degree subdivided into four parts; at the end of which time the diamond, although apparently none the worse, was accidentally broken. A steel point would have suffered in the graduation of only onethird of a single circle upon platinum, so as to have called for additional pressure with the progress of the work, which in so delicate an operation is of course highly objectionable.

ordinary drill-bow and breast-plate for drilling out the hardened steel nipple of a gun, which had been broken short off in the barrel; no material difficulty was experienced, although the stone appeared to be so slenderly held. In collecting the materials for this chapter, I have gratefully to acknowledge the assistance of J. Tennant, Esq., Professor of Mineralogy, King's College, London; Joseph Hall, Esq., Marble and Spar Works, Derby, and some others.

The workshops of the following manufacturers, &c., have also been kindly laid open to my inspection: Messrs. Corotti, Cox, Dallaway, Dumenil, Ellis, Inderwick, Lund, Magnus, &c., by these means I have been able to obtain the most practical information upon the several subjects, the principal difficulty having been to keep *to* matter within tho limits required by the nature of the work.

CHAPTER X. THE METALS. SECT. I.—ARRANGEMENT OF THE SUBJECT.

The numerous materials from the three kingdoms hitherto described, with but three exceptions, namely, clay, horn, and tortoiseshell, are used in the simple natural states in which they are found, without any change of form beyond that of reduction, as the finished works arc in almost every case produced by cutting or chipping away the portions in excess; the works arc consequently smaller than the rough masses from within which they are obtained, and the fragments cannot be re-united without the aid of some artificial means, such as screws, rivets, or cements.

Several of the substances now to be considered, differ greatly from all the foregoing materials, hy possessing in various degrees of excellence, the properties of *fusibility, malleability,* &c. These assist in various ways in the first separation of the metals from the earthy bases in which they are found; also in the perfect combination of the smallest particles, either of one single or of different kinds, into solid masses; and in the conversion of these by the agency of the crucible, the hammer, and various attendant means, to that endless variety of forms in which the metals are used, many of which are merely a state of part-manufacture or what may be called a " transition" stage.

The chemist enumerates forty-one different metals; of these many are entirely confined to the laboratory, part are only used in the chemical arts, and those to which I propose to refer as connected with our subject, are but fourteen, namely, Antimony, Bismuth, Copper, Gold, Iron, Lead, Mercury, Nickel, Palladium, Platinum, Rhodium, Silver, Tin, and Zinc.

Of these, mercury is always fluid in our latitudes, and antimony and bismuth are brittle metals, consequently they are not used alone in construction, and pure nickel although malleable is rarely so employed; subtracting these reduces the number to ten.

182 THE MANUFACTURE OF CAST-IRON.

Palladium, platinum, and rhodium are principally used on account of their infusibility and resistance to acids, for a fewpurposes connected with science; their abstraction reduces the number to seven. Gold and silver are mostly reserved for coin, and articles of luxury, so that taking away these also, the majority of the works of mechanical art fall exclusively on five, namely, copper, iron, lead, tin, and zinc. These five *practical* metals, if the term may be allowed, are again virtually extended by an infinitude of combinations, or alloys, principally amongst themselves, although with the occasional introduction of the metals before named, and some few others.

Of all the metals however, iron is the one to which from its manifold changes and adaptations, and from its abundance, the most importance is to be attached; so much so, that were wc compelled to the choice, it would be doubtless politic to sacrifice all the others for its possession. It is subjected to several of the methods of treatment, common to the other ordinary metals, and to a variety of changes no less important peculiarly its own.

On these several grounds, therefore, the principal share of attention will be devoted to iron and its several modifications, and it is intended as a general illustration, to commence with a slight sketch of the manufacture of iron from the ore, into castiron, wrought iron, blistered, shear, and cast-steel; and its part manufacture for the purpose of commerce, into ingot, bar, sheet and wire. This will be followed by its further preparation, by forging, into some of the elements of machinery and tools; the various changes of hardening and tempering, applied under a variety of circumstances will be then described.

It is proposed subsequently to consider the thirteen other metals, both in their simple and alloyed states, which will lead to a general outline of the methods of casting objects of various forms, and also of the practice of soldering, which is dependent on the fusible property of some of the metals.

SECT. II.— THE MANUFACTURE OF CAST IRON.

The ore having been raised, the first process to which it is subjected is called calcining or roasting; the iron-stone or *rawmine* is intermixed with coal and thrown into heaps, commonly from thirty to sixty feet in length, ten to sixteen feet wide, and about five to ten feet high; the heaps are ignited and allowed to bum themselves out, which takes place in three or four weeks. This calcines the ore, and drives off a portion of the water, sulphur and other volatile matters, after which the ore is said to be terrified; this process is also performed in kilns.

The smelting is generally performed in England and Wales with coke, and therefore another distinct part of the manufacture of iron is the preparation of the coke, which like torrifying the ore is also performed upon an enormous scale either in open heaps or in kilns, more generally the former.

The smelting furnace used in South Wales is represented in fig. 75 ; its height is about forty-five feet, its diameter at the largest part or boshes, from r-U-j twelve to eighteen feet, and it / / terminates at the bottom in the / / hearth,which is originally a cube / / of about a yard on every side, / / but soon becomes of an irregu-/ / 75 lar form from the intensity of / / the heat. *J*

In mountainous countries iN. such as Wales, the furnace is YTi usually built by the side of a / iK hill, upon the summit of which the coke and mine are prepared, so that they, along with the due proportion of limestone, may be wheeled iu barrows along the bridge represented, into the mouth of the furnace. In level countries, the charge has to be dragged to the filling place up an inclined plane, by means of the steam engine; a full barrow proceeds along the upper surface of the rail, arrived at the top it turns over, discharges its contents into the furnace and returns on the lower side, much the same as the buckets of a dredging machine: there are two such barrows, and from their action they are called *tipplers:* the most general plan however, and the best, is to fill the furnace by hand, a man being stationed at the top, on the plane provided for the purpose

The furnace requires in addition to the solid materials, an This wood-cut is reduced from fig. 1, plate 6, of Mr. Mushet's " Papers on Iron and Steel;" the exterior sectional lines are alone copied, to give a general notion of the form of the internal cavity and the thickness of the walls.

184 OPERATION OF THE FURNACE enormous supply of air-, which is driven in by blowing engines of various constructions, either at the ordinary temperature, or in a heated state, and at one, two, or three sides of the cubical hearth through appropriate pipes or *tuyeres;* the fourth or front side of the hearth, being reserved for the *dam* stone, over which the cinder or scoria flows in a fluid state, and for the aperture through which the charge of melted iron is removed.

When the charge arrives at the hottest part of the furnace, the carbon of the fuel is considered to unite with the oxygen of the ore, and to escape in the form of carbonic acid gas, and carbonic oxide. The lime serves as a flux to fuse the clay and silex of the iron-stone into an imperfect glass or scoria; and the particles of the metal now released, ooze out from the iron-stones, mix with some of the carbon of the fuel, fall in drops through the fiery mass, and collect on the bed or hearth of the furnace; whilst the scoria floats on the surface of the fluid metal, and defends it from the air.

When the scoria has accumulated in sufficient quantity to reach a proper aperture in the front of the furnace, it flows away as a constant stream of liquid lava; and the furnace is tapped at intervals, to allow the charge of metal to run out into channels formed in a bed of sand for its reception: it now assumes the name of crude iron, cast-iron, or the pig-iron of commerce, which is a compound more or less pure, of iron and carbon in different proportions.

The choice of the flux depends on the nature of the ore or mine. For argillaceous ores, which are the most common in England, lime is required; and frequently the cinders or slag from the fineries and forge are mixed with the lime. For calcareous ores, like those of the forest of Dean in Gloucestershire, clay is added, in order to establish a similar train of affinities as regards the earthy matters of the ore. One of the main objects being to fuse all the earths into a glass, so fluid as not to detain the globules of metal, from descending through it to the general mass of fluid iron beneath.

When the iron ores are very pure, as the rich haematite ores of Cumberland, clay is introduced into the furnace, but these pure ores are more commonly mixed in small quantities with the poorer of other districts, and are used without being calcined. It is fortunate that in this country nature has generally supplied the three materials, iron-stone, coal, and lime, at the same localities, as otherwise their transport would add materially to the cost of the production of this invaluable metal.

It will be correctly supposed that the quality of the iron is jointly dependent, on the nature of the ore, on the due proportion of the mine, coke, and lime, introduced into the furnace, and on their treatment whilst there. The products are commonly distinguished by the names of the localities at which they are made, as Welsh, Staffordshire, Shropshire, Yorkshire, Scotch Iron, &c., and likewise by the terms, hot, and cold, blast, and the qualities, 1, 2, and 3. But they may be broadly distinguished, as No. 1, dark-grey, or soft foundry iron. No. 2, light-grey, or stronger foundry iron, and No. 3, or mottled iron, called also white iron or forge pig. The No. 1 contains the largest portion of carbon, is the most fusible, and valuable, the No. 3 is just the reverse. Nos. 1 and 2 from various districts are mixed in different proportions by the iron-founder, according to the quality required in the castings. No. 3 is used for ship's ballast, and for the manufacture of wrought iron, to tho slight sketch of which I shall almost immediately proceed.

In speaking of the dimension of blast furnaces, Mr. Mushet says at page 272:— "The most advantageous maximum of capacity varies at different works. Those in South Wales are from 15 to 18 feet in diameter at the boshes, or widest place, and half this diameter at the top or filling place. The largest blast furnaces in South Wales—or perhaps in the world—are those at the Plymouth iron works, at Duffryn near Merthyr, 18 feet diameter in the boshes, and 9 or 10 feet at the top or filling place: the height 40 feet. So that their capacity is equal to at least 7000 cubical feet, and when at work, each must contain at least ISO tons of ignited materials for iron smelting.

"There are of these enormous furnaces 3 in number, into which are discharged per minute, at least 20,000 cubic feet of atmospheric air, under a pressure of 14 lb. to the square inch.

"The furnaces in Staffordshire, Shropshire, and other parts of England, seldom exceed in their widest parts, 13 or 14 feet in diameter and from one-fourth to onethird this diameter at the mouth or filling place."

At page 285 Mr. Mushet states as the result of an experiment continued upon one furnace for 24 hours, that the following was the supply and produce, namely,

"94 barrows of coke.... each 330 lb.... total 31,020 lb.

"94 barrows of iron-stone.... 360 33,840

"94 barrows of limestone.... 120 11,280

'Total weight introduced into the furnace.... 76,140 1b.

"Pig iron obtained 13,680

"Cinders ditto 29,304

"Proportion of iron to cinder 1 lb. to 2 lb.

"Iron produced from iron stone 40, per cent.

"Limestone included... 3Ojfc —

"Iron for each charge... 14341b."

But the general yield may be considered from 25 to 30 per cent of the iron-stone 186 THE MANUFACTURE OP MALLEABLE IRON.

SECTION III. THE MANUFACTURE OF MALLEABLE IRON.

Formerly, wrought-iron was obtained either directly from the ore, or from cast iron, by a process still in extensive operation abroad, in which wood charcoal is required. In 1783 and 17 Mr. Cort took out patents for the processes of puddling and rolling, which have been continued with successive improvements, and are now conducted in the following manner.

The crude cast iron is remelted in quantities of from half a ton to one ton, in a furnace called the chafery, or refinery, blown with blast; it is kept fluid for about half an hour, and then cast into a plate about four inches thick, which is purer, finer in the grain than pig metal, and also much harder and whiter, it is then called *refined metal.* The plate when cold is broken up, and from two to four hundred weight of the fragments, with a certain proportion of lime, are piled on the hearth of the puddling furnace, which is a rcverberatory furnace without blast.

In about half an hour the iron begins to melt, and whilst it is in the semi-fluid state, the workman stirs and turns it about with iron tools; he also throws small ladles full of water upon it from time to time. In this condition the metal appears to ferment, and heaves about from some internal change; this is considered to arise from the escape of the carbon in a volatilised form, which ignites at the surface with spirits of blue flame; in about twenty minutes the pasty condition gives way, and the iron takes a granulated form without any apparent disposition to cohesion; the fire is now urged to the utmost, and before the metal becomes a stiff conglomerated mass, the workman divides it into lumps or balls of about fifty pounds in weight.

These balls are taken out one at a time, and *shingled,* or worked under a massive helve or forge-hammer) that weighs six or eight tons, and is moved by the steam engine, this compresses the ball, squeezes out the loose fluid matter, and converts it into a employed. In reference to the above experiment, (No. 8 of a series of 12,) Mr. Mushet says " that during the 24 hours, in which each of the foregoing experiments and observations were made, the furnace was supplied with a continual discharge of air equal to 2,500 cubical feet per minute, amounting, in the 24 hours, to 3,600,000 cubical feet upon the whole, each cubic foot Weighing 555 grains, we shall find (page 289) the immense total quantity of 1,274 tons of atmospheric air!" or upwards of thirty-seven times the weight of *all* the solid materials introduced into the furnace.

SHINGLING AND ROLLING THE IRON. 187 *bloom,* or short rudely formed bar. The bloom is then raised to the welding heat in a reheating furnace, and again passed under the hammer, or through grooved rollers, or it is submitted to both processes, by which it is elongated into a rough bar called No. 1. The shinglingis sometimes performed by large squeezers, somewhat like huge pliers, or by roughened rollers that also serve to compress the iron; but the ponderous flat-faced helve is considered the more effectively to expel the dross and foreign matters from the bloom, and to weld the same more perfectly at every point of its length. See Note Q, Appendix, Vol. II., p. 958.

The No. 1 bars are next cut into short pieces, and piled in groups of four to six; they are again raised to the welding heat in a reheating furnace, and passed through other rollers to weld them throughout their length, and reduce them to the required sizes, when the bars are called No. 2: and sometimes the processes of cutting and welding are again repeated in the manufacture of still superior kinds called No. 3 Iron.

A similar process of manufacture is still occasionally carried on, partly with wood charcoal, in place of coals and coke; the iron thus manufactured, called charcoal iron, is much purer, but it is also more expensive; it is sometimes, by way of distinction, left in ridges from the hammer, when it is called dented iron.

The rollers or *rolls* of the iron works are turned of a variety of forms, according to the section of the iron that is to be produced; in general one pair is used exclusively for each form of iron required; although in the imaginary sketch, fig. 76, it is supposed that the shaded portion represents the upper edge of the bottom roll; and that the top roll, which is not drawn, almost exactly meets the bottom one, with the exception of the grooves, and which are in general turned partly in each roll, in the manner denoted by the black figures.

One pair will have a series of angular grooves for square iron, gradually less and less, as *a, b, c,* fig. 76, so that the bar may be rapidly reduced without the necessity for altering the adjustment of the rolls, which would lose much valuable time; the flat bars, are prepared square, and then flattened in grooves, such as that *&td;* round, or bolt iron, requires semicircular grooves *e;* but round iron often shows a seam down one side, from the thin waste spread out between the rolls being afterwards laid down without being welded, when the iron is turned one quarter round and sent again through the rollers; therefore the best round works are mostly forged from square bars.

Figs. / and *g* are described as angle, and y iron; these are particularly used in making boilers, the ribs of iron steamvessels; also frames, sashes, and various works, requiring strength with lightness. Plain cylindrical rollers serve for producing plate and sheet iron, which vary in thickness from one inch to that of writing-paper, and rolls turned like fig. *i,* are employed for curvilinear ribbed plates, or the corrugated iron, an elegant application lately patented for roofs. Other rollers composed of two series of steeled discs, placed upon spindles, are used to slit thin plates of iron about six inches wide, into a number of small rods for the manufacture of nails, and similar rods are also made of larger sizes called slit iron; they always exhibit two ragged edges, and from being tied up in small parcels, are also known as bundle iron.

Figs. 77, 78, 79 and 80, represent four amongst numerous other sections of railway iron; these bars are produced in rollers turned with counterpart grooves; as before, the shaded portions represent fragments of the lower rollers, and the upper rollers are supposed to occupy the spaces immediately adjoining the section of the rails. For these also, three, four, or more grooves, varying gradually from that of the roughly prepared bar, to that of the finished rail, are employed, and this in like manner saves the necessity for adjusting the distance between the rollers during the progress of the work.

All the foregoing rolls are supposed to be concentric, and to produce parallel bars and plates of the respective sections; but in making fish-bellied railway bars, (no longer used), taper plates for coach springs, and similar tapered works, the rollers whether plain or grooved are turned eccentrically, so as to make the works respectively thicker or deeper in the middle, as in fig. 81; this requires additional dexterity on the part of the workman to introduce the material at the proper time of the revolution, upon which it is unnecessary to enlarge.

The general effect of the manufacture of malleable iron is to deprive the cast-iron of its carbon, this is done in the puddling furnace; the original crystalline structure gives way to the fibrous, from the working under the hammer and rollers, by which every individual particle or crystal is drawn out as it were into a thread, the multitude of which constitute the fibrous har or metallic rope, to which it has some resemblance except in the absence of twist. The rod may now be bent in any direction without risk of fracture; and the superior kinds, even when cold, may be absolutely tied in a knot, like a rope, when a sufficient force is applied.

Should it however occur that the first operation, or shingling process, were imperfectly performed, the error will be extended in a proportional degree throughout the mass; which will account for the general continuance of any imperfection throughout the bar of iron, or a considerable length of the wire in which the reduction or elongation are further extended; and to which evil all metals and alloys, subjected to these processes of elongation are also liable.

Malleable iron is divided into three principal varieties, first, red-short iron; secondly, cold-short iron; thirdly, iron partaking of neither of these evils; and which may be so far denominated pure malleable iron.

One of these ties may be seen at the Architectural Society's Rooms, Lincoln's Inn, and another at the Institution of

Civil Engineers. The bars of iron are two inches diameter, and the openings in the loops measure only 6J by 3J inches.

The ends of the bars were turned through each other whilst hot, forming circles of about three feet diameter, and when quite cold they were gradually drawn to this small size with the hydrostatic press used in proving chain cables, by the gradual application of a force equal to about twenty-six tons. Considerable heat was evolved, sufficient to melt wax placed on the metal, and the friction at the rubbing surfaces was so great as fairly to brighten the metal, although a plentiful supply of grease was applied.

These curious and fine specimens of iron were manufactured and tied at Messrs. Hartford, Davis & Co.'s Chain Cable Works, Ebbw Vale, South Wales, under the direction of C. Manby, Esq., now Secretary of the Institution of Civil Engineers.

The first kind is brittle when hot, but extremely soft and ductile whilst cold; this is considered to result from the presence of a little carbon. The cold-short withstands the greatest degree of heat without fusion, and may be forged under the heaviest hammers when hot, but it is brittle when cold; this is attributed to the presence of a little silex. The third kind is considered to be entirely free from either carbon or silex, &c., and to be the pure simple metal; but in the general way the characters of iron are intermediate between those described.

From one and a half to two tons of pig-iron, have been used to produce one ton of malleable iron; but the average quantity is now from twenty-six to twenty-seven tons for each twenty tons of produce. The No. 8 cast-iron, called also forge-pig, ballast, and white cast-iron, is the kind principally used, as it contains least carbon, the whole of which should be expelled in the conversion of the cast metal into wrought iron.

It appears to be unnecessary to attempt any minute description of the different marks and qualities of iron; first, as these descriptions have been minutely given in many works, some of which are alluded to at the end of this chapter; and secondly, as in common with most other articles, the quality of iron governs the price. The quantity used by the amateur will be comparatively inconsiderable, he will be therefore disposed to ask for the best article, and to pay the best price.

I will only add, that little can be known of the character of iron from its outside appearance, beyond that of its having been well or ill manufactured, so far as regards its formation into *bars.* The smith is principally guided by the fracture when he *breaks down* the iron, that is, when the bar is nicked on opposite sides with the cold chisel, laid across the anvil upon a strip of iron near to the cut that it may stand hollow, and the blows of the pane of the sledge-hammer are directed upon the cut.

The judgment will be partly formed upon the force thus required in breaking the iron; the weakest and worst kinds will yield very readily, when small, sometimes even to the blow of the chisel alone, and will then show a coarse and brilliant appearance, entirely granular or crystalline; this iron would be called very common and bad. If, on the other hand, the iron breaks with difficulty, and the line of separation, instead of being mode MANUFACTURE OF STEEL.

191 rately flat, is irregular, or presents what may be called a hilly surface, the sides of which have a fibrous structure, and a sort of lead-coloured or a dull grey hue, this kind will have a large proportion of *fibre,* which is the very "thew and siuew" of iron, and it will be called excellent tough iron. Other kinds will be intermediate, and present partly the crystalline and partly the fibrous appearance, and their relative values will depend upon how nearly they approach the one or other character.

Another trial is the extent to which iron, when slightly nicked, may be bent to and fro without breaking; the coarse brittle kind will scarcely bend even once, whereas superior kinds, especially stub, charcoal, and dented irons, will often endure many deflections before fracture, and when nicked on the outside only and doubled flat together, will bend as an arch and partly split open through the center of the bar, somewhere near the bottom of the cut made with the chisel, the entire fracture presenting the beautiful fibrous appearance, and dull leaden hue before described.- See Appendix, Note P, page 958 of Vol. II.

SECTION IV.—MANUFACTURE OF STEEL.

Steel is manufactured from pure malleable iron by the process called cementation; the Swedish iron from the Dannemora mines marked with the letter L in the center of a circle, and called "Hoop L," is almost exclusively used; irons of a few other marks are also used for second-rate kinds of steel. The bars are arranged in a furnace that consists of two troughs, about fourteen feet long, and two feet square; a layer of charcoal powder is spread over the bottom, then a layer of bars, and so on alternately; the full charge is about ten tons, the top is covered over first with charcoal, then sand, and lastly with the waste or slush from the grind-stone trough, applied wet, so as to cement the whole closely down, for the entire exclusion of the air.

A coal fire is now lighted below and between the troughs; and at the end of about seven days, the bars are found to have increased in weight the one hundred and fiftieth part, by an absorption of carbon, and to present when broken, a fracture more crystalline, although less shining than before. The bars, when thus converted, are also covered with blisters, apparently from the expansion of the minute bubbles of air within them; this gives rise to the appellation bhstered-steel.

192 BLISTERED, SHEAR, AND CAST-STEEL.

The continuance of the process of cementation introduces more and more carbon, and renders the bars more fusible, and would ultimately cause them to run into a mass, if the heat were not checked; to avoid this mischief a bar is occasionally withdrawn and broken to watch the progress; and the work is complete when the cementation has extended to the center of the bars; the conversion occupies, with the time for charging and emptying the furnace about fourteen days.

A very small quantity of steel is employed in the blistered state, for welding to iron for certain parts of mechanism, but not for edge tools; the bulk of the blistered steel is passed through one of the two following processes, by which it is made either into shear-steel, or cast-steel.

Shear-steel is produced by piling together six or eight pieces of blistered-steel, about thirty inches long, and securing the ends within an iron ring, terminating in a bar about five feet long by way of a handle. They are then brought to a welding heat iu a furnace, and submitted to the helve or tilt hammer, which unites and extends them into a bar called shear-steel, from its having been much used in the manufacture of shears for cloth mills and also German steel, from having been in former years procured from that country. Sometimes the bars are again cut and welded, and called double-shear steel, from the repetition.

This process of working, as in the manufacture of iron, restores the fibrous character, and retains the property of welding: the shear steel is close, hard, and elastic; it is much used for tools, composed jointly of steel and iron; its superior elasticity also adapts it to the formation of springs, and some kinds are prepared expressly for the same under the name of spring-steel.

We are indebted to Mr. Huntsman for the process of manufacturing the *cast-steel,* for which he took out a patent: he justly attained a celebrated name for its manufacture, and his process is still followed almost without change. In making cast-steel, about twenty-six or twenty-eight pounds of fragments of blistered-steel selected from different varieties, are placed in a crucible made of Stourbridge clay, shaped like a barrel, and fitted with a cover, which is cemented down with a fusible lute that melts after a time the better to secure the joining. Either one or two pots are exposed to a vivid heat, in a furnace like the brassfounders' air-furnace, in which the blistered-steel is thoroughly melted in the course of three or four hours; it is then removed by the workmen in a glowing state, and poured into a mould of iron, either two inches square for bars, or about six by eighteen inches, for rolling into sheet steel. For large ingots the contents of two or more pots are run together in the same mould, but it requires extremely great care in managing the very intense temperature, that it shall be alike in both or all the pots.

The ingots are reheated in an open fire much like that of the common forge, and are passed under a heavy hammer weighing several tons, such as those of iron works; the blows are given gently at first, owing to the crystalline nature of the mass, but as the fibre is eliminated the strength of the blows is increased.

Steel is reduced under the heavy hammer to sizes as small as three-quarters of an inch square. Smaller bars are finished under tilt hammers, which are much lighter than the preceding, move considerably quicker, and are actuated by springs instead of gravity alone; these condense the steel to the utmost. Rollers are also used, especially for steel of round, half-round, and triangular sections, but the tilt hammer is greatly preferred.

Cast-steel is the most uniform in quality, the hardest, and altogether the best adapted to the formation of cutting tools, especially those made entirely of steel; but much of the cast-steel will not endure the ordinary process of welding, but will fly in pieces under the hammer when struck.

Mr. Mushet took out a patent in 1800, for the manufacture of cast-steel, by the direct fusion of malleable iron, charcoal, and other carbonaceous agents, in the respective proportions; and the same process was likewise extended to some of the superior varieties of iron ores, so as to arrive at one step from the ore to cast-steel; but the method appears to have been only applicable to limited and common purposes, and not to have entered into serious competition with the ordinary mode just described.

In respect to steel, the same general remarks offered upon iron may be repeated, namely, that price in a great measure governs quality. Steel when broken does not show the fibrous character of iron, and in general the harder or *harsher* the steel, the more irregular or the less nearly flat will be its fracture.

The blistered-steel should appear throughout its substance of an uniform appearance, namely, crystalline and coarse, much like inferior iron, but with less lustre and less of the bluish tint; 194 APPEARANCES WHEN FRACTURED. when but partially converted, the film of iron will be readily distinguished in the center. The blistered-steel when it has been once passed through the fire and well *hammered,* assumes as may be supposed, a much finer grain, as in fact the operation converts it, (although in the small way,) into shear steel.

Shear steel breaks with a much finer fracture, but the crystalline appearance is still readily distinguished. Cast-steel is in general the finest of all in its fracture, and unless closely inspected, its separate crystals or granulations should be scarcely observable, but the appearance should be that of a fine, light, slaty-grey tint, almost without lustre.

The quality of steel is considerably improved, especially as regards cutting tools, when after being forged it is hammerhardened, or well worked with the hammer until quite cold, as this tends to close the " pores " and to make the material more dense; above all things excess of heat should be avoided, as it makes the grain coarse and shining, almost like that of bad iron, and which deterioration can be only partially restored, by good sound hammering under a peculiar management. The particular degrees of heat at which different samples of iron and steel, bearing the same name, should be worked, can only be found by trial; and it would be hardly possible to describe the shades of difference.

The reader desirous to examine the several conditions of iron, from its state of ore to that of cast-steel, should visit the museum of Economic Geology before referred to, which contains a good series of specimens.

It would have been incompatible with the nature of this work to have entered more largely into the manufacture of iron and steel, or to have attempted the notice of the various alloys of steel which have received many attractive denominations, especially when so much has been already written on the subject by those possessing superior opportunities.

Of all the works published on the manufacture of iron and steel, the one of the most grand importance is the collection of Mr. Mushet's papers, which have appeared in the Philosophical Magazine at various times subsequent to 1798, and were recently collected and published by himself under the title " Papers on Iron and Steel:" the labour and research therein recorded are almost beyond belief.

Of the more brief and popular accounts of this subject, perhaps the best is No. 106 of the Library of Useful Knowledge," On the Manufacture of Iron," published by the Society for the Diffusion of Useful Knowledge. Aikin's Dictionary of Chemistry and Mineralogy, &c.; three volumes on the Manufactures in Metal, in Lardner's Cyclopedia; and Ure's Dictionary of Manufactures and Mines, contain likewise a very large store of information on the metals generally. The reader will also consult with advantage, Aikin's "Illustrations of Arts and Manufactures. " and various articles in the Encyclopedias, &c. &c.

CHAPTER XI. FORGING IRON AND STEEL.

SECT. I.— GENERAL VIEW OF THE PRACTICE OF FORGING

IRON AND STEEL.

In entering upon this subject, which performs so important and indispensable a part in every branch of mechanical industry, it is proposed first to notice some of the general methods pursued, commencing with the heaviest works, and gradually proceeding to those of the smallest proportions. This arrangement is principally adopted that the apparatus, which undergo a corresponding change in their kind and dimensions, may be adverted to.

After this, the management of the fire, and the degrees of heat required for various purposes, will he described; and then the elementary practice of forging will be attempted: those works made principally in one piece will be first treated of, and afterwards such as are composed of two or more parts, and which require the operation of welding.

The heaviest works of all, are generally heated in air furnaces of various descriptions, some of which resemble but greatly exceed in size those employed in the works where iron is manufactured, and in which the process of forging may be truly considered to commence with the very first blow given upon the *ball,* as it leaves the puddling furnace for being converted into a *bloom.*

At these works, in addition to the ordinary manufactures of bar, plate, and hoop iron in all their varieties, the hammer-men are employed in preparing masses, technically called *"uses,"* which mean pieces to be *used* in the construction of certain large works, by the combination or welding of several of these masses A square shaft, to be used at an iron-works in "Wales, was made by laying together sixteen square pieces, measuring collectively about twenty-six inches square, and six feet long. These were 196 LARGEST PADDLE SHAFTS.
bound together, and put into a powerful air furnace, and the ends of the group were welded into a solid mass under the heavy hammer weighing five tons; the weld was afterwards extended throughout the length. The paddle-shafts of the largest steamships are wrought by successive additions at the one end, as follows. A slab or *use* is welded on one side close to the end, and when drawn down to the common thickness, the additional matter becomes thrown into the length; the next *use* is theu placed on the adjoining side of the as yet square shaft, and also drawn into the length, and so on until the full measure is attained.
These ponderous masses are managed with far more facility than might be expected by those who have never witnessed such interesting proceedings. First, the *"heat"* has a loug iron rod attached to it in continuation of its axis, to serve as a *"porter"* or guide rod; the mass is suspended under a traversing crane at that point where it is nearly equipoised, the crane not only serves to swing it round from the fire to the hammer, but the traverse motion also moves the work endways upon the anvil, and small changes of elevation are sometimes effected by a screw adjustment in the suspending chain. The circular form is obtained by shifting the work round upon its axis by means of a cross lever fixed upon the porter, and moved by one or two men, so as to expose each part of the circumference to the action of the helve; this is readily done as the crane terminates in a pulley, around which an endless band of chain is placed, and the work lies within the chain, which shifts round when the work is turned upon the anvil: the precision of the forgings produced by these means is very surprising.

A similar mode of work is adopted on a smaller scale for The above remarks refer to the paddle shafts of engines, such as those built by Messrs. Maudslays for the "Great Western" steamship, consisting of three pieces connected as usual by drag links. The center piece, or *middle length,* was 12 feet long, the two outer, or *the paddle shpftt,* wore 22 feet long each, the largest or the central diameter was 18 inches, and they tapered off to 12 inches at the smallest or the external parts, the bearings being 16 and 15 inches diameter. The collective weight of the three pieces was near 20 tons, and their value upwards of a thousand pounds before they left the forge of Messrs. Acramans of Bristol, at which they were made.

The reader is here referred to the account of Nasmyth's Direct Action Steam Hammer, now principally used in forging heavy works, and to his Pile Driving Engine. *See* Notes Q and R, in the Appendix to vol. ii. pages 958—961.

SMALLER SHAFTS; SCRAP IRON; SMITH'S HEARTH. 197 many of the spindles, shafts, and other parts of ordinary mechanism, which are forged under the great hammer, often of several bars piled together and *faggoted;* a suitable term, as they are frequently made of a round bar in the center, and a group of bars of angular section, called *mitre iron,* around the same, which are temporarily wedged within a hoop, somewhat after the manner of a faggot of wood. Such works are likewise made of scrap-iron, which consists of a strange heterogeneous medley of odd scraps and refuse from a thousand works, scarcely two pieces of which are alike. A number of these fragments are enveloped in an old piece of sheet iron, and held together by a hoop, the mass is raised to the welding heat in a blast or air furnace, and the whole is consolidated and drawn down under the tilt-hammer; one long bar that serves as the porter being welded on by the first blow. The mingling of the fibres in the scrap-iron is considered highly favourable to the strength of the bar produced. The scrap-iron is sometimes twisted during the process of manufacture, to lay all the filaments like a rope, and prevent the formation of *spills,* or the longitudinal dirty seams found on the surface of inferior iron.

Sometimes the formation of the scrap-iron is immediately followed by the production of the shafts and other heavy works for which it is required; at other times the masses are elongated into bars sold under the name of scrap-iron, although it is very questionable if all the iron that is so named is produced in the manner implied.

The long furnaces are particularly well suited to straight works and bars, but when the objects get shorter and of more complex figures, the open fire or ordinary smith's hearth is employed. This, when of the largest kind, is a trough or pit of brickwork about six feet square, elevated only about six inches from the ground; the one side of the hearth is extended into a vertical wall leading to the chimney, the lower end of which terminates in a hood usually of stout plate iron, which serves to collect the smoke from the fire. The back wall of the forge is fitted with a large cast-iron plate, or a *back,* in the center of which is a very thick projecting nozzle also of iron, perforated for admitting the wind used to urge the fire; the aperture is called the *tuyere.* 19S LARGE HKAUTHJ ANCHORS; VERTICAL HAMMEli.

The blast is sometimes supplied from ordinary bellows of various forms; at other times, as at the Woolwich dockyard, by three enormous air-pumps, which lead into a fourth cylinder or regulator, the piston of which is loaded with weights, so as to force the air through pipes all over the smithy, and every fire has a valve to regulate its individual blast; but the more modern and general plan is the revolving fan, also worked by the engine, the blast from which is similarly distributed.

In some cases the cast-iron forge back is made hollow, that a stream of water may circulate through it from a small cistern; the *water-back* is thereby prevented from becoming so hot as the others, and its durability is much increased. In other cases the air, in its passage from the blowing apparatus, flows through chambers in the back plate so as to become heated in its progress, and thus to urge the fire with *hot blast,* which is by many considered to effect a very great economy in the fuel.

Some heavy works of rather complex form, such as anchors, are most conveniently managed by hand forging; many of these require two gangs of men with heavy sledge hammers, each consisting of six to twelve men, who relieve each other at short intervals, as the work is exceedingly laborious. Their hammers are swung round and made to fall upon one particular spot, with an uniformity that might have suggested to the immortal Handel the metre of his " Harmonious Blacksmith," but not certainly the melody; the conductor of this noisy, although dumb concert so far as relates to voice, stands at a respectful distance, and directs the blows of his assistants with a long wooden wand. The Hercules or crane, used for transferring the. work from the fire to the anvil, which is at about the same elevation as the fire itself, is still retained.

The square shanks of anchors are partly forged under a vertical hammer of very simple construction,called a *"monkey."* It consists of a long iron bar running very loosely through an eye or aperture several feet above the anvil, and terminating at foot in a mass of iron, or the ram. The hammer is elevated by means of a chain, attached to the rod, and also to a drum overhead, which is put into gear with the engine, and suddenly released by a simple contrivance, when the hammer has reached the height of from two to five feet, according to circumstances. The ram is made to fall upon any precise spot indicated by the SMALLER WORKS AND HEARTHS. 199 wand of the" foreman, as it has a horizontal range of some twenty inches from the central position, and is guided by two slight gye rods, hooked to the rani and placed at right angles; the gyes are held by two men, who watch the directions given. This contrivance is far more effective than the blows of the sledge hammers, and although now but little used is perhaps more suitable to such purposes than the helve or lift hammer, which always ascends to one height, and falls upon one fixed spot.

The square shank of the anchor, and works of the same section, are readily shifted the exact quarter circle, as the slingchain is made with flat links, each a trifle longer than the side of the square of the work, which therefore bears quite flat upon one link, and when twisted it shifts the chain the space of a link, and rests as before.

Many implements and tools, such as shovels, spades, mattocks, and cleavers, are partly forged under the tilt-hammer; the preparatory processes, called moulding, which include the insertion of the steel, are done by ordinary hand forging. The objects arc then spread out under the broad face of the tilt-hammer, the workman in such cases being sometimes seated on a chair suspended from the ceiling, and by paddling about with his feet, he places himself with great dexterity in front or on either side of the anvil with the progressive changes of the work; the concluding processes are mostly done by hand with the usual tools. A similar arrangement is also adopted in tilting small sized steel.

With the reduction of size in the objects to be forged, the number of hands is also lessened, and the crane required for heavy work is abandoned for a chain or sling from the ceiling, but for the majority of purposes two men only are required, when the work is said to be *two-handed.* The principal or the fireman, takes the management of the work both in the fire and upon the anvil; he directs and assists with a small hammer of from two to four pounds weight; the duty of his assistant is to blow the bellows and wield the sledge-hammer, that weighs from about ten to fourteen pounds although sometimes more, and from which he derives his name of hammer-man.

As the works to be forged become smaller, the hearth is gradually lessened in size, and more elevated, so as to stand about two and a half feet from the ground: it is now built hollow, with an arch beneath serving as the ash-pit to receive 200 ORDINARY FORGE; THE TONGS.
the cinders and clinkers. The single hearths are made about a yard square, and those forges which have two fires under the same hood, measure about two yards by one; a double trough to contain water in the one compartment and coals in the other, is usually added, and the ordinary double bellows are used. In proportion as the hearth is more elevated, so is the anvil likewise, that in ordinary use standing about two feet or two-and-a-half feet from the ground, its weight being from two to four hundred-weight.

Numerous small works are forged at once from the end of the bar of iron, which then also serves the office of the *porter* required for heavy masses; but when the small objects are cut off from the bar, or the pieces are too short to be held in the hand, tongs of different forms are needful to grasp the work. These are made of various shapes, magnitudes, and lengths, according to circumstances; but the annexed figures will serve to explain some of the most general kinds, although variations are continually made in their forms, to meet peculiar cases.

Figs. 82, and 83, are called *flat-bit* tongs; these are either made to fit very close as in fig. 83, for thin works, or to stand more open as in fig. 82, for thicker bars, but always parallel; and a ring or coupler, is put upon the handles or *reins,* to maintain the grip upon the work. Others of the same general form are made with hollow half-round bits; but it is much better they should be angular, like the ends of fig. 84, as then they serve equally well for round bars, or for square bars held upon their opposite angles. Tongs that are made long, and swelled open behind, as in fig. 84, are very excellent for general purposes, and also serve for bolts and similar objects, with the heads placed The common bellows and rocking staff are too familiar to need description; bellows of the same general construction have also been made round and square; and amongst various others, are Street's patent bellows, 1815; Jeffries and Halley's improved blowing machine, rewarded by the Society of Arts in 1819. *See* Trans. Vol. 38, p. 87; but the mode which is now most generally used in large works is the revolving fan, first introduced for iron furnaces and forges, by Messrs. Carmichael of Dundee. *See* Trans. Soc. of Arts, Vol. 48, p. 159. A similar instrument on a small scale is sold under the name of Clark's Blowing Machine.

The notice of the bellows of the ancient Egyptians, the present Indians, Chinese, Neapolitans, and many others, made of goat-skins, bamboos, wood, and other substances, most of which apparatus consist of two distinct bellows worked alternately, some with the hands, others with the feet, (the reservoir of our double bellows not being employed,) might be considered curious but rather misplaced.
TONGS AND GENERAL TOOLS. 201 inwards. The *pincer* tongs, fig. 85, are also applied to similar uses, and serve for shorter bolts.
Figs. 83. 84. 85. 86. 87. 88. 89.

Fig. 86, represents tongs much used at Sheffield, amongst the cutlers; they are called *crook-bit* tongs; their jaws overhang the side, so as to allow the bar of iron or steel to pass down beside the rivet, and the nib at the end prevents the rod from being displaced by the jar of hammering; these are very convenient. Fig. 87, or the *hammer* tongs, are used for managing works punched with holes, such as hammers and hatchets; as the pins enter the holes, and maintain the grasp, they should be made stout and long, so as to admit of being repaired from time to time, as the bits get destroyed by the fire.

Fig. 88, or *hoop* tongs, are very much used by ship-smiths, for grasping hoops and rings, which may be then worked either on the edge, when laid flat on the anvil, or on the side, when upon the beak-iron: and lastly, fig. 89, represents the smith's *pliers,* or light tongs, used for picking up little pieces of iron, or small tools and punches, many of which are continually driven out upon the ground in the ordinary course of work; they are also convenient in hardening small tools.

In addition to the hearth, anvil, and tongs, the smithy contains a number of chisels, punches, and swages or striking tools, called also top and bottom tools, of a variety of suitable forms and generally in pairs; these may be considered as reduced copies of the grooves turned in the rollers, and occasionally made on the faces of the tilt-hammers of the iron-works for the 202 SMALL FORGES AND TOOLS.
production of square, flat, round, T form iron, augle iron, and. railway bars, as referred to. The bottom tools of the ordinary smith's shop, have square tangs to fit the large hole in the anvil, in using them the fireman holds the work upon the bottom tool, and above the work he places the top or rod tool, which is then struck by the sledge-hammer of his assistant.

The smith who works without any helpmate is much more circumscribed as to tools, and he is from necessity compelled to abandon all those used in pairs, unless the upper tools have some mechanical guide to support and direct them. In addition to the anvil he only uses the fixed cutter and heading tools; he may occasionally support the end of the tongs in a hook attached to his apron-string, or suspended from his neck, whilst he applies a hand-chisel, a punch, or a name-mark in the left hand, and strikes with the hammer held in the right. The method is, however, ample for a variety of small works, such as cutlery, tools, nails, and small ironmongery, which are wrought almost exclusively by the hand hammer.

Attempts to work small tilt-hammers with the foot have beeu found generally ineffective, as the attention of the individual is too much subdivided in managing the whole, neither is his strength sufficient for a continued exertion at such work; but the " *Oliver"* described in note S of the Appendix to Vol. II., page 962, is one of the best tools of this class.

For single hand-forging, the fire becomes still further reduced in size, and proportionally elevated above the ground; and this being the scale of work most commonly followed by the amateur, a portable forge of suitable dimensions, and made entirely of iron, is represented in fig. 90; the bellows are placed beneath the hearth and worked by a treadle.

This forge is also occasionally fitted with a furnace for melting small quantities of metal, and with various apparatus for In fitting the hazel rods to the top tools, the rods are alternately wetted in the middle of their length, and warmed over the fire to soften them; that portion is then twisted like a rope, and the rod is wound once round the head of the tool, and retained by an iron ferrule or coupler; a rigid iron handle would jar the hand.

When these tools are used for large works, a square plate of sheet-iron, with a hole punched in the middle of it, is put on the rod towards the tool, to shield the hand of the workman from the heat; and it not unfrequently happens with such large works that the rod catches fire, and the tool is then dipped at short intervals in the sl.ike trough to extinguish it. other applications of heat, such as soldering, either with a small charcoal fire, or a lamp and blow-pipe, which are likewise urged with the bellows. These applications, and also that of hardening and tempering tools, which will be severally returned to at their respective places, are much facilitated by the bellows being worked with the foot, as it leaves both hands at liberty for the management either of the work or fire, with the so-called fireirons, which include a poker, a slice or shovel, and a rake, in addition to the supply of tongs of some of the former shown.

The forge represented is sufficiently powerful for a moderate share of those works which require the use of the sledge hammer; hut when the latter tool is used, the anvil should not fall short It has been considered-unnecessary to represent any of the larger machinery ior forging, many of which have been repeatedly engraved for various popular works.
of one hundred pounds in weight; and the heavier it is, the less it will rebound under the hammer. SECT. II.— MANAGEMENT OF THE FIRE, THE DEGREES OF HEAT.
The ordinary fuel for the smith's forge is coal, and the kinds to be preferred, are such as are dense and free from metallic matters, as these are generally accompanied with sulphur, which is highly detrimental. In London, the Tanfield Moor coals, from the county of Durham, are preferred, and the lumps are broken quite small before use.

The fire is sometimes made open, at other times hollow, or like a tunnel; and the larger the fire is required to be, so much the more distant is it situated from the *tuyere* iron. Before lighting the fire, the useful cinders are first turned back on the hearth, and the exhausted dust or *slack,* is cleared away from the iron back and thrown into the ash-pit; a fair-sized heap of shavings is then lighted, and allowed to burn until the flame is nearly extinguished, when the embers are covered over with the cinders, and the bellows are urged; a dense white smoke first rises, and in two or three minutes the flame bursts forth, unless the fire be choked, when the poker is carefully passed into the mouth of the tuyere. The work is now laid on the fire, and covered over with green or fresh coals, which are beaten around the tuyere and the work, the blast being continued all the while: the whole mass will soon be in a state of ignition. A heap of fresh coals is always kept at the outside wall of the fire, and they are gradually advanced at intervals into the center of the flame, to make up for those consumed.

In making a large hollow fire, after a good-sized fire has been lighted in the ordinary way, the ignited fuel is brought forward on the hearth to expose the tuyere iron, into the central aperture of which the poker is introduced. A mass of small wetted coal is beaten hard round the poker to constitute the *stock,* the magnitude of which will depend on the distance at which the fire is required to stand off", and a second stock is also made opposite the first, the two resembling two hills with the lighted Copper is usually forged in a coke fire, silver and gold in those made of charcoal, but the hearths do not materially differ from those used for iron.

Williams's compressed peat charcoal has been strongly recommended on account of its freedom from sulphur, one of the greatest enemies in nearly all metollurgic operations.
fuel lying between them: the durability of the fire will depend on the stocks being hard rammed, which for large works is often done with the sledge-hammer. The work is now laid in the hollow just opposite the blast-pipe, and covered on its two sides and top with thin pieces of wood, and a heap of wetted coals is carefully banked up around the same, and beaten down with the slice or shovel: when carefully done, the heap is made to assume the smooth form of an embankment of earthwork. The bellows are blown gently all the time, and the work is not withdrawn until the wood is consumed, and the flame peeps through at each end of the aperture, so as to cake the coals well together into a hard mass; after which the work may be removed or shifted about without any risk of breaking down the fire.

Sometimes when a fire is required only for hardening, the centering of the arch is made entirely of wood, either in one or several pieces: and in this manner it may be built of any required form, as angular for knees, circular for hoops, and so on (although such works are usually done iu open fires, which resemble the above in all respects, except the covering-in or roof): small-coal is thrown at intervals into the hollow fire to replace that which is burned, and by careful management, one of these combustible edifices will last half a day, or even the entire day, without renewal. Occasionally, the stock around the tuyere iron will serve with a little repair for a second day, if when the fire is turned back at night, that part is allowed to remain, and the fire is extinguished with water.

When a small hollow fire is required, the same general methods are less carefully followed, and an iron tube, introduced amidst the coals, makes a very convenient muffle or oven for some purposes.

In forging, the iron or steel is in almost every case heated to a greater or less degree, to make it softer and more malleable by lessening its cohesion; the softening goes on increasing with the accession of temperature, until it arrives at a point beyond that which can be usefully employed, or at which the material, whether iron or steel, falls in pieces under the blows of the In localities where wood is scarce, small iron rods are placed around the principal mass, often designated the *heat*; the small rods are first withdrawn when the fire has burned up, to allow room for the removal of the work.
hammer, but which degree is very different with various materials, and even with varieties bearing the same name.

Pure iron will bear an almost unlimited degree of heat, the hot-short iron bears mucb less, and is in fact very brittle when heated; other kinds are intermediate; of steel, the shear-steel will generally bear the highest temperature, the blistered-steel the next, and the cast-steel the least of all; but all these kinds, especially cast-steel, differ very much according to the processes of manufacture, as some cast-steel may be readily welded, but it is then somewhat less certain to harden perfectly.

Without attempting any refined division, I may add, the smith commonly speaks of five degrees of temperature; namely—.

The black-red heat, just visible by daylight:

The low-red heat:

The bright-red heat, when the black scales may be seen: The wbite-heat, when the scales are scarcely visible: The welding-heat, when the iron begins to burn with vivid sparks.

Steel requires on the whole very much more precaution as to the degree of heat than iron; the temperature of cast-steel should not generally exceed a bright-red heat, that of blistered and shear-steel that of a moderate white-heat. Although steel cannot in consequence be so far softened in the fire as iron, and is therefore always more dense and harder to forge, still from its superior cohesion it bears a much greater amount of hard work under the hammer, when it is not over-heated or burned; but the smallest available temperature should be always employed with this material, as in fact with all others.

It has been recommended to try by experiment the lowest degree of heat at which every sample of steel will harden, and in forging, always to keep a trifle below that point. This proposal however is rarely tried, and still less followed, as the usual attempt is to lessen the labour of forging, by softening the steel so far as it is safely practicable.

Iron is more commonly worked at the bright-red and the white-heats, the welding-heat being reserved for those cases in which welding is required; or others in which, from the great extension or working of the iron, there is risk of separating its fibres or laminae, so as to cause the work to become unsound or hollow, from the disrupture of its substance; whereas these same processes being carried on at the welding temperature, the work would be kept sound, as every blow would effect the operation of welding rather than that of separation. The cracks and defects in iron are generally very plainly shown by a difference in colour at the parts when they are heated to a dull-red; this method of trial is often had recourse to in examining the soundness both of new and old forgings.

When a piece of forged work is required to be particularly sound, it is a common practice to subject every part of the material in succession to a welding heat, and to work it well under the hammer, as a repetition of the process of manufacture to ensure the perfection of the iron: this is technically called, *taking a heat over it,* in fact, a heat is generally understood to imply the welding heat. For a two-inch shaft of the soundest quality, two and a half inch iron would be selected, to allow for the reduction in the fire and the lathe; some also twist the iron before the hammering to prevent it from becoming *"spilly."*

The use of sand sprinkled upon the iron is to preserve it from absolute contact with the air, which would cause it to waste away from the oxidation of its surface, and fall off in scales around the anvil. If the sand is thrown on when the metal is only at the full red heat, it falls off without adhering; but when the whiteheat is approached, the sand begins to adhere to the iron, it next melts on its surface, over which it then runs like fluid glass, and defends it from the air; when this point has been rather exceeded, so that the metal nevertheless begins to burn with vivid sparks and a hissing noise like fireworks, the welding temperature is arrived at, and which should not be exceeded. The sparks are however considered a sign of a dirty fire or bad iron, as the purer the iron the less it is subject to waste or oxidation, in tbe course of work.

In welding two pieces of iron together, care must be taken that *both* arrive at the welding heat at the same moment; it may be necessary to keep one of the pieces a little on one side of the most intense part of the fire, (which is just opposite the blast,) should the one be in advance of the other. In all cases a certain amount of *time* is essential, otherwise if the fire be unnecessarily urged, the outer case of the iron may be at the point of ignition before the center has exceeded the red-heat. In welding iron to steel, the latter must be heated in a considerably less degree 208 GENERAL

PRINCIPLES OP FORGING.
than the iron, the welding heat of steel being lower from its greater fusibility, but the process of welding will be separately considered under a few of its most general applications, when the ordinary practice of forging has been discussed, and to which we will now proceed. SECT. III.—ORDINARY PRACTICE OP PORGING.

The general practice of forging works from the bar of iron or steel, are for the most part included in the three following modes; the first two occur in almost every case, and frequently all three together, namely,

By *drawing down,* or reduction: *By jumping* or *up-setting,* otherwise thickening and shortening: By *building-up,* or welding.

When it is desired to reduce the general thickness of the object, both in length and width, then the flat face of the hammer is made to fall level upon the work; but where the length or breadth alone is to be extended, the pane or narrow edge of the hammer is first used, and its blows are directed at right angles to the direction in which the iron is to be spread. To meet the variety of cases which occur,'the smith has hammers in which the panes are made in different ways, either at right angles to the handle, parallel with the same, or oblique.

In order to obtain the same results with more precision and effect, tools of the same characters, but which are struck with the sledge-hammer are also commonly used: those with flat faces are made like hammers, and usually with similar handles, except that for the convenience of reversing them they are not wedged in; these are called *set-hammers;* others which have very broad faces, are called *flatters,* and the top tools with narrow round edges like the pane of the hammer, are called *topfullers;* they all have the ordiuaryhazel rods.

When the sides of the object are required to be parallel, and it is to be reduced both in width and thickness, the flat face of the hammer is made to fall parallel with the anvil, as represented in fig. 91, or oblique for producing taper pieces as in fig. 92, and action and reaction being equal, the lower face of the work receives the same absolute blow from the anvil as that applied above by the hammer itself; it is not requisite BEMARKS O.V THE USE OF THE HAMMER. 209 therefore to present every one of the four sides to the hammer, hut any two, at right angles to each other. In twisting the work round the quarter circle, some practice is called for, in order to retain the rectangular section, and not to allow it to degenerate into the lozenge or rhomboidal form, which error it is difficult to retrace.

This indeed may be considered the first stumbling-block in forging, and one for which it is difficult to provide written rules. Of course in converting a round bar into a square with the hammer, the accuracy will depend almost entirely upon the change of exactly ninety degrees being given to the work, and this the experienced smith will accomplish with that same degree of feeling, or intuition, which teaches the exact distances required upon the fingerboard of a violin, which is defined by habit alone.

In the original manufacture of the iron, the carefully turned grooves, *a, b, c,* of the rollers, page 187, produce the square figure with great truth and facility; and under the tilt-hammer the two opposite sides are sure to be parallel, from the respective parallelism of the faces of the hammer and anvil; and the tilters, from constant practice, apply the work with great truth in its second position. So that under ordinary circumstances the prepared materials are true and square, and the smith has principally to avoid losing that accuracy.

First he must acquire the habit of *feeling* when the bar lies perfectly flat upon the anvil, by holding it slenderly, leaving it almost to rotate in his grasp, or in fact to place *itself.* Next, he must cause the hammer to fall flat upon the work; with which view he will neither grasp its handle close against the head of the hammer, nor at the extreme end of the handle, but at that intermediate point where he finds it comfortably to rebound from the anvil, with the least effort of, or jar to his wrist. And the height of the wrist must also be such as not to This is only true for works of moderate dimensions: iu large masses, such as anchors, the soft doughy state of the metal acts as a cushion, and greatly lessens the recoil of the anvil, and on this account such works are presented to the hammer on all four sides. It is also very injudicious in such cases to continue the exterior finish, or *battcriny-off,* too long, as this extends the outer case of the metal more than the inner part, and sometimes separates the two. When imperfect forginga are broken in the act of being proved, the inner bars are sometimes found not to be even welded together, and the outxide part is a detached sheath, almost like the rind or bark of a tree. See also Note D, page 460 of this volume. Vol L r *no* REMARKS ON THE USE OF THE HAMMER.
allow either the front or hack edge of the hammer face to strike the work first, which would indent it, but it must fall fair and parallel, and without bruising the work.

Figs. 91. 92. 93.
of 15

It would be desirable practice to hammer a bar of cold iron, or still better one of steel, as there would be more leisure for observation; the indentations of the hammer could be easily noticed; and if the work, especially steel, were held too tightly, or without resting fairly on the anvil, it would indicate the error by additional noise and by jarring the wrist: whereas, when hot, the false blows or positions would cause the work to get out of shape, without such monitorial indications.

As to the best form of the hammer, there is much of habit and something of fancy. The ordinary hand-hammer is represented in figs. 91 and 92, but the Sheffield cutlers, and most tool makers, prefer the hammer without a pane, and with the handle quite at the top, the two forming almost a right angle, or from that to about eighty degrees; and sometimes the head is bent like a portion of a circle. Similar but much heavier hand-hammers, occasionally of the weight of twelve or fourteen pounds, are used by the spade-makers for planishing; but the work being thin and cold, the hammer rises almost exclusively by the reaction, and requires little more than guidance. Again, the farriers prefer for sonic parts of their work, a hammer the head of which is almost a sphere; it has two flat faces, one rounded face for the inside of the shoe, and one very stunted pane at right angles to the handle, used for drawing down the clip in front of the horse-shoe; in fact, nearly a small volume might be written upon all the varieties of hammers.

To return to the forging: the flat face of the hammer should not only fall flat, but also centrally upon the work; that is, the center of the hammer, in which point the principal force of the blow is concentrated, should fall on the center of the bar, otherDRAWING DOM, OR REDUCING. 211 wise that edge of the work to which the hammer might lean would be the more reduced, and consequently the parallelism of the work would be lost. It would also be bent in respect to length, as the thinned edge would become more elongated, and thence convex; and when the blows were irregularly scattered, the work would become twisted or put in winding, which would be a still worse error.

I will suppose it required to *draw down* (the technical term for reduction) six inches of the end of a square or rectangular bar of iron or steel; the smith will place the bar across the anvil with perhaps four inches overhanging, and not resting quite flat, but tilted up about a quarter or half-an-inch at the near side of the anvil, as in fig. 92, but less in degree, and the hammer will be made to fall as there shown, except that it will be at a very small angle with the anvil.

Having given one blow, he will, as the only change, twist the work a quarter turn, and strike it again; then he will draw the bar half an inch or an inch towards him, and give it two more similar blows, and so on until he arrives at the extreme end, when he will recommence; but this will be done almost in the time of reading these words. The descent of the hammer, the *drawing* the work towards himself, (whence perhaps the term,) and the quarter turn backwards and forwards, all go on simultaneously and with some expedition. At other times the work is drawn down over the beak-iron, in which casethe curvature of this part of the anvil, makes it less material at what angle the work is held or the blows given, provided the two positions be alike.

In smoothing off the work, the position of fig. 91 is assumed; the work is laid flat upon the anvil, and the hammer is made to fall as nearly as possible horizontally; a series of blows are given all along the work between every quarter turn, the hammer being directed upon one spot, and the work drawn gradually beneath it.

The circumstances are exactly the same as regards the sledgehammer, which is used *up-hand* for light work; the right hand being slid towards the head in the act of lifting the hammer from off the work, and slipped down again as the tool descends; and the conditions are scarcely altered when the smith swings the hammer about in a circle, the signal for which is "*about sledge;*" whereas when, in either case, the blows of the sledge212 SETTING DOWN, OU MAKING SHOULDERS.
hammer are to be discontinued, the fireman taps the anvil with his hand hammer, which is, I believe, an universal language.

In drawing down the tang or taperpoint of a tool, the extreme end of the iron or steel is placed a little beyond the edge of the anvil, as in fig. 92, by which means the risk of indenting the anvil is entirely removed, and the small irregular piece in excess beyond the taper is not cut off until the tang is completed. Tig. 93 shows the position of the chisel in cutting off the finished object from the bar of which it formed a part; that is, the work is placed betwixt the edge of the anvil, and that of the chisel immediately above the same; the two resemble in effect a pair of shears. Sometimes the edge of the anvil alone is used for small objeots, first to indent, and theu to break off the work, but this is likely to injure the anvil, and is a bad practice.

When it is required to make a *set-off,* it is done by placing the intended shoulder at the edge of the anvil: the blows of the hammer will be effective only where opposed to the anvil, but the remainder of the bar will retain its full size and sink down as represented in fig. 91. Should it be necessary to make a shoulder on both sides, a flat-ended set hammer, struck by the sledge, is used for *setting* down the upper shoulder, as in fig. 95, as the direct blows of the hammer could not be given with so much precision. In each of these cases some precaution must be observed, as otherwise the tools, although so much more blunt than the chisel, fig. 93, will resemble it in effect, and cripple or weaken the work in the corner; on this account the smith's tools are rarely quite sharp at the angles: this mischief is almost removed when the round fullers, fig. 96, are used for reducing the principal bulk, and the sharper tools are only employed for trimming the angles with moderate blows.

When the iron is to be set down, and also spread laterally, as in fig. 97, it is first nicked with a round fuller as upon the MAKING BOLTS; FIRST AND SKCOND MODE.
213 dotted line at *a,* and the piece at the end is spread by the same tool, upon the short lines of the object, or parallel with the length of the bar: the first notch greatly assists in keeping a good shoulder at the bottom of the part set down, and the lines are supposed to represent the rough indentations of the round fuller before the work is trimmed up.
There is often considerable choice of method in forging, and the skilful workman selects that method of proceeding which will produce the result with the least portion of manual labour. Thus an ordinary screw-bolt, that I will suppose to measure five-eighths of an inch in diameter in the stem, and one inch square in the head, may be made in either of the three following ways adverted to in the outset:

First, by drawing-down:—A bar of iron is selected one inch square, or of the size of the *head* of the bolt, and a short portion of the same is set down, according to fig. 9G, by a pair of fullers that are convex in profile as shown, and also slightly concave upon the line at right angles to the paper; this prepares the shoulder or joining of the two dimensions; the bolt is made cylindrical, and of proper diameter between the rounding tools, fig. 98; and lastly, it is cut off with the chisel, as in fig. 93, p. 210, so much of the original square bar as suffices for the thickness of the head being allowed to remain.

Secondly, by jumping:—A piece of bolt-iron of five-eighths of an inch diameter, or of the size of the *stem* of the bolt, is cut off somewhat longer than the intended length; "*a short heat*" is taken upon it, that is, the extreme end alone is made whitehot, then placed perpendicularly upon the anvil, and the cold end is struck with the hammer as in driving in a nail; this thickens the metal or upsets it, and makes a thick conical button. The head is completed, by driving the bolt into a heading tool with a circular hole of five-eighths diameter; the thickened part of the head prevents the piece from passing through, and the lump is flattened out by the hammer into an irregular button or disk, 214 BOLTS, tTC. J THIRD MODE.

which is afterwards beaten square to complete the bolt. Figs. 99, 100, and 101, explain these processes; the latter is a single tool, hut the heading tool, fig. 102, with several holes, is also used.

In upsetting the end of the work, if more convenient, it may be held horizontally across the anvil, and struck on the heated extremity with the hand hammer; or it can be jumped forcibly upon the anvil, when its own weight will supply the required momentum. If too considerable a portion of the work is heated, it will either bend, or it will swell generally; and therefore to limit the enlargement to the required spot, should the heat be too long, the neighbouring part is partially cooled by immersing it in the water-trough, as near to the heat as admissible.

Thirdly, the same bolt may be made by building up or welding:—An eye is first made at the end of a small rod of square or flat iron; by bending it round the beak iron, as in fig. 108, it is placed around the rod of five-eighths round iron, and the curled end is cut off with the chisel, as in fig. 104, enough iron being left in the ring, which is afterwards *welded* to the five-eighths inch rod to form the head of the bolt, by a few quick light blows given at the proper heat; the bolt is then completed by any of the tools already described, that may be preferred. A swage at the angle of sixty degrees, fig. 105, will be found very convenient in forming hexagonal heads, as the horizontal blow of the hammer completes the equilateral triangle, and two positions operate on every side of the hexagon; fig. 105 is essential likewise in forging triangular files and rods.

Of these three modes of making a bolt, and which will apply LONGER BUT SIMILAR WORKS; CORKINGS, HOLES. 215 to a multitude of objects somewhat analogous in form, the first is the most general for small and short bolts; the second for small but longer kinds; and the third is perhaps the most common for large bolts, although the least secure; it is used for bolts for ordinary building purposes, but is less generally employed for the parts of mechanism.

For works of the same character, in which a considerable length of two different sections or magnitudes of iron are required, the method by drawing down from the large size would be too expensive; the method by upsetting would be impracticable; and therefore a more judicious use is made of the iron store, and the object is made in two parts, of bars of the exact sections respectively. The larger bar is reduced to the size of the smaller, generally upon the beak iron with top fullers, and with a gradual transition or taper extending some few inches, as represented in fig. 106; the two pieces are *scarfed* or prepared for welding, but which part of the subject is for the present deferred, in order that the different examples of welding may be given together.

The figure 106 is also intended to explain two other proceedings very commonly required in forging. Bars are bent down at right angles as for the short end or corking of the piece, fig. 106, by laying the work on the anvil, and holding it down with the sledge hammer, as in fig. 107; the end is then bent with the hand hammer, and trimmed square over the edge of the anvil; or when more precision is wanted, the work is screwed fast in the tail-vice, which is one of the tools of every smith's shop, and it is bent over the jaws of the vice. When the external angle, as well as the internal, is required to be sharp and square, the work is reduced with the fuller from a larger bar to the form of figure 108, to compensate for the great extension in length that occurs at the outer part, or *heel* of the bend, of which the inner angle forms as it were the center.

The holes in fig. 106, for the cross bolts, are made with a rodpunch, which is driven a little more than half way through from the one side whilst the work lies upon the anvil, so that when turned over, the cooling effect of the punch may serve to show the place where the tool must be again applied for the completion of the hole; the little bit or *burr* is then driven out, either through the square hole in the anvil that is intended for the 216 SQUARE AND HEXAGON NUTS, ETC.; PUNCHING.

bottom tools, or else upon the bolster, fig. 109, a tool faced with steel, and having an aperture of the same form and dimensions as the face of the punch.

Figs. 107. 106.

109, C=::= ==*i* n,

Fig. 112 shows the ordinary mode of making the square nuts for bolts. A flat bar is first nicked on the sides with the chisel, then punched, and the rough nuts if small, are separated and strung upon the end of the poker (a slight round rod bent up at the end), for the convenience of managing them in the fire, from which they are removed one at a time when hot, and finished on tlic triblet, fig. 113, which serves both as a handle, and also as the means of perfecting the holes.

For making hexagon nuts, the flat bar is nicked on both edges with a narrow round fuller; this gives a nearer approach to the hexagon: the nuts are then flattened on the face, punched, and dressed on the. triblet within the angular swage, fig. 105. page 214, before adverted to. Thick circular collars are made precisely in the same way, with the exception that they are finished externally with the hammer, or between top and bottom rounding tools of corresponding diameter.

It is usual in punching holes through thick pieces, to throw a little coal-dust into the hole when it is partly made, to prevent the punch sticking in so fast as it otherwise would: the punch generally gets red-hot in the process, and requires to be immediately cooled on removal from the hole.

In making a socket, or a very deep hole in the one end of a bar, some difficulty is experienced in getting the hole in the axis of the bar, and in avoiding to burst open the iron; such holes HOLES; MORTICES; PIECE WITH THREE TAILS.
217 are produced differently, by sinking the hole as a groove in the center of a flat bar by means of a fuller; the piece is cut nearly through from the opposite side, folded together lengthways, and welded. The hole thus formed will only require to be perfected by the introduction of an appropriate punch, and to be worked on the outside, with those tools required for dressing off its exterior surface, whilst the punch remains in the hole to prevent its sides from being squeezed in: this method is very good.

For punching square holes, square punches and bolsters are used, and fig. 110, the split bolster, is employed for cutting out long rectangular holes or mortises, which are often done at two or more cuts with an oblong punch.

Mortises, when of still greater length, are usually made by punching a hole of their full width at each end, and cutting out a strip of metal between them, by two long incisions made with the rod-chisel; at other times one cut only is made, and the mortise is *opened out;* this retains all the iron, but makes the ends narrower than the middle. In finishing a mortise, a parallel plate or drift is inserted in the slit, the drift is laid across the chaps of the vice, whilst the bar of iron lies partly between its jaws; in order that the blows of the hammer may be effective, on the upper and under surfaces of the one rib at the same time. The drift serves as a temporary anvil, the other rib is completed in the same manner, and the work is finally closed to its true width upon the anvil, the drift still lving in the mortise.

When a thick lump is wanted at the end of a bar, it is often made by cutting the iron nearly through and doubling it backwards and forwards, as in fig. Ill; the whole is then welded into a solid mass as the preparatory step.

A piece with three tails, such as fig. 114, is made from a large square bar; an elliptical hole is first punched through the bar, and the remainder is split with a chisel, as in fig. 115, the work at the time being laid upon a soft-iron cutting plate in order to shield the chisel from being driven against the hardened 218 SLING CONCLUDED; HOOK WEENCH Oil SET.
steel face'of the anvil; the end is afterwards opened into a fork, and moulded into shape over the beak-iron, as indicated by the dotted lines.

The concave lines about the object are principally worked with the fuller, or half-round set-hammer; and in making all the holes, narrow oval punches are used as described at the commencement, and the slits are enlarged into circular holes by conical mandrels; these bulge the metal out, and the holes are more judiciously formed in this manner than if the metal were wasted by cutting out great circular holes, which would sever a large quantity of the fibres and reduce the strength.

The mandrels are left in the holes whilst the parts around them are finished, which tends to the perfection of both parts; as the holes more closely copy the mandrels, and the marginal parts are better finished when the apertures are for the time rendered solid. Supposing a hole to be wanted in the cylindrical part of the work that should be finished between the rounding tools; the mandrel could not be allowed to remain in; and therefore a short piece of iron is forged or *drawn down* to the size of the hole, cut off in length to the diameter of the part, and inserted in the hole to preserve it from being compressed, yet without interference with the completion of the cylindrical portion; which accomplished, this little bit, called by the un-mechanical name of a *devil,* is driven out, unless by a very careless use of the welding temperature it should have been permanently fastened in. Towards the conclusion a long mandrel is passed through the two holes in the fork of fig. 114, to show whether their common axis is at right angles to the main rod, otherwise the one or other arm is drawn out, or upset, according as the work may err in respect to deficiency or excess of length. Such a piece as fig. 114, if of large dimensions, would be made in two separate parts, and welded through the central line or axis.

Should it happen the two arms are not quite parallel, that is, when viewed edgeways should they stand oblique to each other, or to the central bar, an error that could scarcely be corrected by the hammer alone; the work would be fixed in the vice with the two tails upwards, and the one or other of these would be twisted to its true position by a *hook wrench* or *set,* made like the three sides of a square, but the one very long to serve as a lever; it is applied exactly in the manner of a key, spanner, or screw wrench, in turning round a bolt or screw. The hook,wrench is constantly used for taking the twist out of work, or the error of winding, as the hammer can only be successfully employed for correcting the curvatures of length.

Some bent objects, such as cranks and straps, are made from bar-iron, bent over specific moulds, which are sometimes made in pairs like dies, and pressed together by screw contrivances. When the moulds are single, the work is often retained in contact with the same, at some appropriate part, by means of straps and wedges; whilst the work is bent to the form of the mould by top tools of suitable kinds.

Objects of more nearly rectilinear form are cut out of large plates and bars of iron with chisels; for example, the cranks of locomotive engines are fagoted up of several bars or uses laid together, and pared to the shape: they are sometimes forged in two separate parts, and welded between the cranks, at other times they are forged out of one parallel mass, and afterwards twisted with a hook-wrench, in the neck between the cranks, to place the latter at right angles. The notches are sometimes cut out on the anvil whilst the work is red-hot; or otherwise by machinery when in the cold state.

A very different method of making rectangular cranks and similar works is also recommended, by bending one or more straight bars of iron to the form; the angles, which are at first rounded, are perfected by welding on outer caps. In this case the fibre runs round the figure, whereas when the gap is cut out, a large proportion of the fibres are cut into short lengths, and therefore a greater bulk must be allowed for equal strength: this method is however seldom used.

All kinds of levers, arms, brackets and frames, are made after these several methods, partly by bending and welding, and partly by cutting and punching out; and few branches of industry present a greater variety in the choice of methods, and which call the judgment of the smith continually into requisition.

SECT. IV.—GENERAL EXAMPLES OP WELDING.

The former illustrations of forging have been principally descriptive of such works as could be made from a single bar of iron, on purpose that the examples to be advanced in welding or joining together two pieces of iron by heat, technically called 220 EXAMPLES OP WELDING.— SCAUP JOINT.

"shutting together," or *"shutting up,"* might be collected at one place.

There are several ways of accomplishing this operation, and which bear some little analogy to the joints employed in carpentry; more particularly that called scarfing, used in the construction of long beams and girders by joining two shorter pieces together endways, with sloping joints, which in carpentry are interlaced or mortised together in various ways, and then secured by iron straps or bolts. In smith's work likewise, the joinings are called *scarfs,* but from the adhesive nature of the iron when at a suitable temperature, the accessories called for in carpentry, such as glue, bolts, straps and pins, are no longer wanted. It would be desirable that the remarks on the fire and degrees of heat, page 206, should be again perused before the reader attempts the process of welding.

The example, fig. 106, page 216, was left unfinished, but wc will proceed to show the mode of joining the two cylindrical ends of the work. The scarfs required for the " *shut,"* are made by first upsetting or thickening the iron by blows upon its extremity, to prepare it for the loss it will sustain from scaling off, both in the fire and upon the anvil, and also in the subsequent working upon the joint. It is next rudely tapered off to the form of a flight of steps, as shown in figs. 110 and 117, and the sides are slightly bevelled or pointed, as in fig. 117, the proportions being somewhat exceeded to render the forms more apparent.

The two extremities are next heated to the point of ignition; and when this is approached, a little sand is strewed upon each part, which fuses and spreads something like a varnish, and partially defends them from the air; the heat is proper when, notwithstanding the sand, the iron begins to burn away with vivid sparks. The two men then take each one piece, strike them forcibly across the anvil to remove any loose cinders, place them in their true positions, exactly as in fig. 116, and two or three blows of the small hammer of the principal or fireman Btick them together; the assistant then quickly joins in with the sledge hammer, and the smoothing oft' and completion of the work are soon accomplished.

It is of course necessary to perform the work with rapidity, and literally " to strike whilst the iron is hot;" the smith afterwards jumps the end of the rod upon the anvil, or strikes it endways with the hammer; this proves the soundness of the joint, but it is mostly done to enlarge the part, should it during the process have become accidentally reduced below the general size. The sand appears to be quite essential to the process of welding, as although the heat might be arrived at without its agency, the surfaces of the metal would become foul and covered with oxide when unprotected from the air; at all events common experience shows that it is always required. The scarf joint, shown in figs. 116 and 117, is commonly used for all straight bars, whether flat, square or round, when of medium size.

In very heavy works the welding is principally accomplished within the fire: the two parts are previously prepared either to the form of the *tongue* or *split* joint, fig. 118, or to that of the *butt* joint, fig. 119, and placed in their relative positions in a large hollow fire. When the two parts are at the proper heat, they are jumped together endways, which is greatly facilitated by their suspension from the crane, and they are afterwards struck on the ends with sledge hammers, a heavy mass being in some cases held against the opposite extremity to sustain the blows; the heat is kept up, and the work is ultimately withdrawn from the fire, and finished upon the anvil.

The butt joint, fig. 119, is materially strengthened, when, as it is usually the case for the paddle shafts of steam-vessels and similar works, the joint whilst still large is notched in on three or four sides, and pieces called *stick-in* pieces, *dowels,* or *charlins,* one of which is represented by the dotted lines, are prepared at anotherfire, and laid in the notches; the whole, when raised to the welding heat, is well worked together and reduced to the intended size; this mingles all the parts in a very substantial manner. For the majority of works however, the scarf joint, fig. 116, is 222

T FOIIM AND CORNER JOINTS.

used, but the stick-in pieces are also occasionally employed, especially when any accidental deficiency of iron is to be feared.

When two bars are required to form a T joint, the transverse piece *is thinned down* as at *a,* in fig. 120; for an angle or corner the form of *b* may be adopted; but *c,* in which each part is cut off obliquely is to be preferred. The pieces *a, b, c,* are represented upside down, in order that the ridges set down on their lower surfaces may be seen. In most cases when two separate bars are to be joined, whatever the nature of the joint, the metal should be first upset, and then set down in ridges on the edge of the anvil, or with a set hammer, as the plain chamfered or sloping surfaces are apt to slide asunder when struck with the hammer, and prevent the union. When a T joint is made of square or thick iron, the one piece is upset, and moulded with the fuller much in the form of the letter; it is then welded against the flat side of the bar: such works are sometimes welded with dowel or tenon joints, but all the varieties of method cannot be noticed.

There are many works in which the opposite edges, or the ends of the same piece, require to be welded; in these the risk of the two parts sliding asunder scarcely exists, and the scarfs are made with a plain chamfer, or simply to overlap or fold together without any particular preparation.

Of the last kind, fig. Ill, page 216, may be taken as an example, in which the parts have no disposition to separate; in this and similar cases the smith often leaves the parts slightly open, in order that the very last process before welding may be the striking the whole edgeways upon the anvil, to drive out any loose scales, cinders or sand, situated between the joints; which if allowed to remain would be cither inclosed amidst the sound parts of the work, or would partially prevent the union.

In works that have accidentally broken in the welded part, the fracture will be frequently seen to have arisen from some dirty matter having been allowed to remain between them, on which account, *shuts* or welded joints extending over a large surface are often less secure than those of smaller area, from the greater risk of their becoming foul. In fact, throwing a little small coal between the contiguous surfaces of work not intended to be united, is a common and sometimes a highly essential precaution to prevent them from becoming welded.

WELDING SOCKETS, HINGES, MUSKET BARRELS. 223

The conical sockets of socket chisels, garden spuds, and a variety of agricultural implements, are formed out of a bar of flat iron, which is spread out sideways or to an angle, with the pane of the hammer, and then bent within a semicircular bottom tool also, by the pane of the hammer, to the form of fig. 121; after which the sockets are still more curled up by blows on the edges, and "are" perfected upon a taper pointed mandrel, so that the two edges slightly overlap at the mouth of the socket, and meet pretty uniformly elsewhere, as in fig. 122; and lastly, about an inch or more at the end is welded. Sometimes the welding is continued throughout the length, but more commonly only a small portion of the extremity is thus joined, and the remainder of the edges are drawn together with the pane of the hammer.

In making wrought-iron hinges, two short slits are cut lengthways and nearly through the bar, towards its extremity; the iron is then folded round a mandrel, set down close in the corner, and the two ends are welded together. To complete the hinge, it only remains to cut away transversely, either the central piece or the two external pieces to form the knuckles, and the addition of the pin or pivot finishes the work.

Musket barrels, when made entirely by hand, were forged in the form of long strips about a yard long and four inches wide, but taper both in length and width, which were bent round a cylindrical mandrel until their edges slightly overlapped; they were then welded at three or four heats, by introducing the mandrel within them instantly on their removal from the fire at the proper heat, in order to prevent the sides of the tube from being pressed together by the blows of the hammer.

They have been subsequently, and are now almost universally welded by machinery at one heat, and whilst of the length of only one foot, as on removal from the fire the mandrel is quickly introduced, and the two are passed through a pair of grooved rollers: they are afterwards extended to the full length by similar means, but at a lower temperature, so that the iron is not so much injured as when thrice heated to the welding point.

The twisted barrels are made out of long ribands of iron *22i* TWISTED BARRELS; DAMASCUS BARRELS; TUBES. wound spirally around a mandrel, and welded on their edges by jumping them upon the ground or rather on an anvil embedded therein. The plain stub barrels are made in this manner from iron manufactured from a bundle of stubnails, welded together and drawn out into ribands to ensure the possession of a material most thoroughly and intimately worked. The Damascus barrels are made from a mixture of stub-nails and clippings of steel in given proportions, puddled together, made into a bloom, and subsequently passed through all the stages of the manufacture of iron already explained; to obtain an iron that shall be of unequal quality and hardness, and therefore display different colours and markings when oxidised or browned.

Other twisted barrels are made in the like manner, except that the bars to form the ribands are twisted whilst red-hot like ropes, some to the right, others to the left, and which are sometimes again laminated together for greater diversity; they are subsequently again drawn into the ribands and wound upon the mandrel, and frequently two or three differently prepared pieces are placed side by side to form the complex and ornamental figures for the barrels of fowling-pieces, described as "*stub-tuisl, wire-twist, Damascus-twist,*" &c., which processes are minutely explained and figured by Mr. Greener.

All these matters are also explained in Mr. Wilkinson's recent and interesting work,f which likewise treats of one method amongst others of the formation of the Damascus gun barrels; by arranging twenty-five thin bars of iron, and mild steel in alternate layers, welding the whole together, drawing it down small, twisting it like a rope, and again welding three such ropes, for the formation of the riband which is then spirally twisted to, form a barrel, that exhibits, when finished and acted upon by acids, a diversified laminated structure, resembling when properly managed an ostrich feather.

Let me now turn from these engines of destruction to a modification of one of them to a happier purpose. When the illumination by gas was first introduced in the large way, by Aaron Manby, Esq. , then of the Horsley Iron Works, the old musketbarrels, laid by in quiet retirement from the fatigues of the last war, were employed for the conveyance of gas; and by a curious Greener " Ou the Gun," 1835; and Greener " On Gunnery," 1341.

t "Wilkinson's Engines of War," 1841, pp. 89—98.

TUBES FOR GAS.—RUSSELL'S PATENT TUBE.

225 coincidence, various iron foundries desisted in a great measure from the manufacture of iron ordnance, and took up the peaceful employment of casting pipes for gas and water.

The hreech ends of the musket-barrels were broached and tapped, and the muzzles were screwed externally, to connect the two without detached sockets. From the rapid increase of gas illumination, the old gun-barrels soon became scarce, and new tubes with detached sockets, made by the old barrel-forgers, were first resorted to. This led to a series of valuable contrivances for the manufacture of the wrought-iron tubes, commencing with Russell's patent, in 1824, under which the tubes were first bent up by hand hammers and swages, to bring the edges near together; and they were welded between semicircular swages, fixed respectively in the anvil, and the face of a small tilt-hammer worked by machinery, by a series of blows along the tube, either with or without a mandrel. The tube was completed on being passed between rollers with half-round grooves, which forced it over a conical or egg-shaped piece at the end of a long bar, to perfect the interior surface.

Various steps of improvement have been since made; for instance, the skelps were bent at two squeezes, first to the semicylindrical, and then to the tubular form (preparatory to welding), between a swage-tool five feet long worked by machinery. The whole process was afterwards carried on by rollers, but abandoned on account of the unequal velocity at which the greatest and least diameters of the rollers travelled.

In the present method of manufacturing the patent welded tube, the end of the skelp is bent to the circular form, its entire length is raised to the welding heat in an appropriate furnace, and as it leaves the furnace almost at the point of fusion, it is dragged by the chain of a draw-bench, after the manner of wire, through a pair of tongs with two bell-mouthed jaws; these are opened at the moment of introducing the end of the skelp, which is welded without the agency of a mandrel.

By this ingenious arrangement, wrought-iron tubes may be made from the diameter of six inches internally and about oneeighth to three-eighths of an inch thick, to as small as one-quarter inch diameter and one-tenth bore; and so admirably is the joining effected in those of the best description, that they will withstand the greatest pressures of gas, steam, or water, to which *you* i. li they have been subjected, and they admit of being bent both in the heated and cold state almost with impunity. Sometimes the tubes are made one upon the other when greater thickness is required; but these stout pipes, and those larger than three inches are comparatively but little used. See Note T, Appendix, Vol. II., pp. 963—969.

Various articles with large apertures are n;;tde,not by punching or cutting out the holes, but by folding the metal around the beak iron, and finishing them upon a triblet of the appropriate figure; thus the complete smithy is generally furnished with a series of cones turned in the lathe, for making rings the ends of which are folded together and welded, such as fig. 123, page 228. The same rings when made of such cast-steel as does not admit of being welded, are first punched with a small hole, and gradually thinned out by blows around the margin, until they reach the diameter sought; but this, like numerous other works, requires considerable forethought to proportion the quantity of the material to its ultimate form and bulk, so that the work may not in the end become either too slight or too heavy.

Chains may be taken as another familiar example of welding; in these the iron is cut off with a plain chamfer, as from the annular form of the links their extremities cannot slide asunder when struck; every succeeding link is bent, introduced, and finally welded. In some of these welded chains the links are no more than half an inch long, and the iron wire one-eighth of an inch diameter; several inches of such chain are required to weigh one pound: these are made with great dexterity by a man and a boy at a small fire. The curbed chains are welded in the ordinary form and twisted afterwards, a few links being made red-hot at a time for the purpose.

The massive cable-chains are made much in the same manner, although partly by aid of machinery: the bar of iron, now one, A piece of tube of the smallest dimensions,and fourteen feet long, which has been bent cold almost into the form of the Gordian knot, may be seen at the Institution of Civil Engineers. The wrought-iron tubes of hydrostatic presses, which measure about half an inch internally, and one-fourth to three-eighths of an inch thick in the metal, are frequently subjected to a pressure equal to *four tout* on each square inch. Pipes prov. d to the same degree are also used in Mr. Perkins' patent apparatus for warming buildings, and in his patented steam-boiler. The safety of each of these is entirely secured by a fusible plug, which melts and allows the water to escape into the fire when its temperature exceeds any predetermined degree, namely, from about 300 to 600 F., generally the former.

one and a half, or even two inches diameter, is heated, and the scarf is made as a plain chamfer by a cutting machine; the link is then formed by inserting the end of the heated bar within a loop in the edge of an oval disk, which may be compared to a chuck fixed on the end of a lathe mandrel. The disk is put in gear with the steam-engine; it makes exactly one revolution, and throws itself out of motion; this bends the heated extremity of the iron into an oval figure, afterwards it is detached from the rod with a chamfered cut by the cutting machine, which at one stroke makes the second scarf of the detached link, and the first of that next to be curled up.

The link is now threaded to the extremity of the chain, closed together, and transferred to the fire, the loose end being carried by a traverse crane; when the link is at the proper heat, it is returned to the anvil, welded, and dressed off between top and bottom tools, after which the cast-iron transverse stay is inserted, and the link having been closed upon the stay, the routine is recommenced. The work commonly requires three men, and the scarf is placed at the side of the oval link, and flatway through the same. In similar chains made by hand it is perhaps more customary to weld the link at the *crown* or small end.

The succeeding illustration of the practice of forging will be that of the formation of a hatchet, figs. 124 and 125, which like many similar tools is made by doubling the iron around a mandrel, to form the *eye* of the tool: it will also permit the The tires of wrought-iron wheels for locomotive engines and carriages, are in general bent to the circle by somewhat analogous means to those employed in chain-making, as are likewise the skelps for the twisted barrels of guns; the latter only require a mandrel or spindle with a winch handle at the one extremity, and a loop for the nd of the skelp, which is wound in contact with the mandrel by means of a fixed bar placed near the same; such barrels are coiled up in three lengths, which are joined together after the spirals are welded.

Wheels for railways display many curious examples of smithing: thus some except the nave, are made entirely by welding; others are partly combined with rivets; in all the nave or boss is a mass of cast-iron usually poured around the ends of the spokes, with the exception of Bourne aud Bartley's patent wheel, in which the nave, spokes, and periphery, are made entirely of wrought-iron and welded together.

The common practice of welding the tiles of railway wheels is now as follows: the tires are cut off with ridges in the center, so ai in meeting to form two angular notches, into which two thin iron wedges are subsequently welded radially; the four parts thus united together in the form of a cross, make a very secure joint without the necessity for upsetting the iron, which would distort the form of the tire.

228 MAKING HATCHETS. description of some other general proceedings, and likewise the introduction of the steel for the cutting edge.

In making the hatchet, a piece of flat iron is selected of the width of A E, and twice the length of A D; it is thinned and extended sideways before it is folded together, to form the projections near B and F, by blows with the pane of the hammer or a round-edged fuller, on the lines A B to E F, but the metal must be preserved of the full thickness at the part A E, to form the poll of the hatchet, although a piece of steel is frequently welded on at that part as a previous step. The work is then bent round a mandrel, figs. 126 and 127, exactly of the section of the eye as seen in fig. 125, and the work is welded across the line B F; the mandrel is again introduced, and the eye is perfected.

A slip of shear-steel, equal in length to D H, is next iuserted between the two tails of the iron, as yet of their original size, up to the former weld, and all three are welded together between C, G, D, H: the combined iron and steel are now drawn out sideways, by blows of the pane of the hammer on and between C D and G H, to extend them together to I J. The tool is then flattened and smoothed with the face of the hammer, and the edges are pared with straight or circular chisels to the particular pattern, and trimmed with a round-faced hammer, or a top fuller.

In smoothing off the work, the smith pursues his common method of first removing with a file the hard black scales that appear like spots when the work is removed from the fire; he then dips the hammer in the slake trough, and lets fall upon the anvil a few drops of the water it picks up, the explosion of which when the red-hot metal is struck upon it, makes a smart report and detaches the scales that would be otherwise indented in the work. It should be observed that the mandrel, fig. 126, is purposely made very taper, and is introduced into the hole from both sides, so that the eye may be smaller in the middle: when therefore the handle of the tool is carefully fitted and wedged in, the handle is as it were dove-tailed, and the tool can neither fly off or slip down the handle; the same mode is also adopted for the heads of hammers.

In spades, and many similar implements, the steel is introduced between the two pieces of iron of which the tools are made; in others, as plane irons anil socket chisels, it is laid on the outside, and the two are afterwards extended in length or width to the required size. The ordinary chisel for the smith's shop is made by inserting the steel in a cleft, as in fig. 118, page 220, and so is also the *pane* of a hammer; but the flat *face* of the hammer is sometimes stuck on whilst it continues at the extremity of a flat bar of steel; it is then cut off, and the welding is afterwards completed. At other times the face of the hammer is prepared like a nail, with a small spike and a very large head, so as to be driven into the iron to retain its position, until finally secured by the operation of welding.

In putting a piece of steel into the end of an iron rod to serve for a center, the bar is heated, fixed horizontally in the vice, and punched lengthways with a sharp square punchy for the reception of the steel, which is drawn down like a taper tang or thick nail, and driven in; the whole is then returned to the fire, and when at the proper heat united by welding, the blows being first directed as for forming a very obtuse cone to prevent the piece of steel from dropping out.

For some few purposes the blistered steel is used for welding, either to itself or to iron; it is true the first working inder the hammer in a measure changes it to the condition of shear-steel, but less efficiently so than when the ordinary course of manufacture is pursued, as the hammering is found to improve steel in a remarkable and increasing degree.

For the majority of works in which it is necessary to weld steel to iron, or steel to steel, the shear, or double shear, is exceedingly suitable: it is used for welding upon various cutting tools, as the majority of the cast-steel will not endure the heat wit out crumbling under the hammer. Shear-steel is also used
230 CONCLUDING REMAUKS ON rOBGING.
for various kinds of springs, and for some cutting tools requiring much elasticity.

It is more usual to reserve the cast-steel for those works in which the process of welding is not required, although of late years mild cast-steel, or welding cast-steel, containing a smaller proportion of carhon has heen rather extensively used; but in general the harder the steel the less easily will it admit of welding, and not unfrequently it is altogether inadmissible.

The hard or *harsh* varieties of cast-steel, are somewhat more manageable when fused borax is used as a defence instead of sand, cither sprinkled on in powder or rubbed on in a lump: and caststeel otherwise intractable may be sometimes welded to iron by first heating the iron pretty smartly, then placing the cold steel beside it in the fire, and welding them the moment the steel has acquired its maximum temperature, by which time the irou will be fully up to the welding heat. When both are put into the fire cold alike, the steel is often spoiled before the iron is nearly hot enough, and therefore it is generally usual to heat the iron and steel separately, and only to place them in contact towards the conclusion of the period of getting up the heat. In forging works either of iron or steel, the *uniformity* of the hammering tends greatly to increase and equalize the strength of each material: and in steel, judicious and equal forging greatly lessens also the after-risk in hardening.

SECT. V.—CONCLUDING REMARKS ON FORGING AND THE APPLICATIONS OF HEADING TOOLS, SWAGE TOOLS, PUNCHES, ETC.

With the utmost care and unlimited space, it would have been quite impossible to have conveyed the instructions called for, in forging the thousand varieties of tools, and parts of mechanism When cast-steel has been spoiled by overheating, it may be partially recovered by four or five reheatings and quenchings in water, each carried to an extent a little less and less than the first excess; and lastly, the steel must have a good hammering at the ordinary red heat. Some go so far as to prefer for cutting toote the steel thus recovered, but this seems a most questionable policy, although the change wrought by this treatment is really remarkable; as the fragment broken off from the bar in the spoiled state, and another from the same bar after part restoration and hardening, will exhibit the extreme characters of coarse and fine.

The hammering I suspect to be the principal requisite, and in superior tools it should be continued until the work is nearly cold, to produce the maximum amount of condensation before hardening; but no hammering will restore the loss of tenacity consequent upon the over-heating, or even the two frequent heating of eteel, without excess.
HEADING TOOLS: SWAGE BLOCK. 231
the smith is continually called upon to produce; and all that could be reasonably attempted in this place, was to convey a few of the general features and practices of this most useful and interesting branch of industry. It is hoped, that such combinations of these methods may be readily arrived at as will serve for the majority of ordinary wants. The smith in all cases selects or prepares that particular form and magnitude of iron, and also adopts that order of proceeding, which experience points out as being the most exact, sound, and economical. In this he is assisted by a large assortment of various tools and moulds for such parts of the work as are often repeated, or that are of a character sufficiently general to warrant the outlay, and to some of which I will advert.

The heading tools, figs. 101 and 102, page 214, are made of all sizes and varieties of form; some with a square recess to produce a square beneath the head, to prevent the bolt from being turned round in the act of tightening its nut; others for countersunk and round-headed bolts, with and without square shoulders; many similar heading tools are used for all those parts of work which at all resemble bolts, in having any sudden enlargement from the stem or shaft. The holes in the swage block, fig. 128, are used after the manner of heading tools for large objects; the grooves and recesses around its margin, also serve in a variety of works as bottom swages beyond the size of those fitted to the anvil. At the opposite extreme of the heading tools, as to size, may be noticed those constantly employed in producing the smallest kinds of nails, brads and rivets, of various denominations, some of which heading tools divide in two parts like a pair of spring forceps to release the nails after they have been forged. These kinds are called *wrought nails* and brads, in The forge used by the nail-makers is built as a circular pedestal with the fire *in* the center and the chimney directly over it; the rock-staff of the bellows extends entirely around the forge, so that one of the four or five persons who work at the same fire is continually blowing it, whence the fire is always at the heat proper for welding, and which keeps the nails sound and good.
232 TOP AND BOTTOM SWAOE TOOLS.
contradistinction to similar nails cut out of sheet-iron by various processes of shearing and punching, which latter kinds are known as *ait brads* and nails, and will be adverted to in the second volume.
The top and bottom rounding tools, fig. 9S, p. 214, are made of all diameters for plain cylindrical works: and when they are used for objects the different parts of which are of various diameters, it requires much care to apply them equally on all parts of the work, that the several circles may be concentric and true one with the other, or possess one axis in common. To ensure this condition some

of these rounding tools arc made of various and specific forms, for the heads of screws, for collars, flanges or enlargements, which are of continual Occurrence in machinery; for the ornamental swells or flanges about the iron work of carriages, and other works. Such tools, like the pair represented in figs. 129 and 130, are called swage or collar tools; they save labour in a most important degree, and are thus made. A solid mould, core or striker, exactly a copy of the work to be produced, is made of steel by haud-forgiug, and then turned in the lathe to the required form, as shown in fig. 131.

The top tool is first moulded to the general form in an appropriate aperture in the swage block, fig. 12S; it is faced with steel like a hammer, and the core,'fig. 131, is indented into it; the blows of the sledge hammer not being given directly upon the core, but upon some hollow tool previously made; otherwise the core must be filed partly flat to present a plane surface to the hammer. The bottom tool, which is fitted to the anvil, is made in a similar manner, and sometimes the two are finished at the same time whilst hot, with the cold striker between them; their edges are carefully rounded with a file so as not to cut the work, and lastly they are hardened under a stream of water.

In preparing the work for the collar tools, when the projection is inconsiderable, the work is always drawn down rudely to the form between the top and bottom fullers, as in fig. 96, p. 212: but for greater economy, large works in iron are sometimes made by folding a ring around them as in fig. 104, p. 214. The metal for a large ring is occasionally moulded in a bottom tool, like fig. 132, and coiled up to the shape of fig. 133, after which it is closed upon the central rod between the swages, and then welded within them. The tools are slightly greased, to prevent the work from hanging to them, and from the same motive their surfaces are not made quite flat or perpendicular, but slightly conical, and all the augles are obliterated and rounded.

The spring swage tool, represented in fig. 134, is used for some small manufacturing purposes; it differs in no respect from the former, except in the steel spring which connects the two parts; it is employed for light single hand-forgings. Other workmen use swage tools, such as fig. 135, in which there is a square recess in the bottom tool to fit the margin of the top-tool so as to guide it exactly to its true position: this kind also may be used for single hand works, and is particularly suited to those which are of rectangular section, as the shoulders of table-knives; these do not admit of being twisted round, which movement furnishes the guide for the position of the top-tool in forging circular works.

The smith has likewise a variety of punches of all shapes and sizes, for making holes of corresponding forms; and also drifts or mandrels, used alone for finishing them, many of which, like the turned cones, are made from a small to a large size to serve for objects of various sizes. Two examples of the very dexterous use of punches, are in the hands of almost every person, namely, ordinary scissors and pliers.

The first are made from a small bar of flat steel; the end is flattened and punched with a small round hole, which is gradually opened upon a beak-iron, fig. 136, attached to the square hole of the anvil; the beak-iron has a shallow groqve (accidentally In practice the recess in the bottom tool would be deeper, and taper or larger above to guide the tool more easily to its place; but if so drawn the figure would have been less distinct.

omitted) for rounding the inside of the bows. The remaining parts of the scissors are moulded jointly by the hammer, and bottom swage tools; but the bows are mostly finished by the eye alone.

In the Lancashire pliers, the central half of the joint is first made; the aperture in the other part is then punched through sideways, and sufficiently bulged out to allow the middle joint to be passed through, after which the outsides are closed upon the center. This proceeding exhibits, in the smallest kinds especially, a surprising degree of dexterity and dispatch, only to be arrived at by very great practice; and which in this and numerous other instances of manufacture could be scarcely attained but for the enormous demand, which enables a great subdivision of labour to be successfully applied to their production.

The reader will derive pleaaure from the perusal of Mr. Cornelius Varley's paper "On working Iron and Steel," in vol. xlviii. part ii. page 250, of the "Transactions of the Society of Arts." The remarks on forging iron are perhaps rather speculative than practical: two points are principally insisted upon; the one, the necessity for introducing between welded surfaces some matter containing carbon, as the filings of steel or cast-iron, to avoid the oxidation of the surfaces in the fire, as if oxidized they would be prevented from becoming soundly welded. The other point is, the propriety of forging works, such as anchors in moulds, by which the condensation of the metal would be uniformly effected, without the power of its stretching unequally at different parts, by which Mr. Varley supposes some of the layers or fibres are strained almost to breaking, and therefore deprive the mass of Bo much of its apparent strength.

The remarks on steel also refer to the necessity of good primary forging and hammering to produce homogeneity; and also to many of the other points generally admitted by practical men as being conducive to the success of hardening.

The reader is also referred to the Appendix to this volume, Notes C. D. E. , pages 460 to 462; and likewiso to the Appendix of the second volume, Notes Q and R, pages 958—962, at which latter place are given some particulars of Nasmyth's Patent Direct-action Steam Hammer and also of his Patent Steam Pile-Driving Engine.

CHAPTER XII. -HARDENING AND TEMPERING.

SECT. I.—GENERAL VIEW OF THE SUBJECT.

When the malleable metals are hammered, or rolled, they generally increase in hardness, in elasticity, and in density or specific gravity; which effects are produced simply from the closer approximation of their particles, and in this respect steel may be perhaps considered to excel, as the process called hammer-hardening, which simply means hammering without heat, is frequently employed as the sole means of hardening some kinds of steel springs, and for which it answers remarkably well.

After a certain degree of compression, the malleable metals assume their closest and most condensed states; and it then becomes necessary to discontinue the compression or elongation, as it would cause the disunion or cracking of the sheet or wire, or else the metal must be softened by the process of annealing.

The metals, lead, tin, and zinc, are by some considered to be perceptibly softened by immersion in boiling water; but such of the metals as will bear it are generally heated to redness, the cohesion of the mass is for the time reduced, and the metal becomes as soft as at first, and the working and annealing may be thus alternately pursued, until the sheet metal, or the wire, reaches its limit of tenuity.

The generality of the metals and alloys suffer no very observable change, whether or not they are suddenly quenched in water from the red heat. Pure hammerediron, like the rest, appears after annealing, to be equally soft whether suddenly or slowly cooled; some of the impure kinds of malleable iron harden by immersion, but only to an extent that is rather hurtful than useful, and which may be considered as an accidental quality.

Steel however receives by sudden cooling that extreme degree of hardness combined with tenacity, which places it so incalculably 236 IMPORTANCE OF STEEL.
beyond every other material for the manufacture of cutting tools; especially as it likewise admits of a regular gradation from extreme hardness to its softest state, when subsequently re-heated or *tempered.* Steel therefore assumes a place in the economy of manufactures unapproachable by any other material; consequently we may safely say that without it, it would be impossible to produce nearly all our finished works in metal and other hard substances; for although some of the metallic alloys are remarkable for hardness, and were used for various implements of peaceful industry, and also those of war, before the invention of steel; yet in point of absolute and enduring hardness, and equally so in respect to elasticity and tenacity, they fall exceedingly short of hardened steel.
Hammer-hardening renders the steel more fibrous and less crystalline, and reduces it in bulk; on the other baud, fire hardening makes steel more crystalline, and frequently of greater bulk; but the elastic nature of hammer-hardened steel, will not take so wide nor so efficient a range as that which is fire-hardened.

If we attempt to seek the remarkable difference between pure iron and steel in their chemical analysis, it appears to result from a minute portion of carbon; and cast-iron, which possesses a much larger share, presents, as we should expect, somewhat similar phenomena. Moreover, as the hard and soft conditions of steel may be reversed backwards and forwards without any rapid chemical change in its substance, it has been pronounced

For the mode of analysis for ascertaining the quantity of carbon in cast-iron and steel, invented by M. V. Regnault,.Alining Engineer,.SV *Annates de Chimic et de Physique,* for January 1339; also Journal of the Franklin Institute, vol. 25, p. 327. It is stated that the analysis is very easy and exact, and may be completed in half an hour.
to result from internal arrangement or crystallization, which may be in a degree illustrated and explained by similar changes observed in glass.

A wine-glass, or other object recently blown, and plunged whilst red-hot into cold water, cracks in a thousand places, and even cooled in warm air it is very brittle, and will scarcely endure the slightest violence or sudden change of temperature; and visitors to the glass-house are often shown that a wine-glass or other article of irregular form, breaks in cooling in the open air from its unequal contraction at different parts. But the objects would have become useful, and less disposed to fracture, if they had been allowed to arrange their particles gradually, during their very slow passage through the long annealing oven or *leer* of the glass-house, the end at which they enter being at the red heat, and the opposite extremity almost cold.

To perfect the annealing, it is not unusual with lamp-glasses, tubes for steam-gages, and similar pieces exposed to sudden transitions of heat and cold, to place them in a vessel of cold water, which is slowly raised to the boiling temperature, kept for some hours at that heat, and then allowed to cool very slowly: the effect thus produced is far from chimerical. For such pieces of flint glass intended for cutting, as are found to be insufficiently annealed, the boiling is sometimes preferred to a second passage through the leer: lamp-glasses are also much *less* exposed to fracture when they have been once used, as the heat if not too suddenly applied or checked, completes the annealing.

Steel in like manner when suddenly cooled is disposed to crack in pieces, which is a constant source of anxiety; the danger increases with the thickness in the same way as with glass, and the more especially when the works are unequally thick and thin.

Another ground of analogy between glass and steel, appears to exist in the pieces of unannealed glass used for exhibiting the phenomena, formerly called double refraction, but now polarization of light; an effect distinctly traced to its peculiar crystalline structure.

In glass it is supposed to arise from the cooling of the external crust more rapidly than the internal mass; the outer crust is therefore in a state of tension, or restraint, from an attempt to squeeze the inner mass into a smaller space than it seems to require; and from the hasty arrangement of the unannealed glass, the natural positions of its crystals are in a measure disturbed or dislocated.

In the philosophical toy, the Prince Rupert's drop, this disruption is curiously evident to the sight, as the inner substance is cracked and divided into a multitude of detached parts, held together by the smooth external coat. The unannealed glass, when cautiously heated and slowly cooled, ceases to present the polarizing effect, and the steel similarly treated ceases to be hard, and may we not therefore indulge in the speculation, that in both cases a peculiar crystalline structure is consequent upon the unannealed or hardened state?.

In the process of hardening steel, water is by no means essential, as the sole object is to extract its heat rapidly, and the following are examples, commencing with the condition of extreme hardness, and ending with the reverse condition.

A thin heated blade placed between the cold hammer and anvil, or other good conductors of heat, becomes perfectly hard. Thicker pieces of steel, cooled by exposure to the air upon the anvil, become rather hard, but readily admit of being filed. They become softer when placed on the cold cinders, or other bad conductors of heat. Still more soft when placed in hot cinders, or within the fire itself, and cooled by their gradual extinction. When the steel is encased in close boxes with charcoal powder, and it is raised to a red-heat and allowed to cool in the fire or furnace, it assumes its softest state; unless lastly, we proceed to its partial decomposition. This is done by enclosing the steel with iron turnings or filings, the scales from the smith's anvil, lime, or other matters that will abstract the carbon from its surface; by this mode it is superficially decarbonized, or reduced to the condition of pure soft iron, in the manner practised by Mr. Jacob Perkins, in his most ingenious and effective combination of processes, employed for producing in unlimited numbers, absolutely identical impressions of bank notes and cheques, for the prevention of forgery. These methods of treating steel will be hereafter noticed.

Mr. Pellatt has shown experimentally, that a re arrangement of the particles of glass occurs in the process of annealing, as of two pieces of the same tube each 40 inches long, the one sent through the leer, contracted one-sixteenth of an inch more than the other, which was cooled as usual in the open air. Tubes for philosophical purposes are not annealed, as their inner surfaces are apt to become soiled with the sulphur of the fuel: they are in consequence very brittle and liable to accident. For other interesting particulars concerning flint-glass, are Minutes of Conversation, lust. Civ. Eng, 1840, pp. 37 to 41. PRACTICE OF HARDENING. AND TEMPERING STEEL. 239

A nearly similar variety of conditions might be referred to as existing in cast-iron in its ordinary state, governed by the magnitude, quality, and management of the castings; independently of which, by one particular method, some cast-iron may be rendered externally as hard as the hardest steel; such are called *chilled-iron castings;* and, as the opposite extreme, by a method of annealing combined with partiahdecomposition, *malleable-iron castings* may he obtained, so that cast-iron nails may be clenched.

Again, the purest iron, and most varieties of cast-iron, may, by another proceeding, be superficially converted into steel, and then hardened, the operation being appropriately named *case-hardening.* I therefore propose to illustrate these phenomena collectively, under three divisions: first the hardening and tempering of steel; secondly, the hardening and annealing of cast-iron; and thirdly, the process of case-hardening.

SECT. II.—PRACTICE OF HARDENING AND TEMPERING STEEL.

It may perhaps be truly said, that upon no one subject connected with mechanical art does there exist such a contrariety of opinion, not unmixed with prejudice, as upon that of hardening and tempering steel; which makes it often difficult to reconcile the practices followed by different individuals in order to arrive at exactly similar ends. The real difficulty of the subject occurs in part from the mysteriousness of the change; and from the absence of defined measures, by which either the steps of the process itself, or the value of the results when obtained, may be satisfactorily measured; as each is determined almost alone by the unassisted senses of sight and touch, instead of by those physical means by which numerous other matters may be strictly tested and measured, nearly without reference to the judgment of the individual, which in its very nature is less to be relied upon.

The excellence of cutting tools, for instance, is pronounced upon their relative degrees of endurance, but many accidental circumstances here interfere to vitiate the strict comparison: and in respect to the measure of simple hardness, nearly the only test is the resistance the objects offer to the file, a mode in two ways defective, as the files differ amongst themselves in hardness; and they only serve to indicate in an imperfect manner to the touch of the individual, a general notion without any 240 GENERAL OBSERVATIONS ON HARDENING.

distinct measure, so that when the opinion of half a dozen persons may be taken, upon as many pieces of steel differing but slightly in hardness; the want of uniformity in their decisions will show the vague nature of the proof.

Under these circumstances, instead of recommending any particular methods, I have determined to advance a variety of practical examples derived from various sources, which will serve in most cases to confirm, but in some to confute one another; leaving to every individual to follow those examples which may be the most nearly parallel with his own wants. There are however some few points upon which it may be said that all are agreed; namely.

The temperature suitable to forging and hardening steel differs in some degree with its quality and its mode of manufacture; the heat that is required diminishes with the increase of carbon:

In every case *the lowest available temperature should* beemployed in each process, the hammering should be applied in *the most equal manner throughout,* and for cutting tools it should be continued until they are nearly cold:

Coke or charcoal is much better as a fuel than fresh coal, the sulphur of which is highly injurious:

The scale should be removed from the face of the work to expose it the more uniformly to the effect of the cooling medium:

Hardening a second time without the intervention of hammering is attended with increased risk; and the less frequently steel passes through the fire the better.

In hardening and tempering steel there are three things to be considered; namely, the means of heating the objects to redness, the means of cooling the same, and the means of applying the heat for tempering or letting them down. I will speak of these separately, before giving examples of their application.

The smallest works are heated with the flame of the blowpipe and are occasionally supported upon charcoal; but as the blowpipe is used to a far greater extent in soldering, its management will be described in the chapter devoted to that process.

For objects that are too large to be heated by the blowpipe, and too small to be conveniently warmed in the naked fire, various protective means are employed. Thus an iron tube or sheet-iron box inserted in the midst of the ignited fuel is a safe and cleanly way; it resembles the muffle employed in chemical works. The work is then managed with long forceps made of steel or iron wire, bent in the form of the letter U, and flattened or hollowed at the ends. A crucible or an iron pot about four to six inches deep, filled with lead and heated to redness, is likewise excellent, but more particularly for long and thin tools, such as gravers for artists, and other slight instruments; several of these may be inserted at once, although towards the last they should be moved about to equalise the heat; the weight of the lead makes it desirable to use a bridle or trevet for the support of the crucihle. Some workmen place on the fire a pan of charcoal dust, and heat it to redness.

Great numbers of tools, both of medium and large size, are heated in the ordinary forge fire, which should consist of cinders rather than fresh coals; coke and also charcoal are used, but far less generally; recourse is also had to hollow fires, the construction of which M as explained at page 204; but the bellows should be very sparingly used, except in blowing up the fire before the introduction of the work, *which should be allowed ample time to get hot,* or as it is called, to " *soak.* "

Which method soever may be resorted to for heating the work, the greatest care should be given to communicate to all the parts requiring to be hardened a *uniform* temperature, and which is only to be arrived at by cautiously moving the work to and fro to expose all parts alike to the fire; the difficulty of accomplishing this of course increases with long objects, for which fires of proportionate length are required.

It is far better to err on the side of deficiency than of excess of heat j the point is rather critical, and not alike in all varieties of steel. Until the quality of the steel is familiarly known, it is a safe precaution to commence rather too low than otherwise, as then the extent of the mischief will be the necessity for a repetition of the process at a higher degree of heat; but It is a common and excellent practice amongst the Sheffield workmen to use coke both in forging and hardening steel goods. They frequently prepare it for themselves, either upon the forge hearth or in a heap in the open yard. In that celebrated town a decided preference is given, for the purpose of hardening, to the light coke of the Deepcar coal, which is obtained about eight miles N.W. of Sheffield, although for ordinary use and for forging, this is considered an inferior kind of coal, and of light quality; other workmen prefer chaicoal for hardening. VOL. I. R the steel if burned or overheated will be covered with scales, and what is far worse, its quality will be permanently injured; a good hammering will, in a degree, restore it; but this iu finished works is generally impracticable.

Less than a certain heat fails to produce hardness, and in the opinion of some workmen has quite the opposite effect, and they consequently resort to it as the means of rapid annealing, not, however, by plunging the steel into the water and allowing it to remain until cold; but dipping it quickly, holding it in the steam for a few moments, dipping it again and so on, reducing it to the cold state in a hasty but intermittent manner.

There is another opinion prevalent amongst workmen, that steel which is *"pinny,"* or as if composed of a bundle of hard wires, is rendered uniform in its substance if it is first hardened and then annealed.

Secondly, the choice of the cooling medium has reference mainly to the relative powers of conducting heat they severally possess: the following have been at different times resorted to with various degrees of success: currents of cold air; immersion It is argued by some, that by heating pieces of steel to different degrees, before plunging them into the water, the one piece attains full hardness, the next the temper of a tool fit for metal, another of a tool fit for wood, a fourth that of a spring, ami Bo on. That this view is not altogether without foundation, appears in the fact that if the end of a piece of steel be made entirely hard, the transition is not quite immediate from the hard to the soft part; and Mr. Ross, iu making tho dividing-points, fur his dividing-engine, hardeus the end of a longer piece of steel than is required, and forms the point upon the grindstone, exactly at the part where the temper suits, without the steel being let down at all; a practice first employed by Mr. Stancliffe, a workman formerly employed by the celebrated Ratusden. In hardening by this method however without tempering, the scale of proper hardness is confined within such extremely narrow limits, as to be nearly useless; thus, it frequently happens that in a number of tools heated as nearly alike as the workman could judge, some few will be found too soft for any use, although they were all intended to receive the ordinary hardness, so as to require letting down, as usual with those tools exposed to violent strains or blows, such as screw-taps, cold chisels and hatchets, although many tools for metal, used with quiet and uniform pressure, are left of the full hardness for greater durability.

With the *excess* of heat, beyond *the lowest that will suffice,* the brittleness rather than the useful hardness of tools is increased; and when *no excess* of heat ii employed beyond that *absolutely requisite* for hardening in the usual manner, the steel does not appear to be injured, aud the colours on its brightened surface that occur in tempering are an excellent, and in general, sufficiently trustworthy kulcx of the inferior degrees of hardness proper for various uses.
iu water in various states, in oil or wax, and in freeziug mixtures; mercury, and flat metallic surfaces have been also used. Mr. Stodart recommended as the result of his experiments, plain water at a temperature of 40 Fahrenheit. On the whole, however, there appears to be an opinion that mercury gives the greatest degree of hardness; then cold salt and water, or water mixed with various "astringent and acidifying matters;" plain water follows; and lastly, oily mixtures.

A so-called natural spring is made by a vessel with a true and a false bottom, the latter perforated with small holes; it is filled with water, and a copious supply is admitted beneath the partition; it ascends through the holes, and pursues the same current as the heated portions, which also escape at the top. This was invented by the late John Oldham, Esq. , Engineer to the Bank of England, and was used by him in hardening the rollers for transferring the impressions to the steel-plates for bank-notes.

Sometimes when neighbouring parts of works are required to be respectively hard and soft, metal tubes or collars are fitted tight upon the work, to protect the parts to be kept soft from the direct action of the water, at any rate for so long a period as they retain the temperature suitable to hardening.

The process of hardening is generally one of anxiety, as the I find but one person who has commonly used the mercury; many presume upou the good conducting power of the metal, and the nonformation of steam, which causes a separation betwixt the steel and water when the latter is employed as the cooling medium. I have failed to learn the *reaton* of the advantage of salt and water, unless the fluid have, as well as a greater density, a superior conducting power. The file-makers medicate the water in other ways, but this is one of the questionable mysteries which is never divulged; although it is supposed that a small quantity of white arsenic is generally added to water saturated with salt. One thing however, may be noticed, that articles hardened in salt and water are apt to rust, unless they are laid for a time in lime-water, or some neutralising agent.

With plain water au opinion very largely exists in favour of that which has been used over and over again even for years, provided it is not greasy; and when the steel is very harsh, the chill is taken off plain water to lessen the risk of cracking it; oily mixtures impart to *thin* articles, such as springs, a sufficient and milder degree of hardness, with less danger of cracking, than from water; and in some cases a medium course is pursued by covering the water w ith a thick film of oil, which is said to be adopted occasionally with scythes, reaping-hooks, and thin edge-tools.

Having experimented upon all these means, I am induced fully to acquiesco in Mr. Stodart's recommendation of plain cold water for general purposes; except in the case of thin elastic works, for w hkh oil, or oily compositions ai e certainly more proper, and some of these are described iu page 249.

RISK OF CRACKS AND DISTORTIONS.
sudden transition from heat to cold often. causes the works to become greatly distorted if not cracked. The last accident is much the most likely to occur with thick massive pieces, which are as it were hardened in layers, as although the external crust or shell may be perfectly hard, there is almost a certainty that towards the center the parts are gradually less hard; and when broken the inner portions will sometimes admit of being readily filed.
When in the fire the steel becomes altogether expanded, and in the water its outer crust is suddenly arrested, but with a tendency to contract from the loss of heat, which cannot so rapidly occur at the central part; it may be therefore presumed that the inner bulk continues to contract after the outer crust is fixed, and which tends to tear the two asunder, the more especially if there be any defective part in the steel itself. An external flake of greater or less extent not unfrequently shells off in hardening; and it often happens that works remain unbroken for hours after removal from the water, but eventually give way and crack with a loud report, from the rigid unequal tension produced by the violence of the process of hardening.

The contiguity of thick and thin parts is also highly dangerous, as they can neither receive, nor yield up heat, in the same times; the mischief is sometimes lessened by binding pieces of metal around the thin parts with wire, to save them from the action of the cooling medium. Sharp angular notches are also fertile sources of mischief, and where practicable they should be rejected in favour of curved lines.

As regards both cracks and distortions, it may perhaps be generally said, that their avoidance depends principally upon *manipulation, or the successful management of every step:* first the original manufacture of the steel, its being forged and wrought, so that it may be equally condensed on all sides with the hammer, otherwise when the cohesion of the mass is lessened from its becoming red-hot, it recovers in part from any unequal state of density in which it may have been placed.

"Whilst red-hot, it is also in its weakest condition; it is therefore prone to injury either from incautious handling with the tongs, or from meeting the sudden cooling action irregularly, and therefore it is generally best to plunge works vertically, as all parts are then exposed to equal circumstances, and less disturbance is risked than when the objects are immersed obliquely or side ways into the water; although for swords, and objects of similar form, it is found the best to dip them exactly as in making a vertical downward cut with a sabre, which for this weapon is its strongest direction.

Occasionally objects are clamped between stubborn pieces of metal, as soft iron or copper, during their passage through the fire and water. Such plans can be seldom adopted and are rarely followed, the success of the process being mostly allowed to depend exclusively upon good general management. *In all cases* the thick unequal scale left from the forge should be ground off before hardening, in order to expose a clean metallic Burface, otherwise the cooling medium cannot produce its due and equal effect throughout the instrument. The edges also should be left thick, that they may not be burned in the fire; thus it will frequently happen that the extreme end or edge of a tool is inferior in quality to the part within, and that the instrument is much better after it has been a few times ground; in urging this point, I cannot do better than quote the couplet inserted in Moxon's Mechanick Exercises, and which he describes as having been very old and familiar to smiths even in his day—
"He that will a good Edge win.
Must Forge thick and Grind thin."

Thirdly, the heat for tempering or letting down. Between the extreme conditions of hard and soft steel there are many intermediate grades, the common index for which is the oxidation of the brightened surface, and it is quite sufficient for practice. These tints, and their respective approximate temperatures, were tabulated by Mr. Stodart.
1. Very pale straw yellow....480 deg. 1 Too,g for 2. A shade of darker yellow.. .. 450 „ *1* 3. Darker straw yellow.. 470 „) Tools for wood and 4. Still darker straw yellow.... 490 „) screw taps, &c. 5. A brown yellow 500 „ i Hatchets, chipping chi 6. A yellow, tinged slightly with purple. 520 „ *(* sels, and other percus 7. Light purple..... 530 „ ' sive tools, saws, &c. 8. Dark purple 550 „) „. *r r* f Springs. 9. Dark blue 570 „ 10. Paler blue 590 „) m„ ii o«-ii I vi »,n ' Too soft for the above 11. Still paler blue 610 „ *I* 12. Still paler blue, with a tinge of green. 630 „ ' PurP0ses Mr. Jordan informs me that in making the magnets for Fox's dipping Needles, which are about ten inches long, one-fourth of an inch wide, and the two-hundredth
The first tint arrives at about 430 F., but it is only seen by comparison with a piece of steel not heated: the tempering colours differ slightly with the various qualities of steel.

The heat for tempering being moderate, it is often supplied by the part of the tool not requiring to be hardened, and which is not therefore cooled in the water. The workman first hastily tries with a file whether the work is hard; he then partially brightens it at a few parts with a piece of grindstone or an emery stick, that he may be enabled to watch for the required colour; which attained, the work is usually cooled in any convenient manner, lest the body of the tool should continue to supply heat. But when, on the contrary, the colour does not otherwise appear, partial recurrence is had to the mode in which the work was heated, as the flame of the candle, or the surface of the clear fire applied if possible, a little below the part where the colour is to be observed, that it may not be soiled by the smoke.

A very convenient and general manner of tempering small objects, is to heat to redness a few inches of the end of a flat bar of iron about two feet long; it is laid across the anvil, or fixed by its cold extremity in the vice; and the work is placed on that, part of its surfsfte which is found by trial to be of the suitable temperature, by gradually sliding the work towards the heated extremity. In this manner many tools may be tempered at once, those at the hot part being pushed off into a vessel of water or oil, as they severally show the required colour, but it requires dexterity and quickness in thus managing many pieces.

Vessels containing oil or fusible alloys carefully heated to the required temperatures have also been used, and I shall have to describe a method called *"blazing off,"* resorted to for many articles such as springs and saws, by heating them over the naked fire until the oil, wax, or composition in which they have been hardened ignites; this can only occur when they part of an inch thick, this precaution entirely failed; and the needles assumed all sorts of distortions when released from between the stiff bars within which they were hardened. The plan was eventually abandoned, and the magnets were heated in the ordinary way within an iron tube, and were set straight with the hammer after being let down to a deep orange or brown colour. Steel however is in the best condition for the formation of good permanent magnets when perfectly hard.
The knife edges, for Captain Eater's experimental pendulum, were very carefully hardened and tempered in a bath heated to 430 deg.; being then found too soft they were re-hardened, and tempered, at only the heat of boiling water, after which they were considered admirably suited to their purpose. respectively reach their boiling temperatures and are evaporated in the gaseous form.

The period of letting down the works is also commonly chosen for correcting, by means of the hammer, those distortions which so commonly occur in hardening; this is done upon the anvil, either with the thin pane of an ordinary hammer, or else with a *hack-hammer,* a tool terminating at each end in an obtuse chisel edge which requires continual repair on the grindstone.

The blows are given on the hollow side of the work, and at right angles to the length of the curve; they elongate the concave side, and gradually restore it to a plane surface, when the blows are distributed consistently with the positions of the erroneous parts. The hack-hammer unavoidably injures the surface of the work, but the blows should not be violent, as they are then also more prone to break the work, the liability to which is materially lessened when it is kept at or near the tempering heat, and the edge of the hack-hammer is slightly rounded.

SECT. III. COMMON EXAMPLES OF HARDENING AND TEMPERING STEEL.

Watchmaker's drills of the smallest kinds, are heated in the blue part of the flame of the candle; larger drills arc heated with the blowpipe flame, applied very obliquely, and a little below the point; when very thin they may be whisked in the air to cool them, but they are more generally thrust into the tallow of the candle or the oil of the lamp; they are tempered either by their own heat, or by immersion in the flame below the point of the tool.

For tools between those suited to the action of the blowpipe, and those proper for the open fire, there are many which require either the iron tube, or the bath of lead or charcoal described at page 241, but the greater number of works are hardened in the ordinary smith's fire, without such defences.

Tools of moderate size, such as the majority of turning tools, carpenters' chisels and gouges, and so forth, are generally heated in the open fire; they require to be continually drawn backwards and forwards through the fire, to equalise the temperature applied; they are plunged vertically into the water and then moved about sideways to expose them to the cooler portions of the fluid. If needful, they are only dipped to a certain depth, the remainder being left soft.

Some persons use a shallow vessel filled only to the height of the portion to he hardened, and plunge the tools to the bottom; but this strict line of demarcation is sometimes dangerous, as the tools are apt to become cracked at the part, and therefore a small vertical movement is also generally given, that the transition from the hard to the soft part may occupy more length.

Razors and penknives are too frequently hardened without the removal of the scale arising from the forging; *this practice, which is not done with the best works, cannot be too much deprecated.* The blades are heated in a coke or charcoal fire, and dipped into the water obliquely. In tempering razors, they are laid on their backs upon a clear fire, about half-a-dozen together, and they are removed one at a time, when the edges, which are as yet thick, come down to a pale straw-colour; should the backs accidentally get heated beyond the straw-colour, the blades arc cooled in water, but not otherwise. Penknife blades are tempered, a dozen or two at a time, on a plate of iron or copper about twelve inches long, three or four wide, and about a quarter of an inch thick; the blades are arranged close together on their backs, and lean at an angle against each other. As they come down to the temper, they are picked out with small pliers aud thrown into water, if necessary; other blades are then thrust forward from the cooler parts of the plate to take their place.

Hatchets, adzes, cold chisels, and numbers of similar tools, in which the total bulk is considerable compared with the part to be hardened, are only partially dipped; they are afterwards let down by the heat of the remainder of the tool, and when the colour indicative of the temper is attained, they are entirely quenched. With the view of removing the loose scales, or the oxidation acquired in the fire, some workmen rub the objects hastily in dry salt before plunging them in the water, in order to give them a cleaner and whiter face.

In hardening large dies, anvils, and other pieces of considerable size, by direct immersion, the rapid formation of steam at the sides of the metal prevents the free access of the water for the removal of the heat with the required expedition; in these cases, a copious stream of water from a reservoir above is allowed to fall on the surface to be hardened. This contrivance is frequently called a "float," and although the derivation of the name is not very clear, the practice is excellent, as it supplies an abundance of cold water; and which, as it falls directly on the center of the anvil, is sure to render that part hard. It is, however, rather dangerous to stand near such works at the time, as when the anvil face is not perfectly welded, it sometimes in part flies off with great violence and a loud report.

Occasionally the object is partly immersed in a tank beneath the fall of water, by means of a crane and slings; it is ultimately tempered with its own heat, and dropped in the water to become entirely cold.

Oil, or various mixtures of oil, tallow, wax and resin, are used for many thin and elastic objects, such as needles, fish-hooks, steel pens and springs, which require a milder degree of hardness than is given by water.

For example, steel pens are heated in large quantities in iron trays within a furnace, and are then hardened in an oily mixture; generally they are likewise tempered in oil, or a composition the boiling point of which is the same as the temperature suited to letting them down. This mode is particularly expeditious, as the temper cannot fall below the assigned degree. The dry heat of an oven is also used, and both the oil and oven may be made to serve for tempers harder than that given by boiling oil; but more care and observation are required for these lower temperatures.

Saws and springs are generally hardened in various compositions of oil, suet, wax and other ingredients, which, however, The composition used by an experienced saw-maker is two pounds of suet and a quarter of a pound of beeswax to every gallon of whale-oil; these are boiled together, and will serve for thin works and most kinds of steel. The addition of black resin, to the extent of about one pound to the gallon, makes it serve for thicker pieces and for those it refused to harden before; but the resin should be added with judgment, or the works will become too hard and brittle. The composition is useless when it has been constantly employed for about a month: the period depends, however, on the extent to which it is used, and the trough should be thoroughly cleaned out before new mixture is placed in it.

The following recipe is recommended by Mr. Gill:

"Twenty gallons of spermaceti oil;

"Twenty pounds of beef suet rendered;

"One gallon of neats-foot oil;

"One pound of pitch;

"Three pounds of black resin.

"The«e two last articles must be previously melted together, and then added to the other ingredients; when the whole must be heated in a proper iron vessel, with a close cover lit ted to it, until the moisture is entirely evaporated, aud the compo250 SPRINGS AND SAWS; MAIN-

SFRINOS FOR WATCHES.
lose their hardening property after a few weeks' constant use; the saws are heated in long furnaces, and then immersed horizontally and edgeways in a long trough containing the composition; two troughs are commonly used, the one until it gets too warm, then the other for a period, and so on alternately. Part of the composition is wiped off the saws with a piece of leather, when they are removed from the trough, and they are heated one by one over a clear coke fire, until the grease inflames; this is called "*blazing off.*" When the saws are wanted to be rather hard, but little of the grease is burned off; when milder, a larger portion; and for a spring temper, the whole is allowed to burn away. When the work is thick, or irregularly thick and thin, as in some springs, a second and third dose is burned off, to ensure equality of temper at all parts alike. Springs and saws appear to lose their elasticity, after hardening and tempering, from the reduction and friction they undergo in grinding and polishing. Towards the conclusion of the manufacture, the elasticity of the saw is restored principally by hammering, and partly by heating it over a clear coke fire to a straw colour; the tint is removed by very diluted muriatic acid; after which the saws are well washed iu plain water and dried.

Watch springs are hammered out of round steel wire, of suitable diameter, until they fill the gage for width, which at the same time ensures equality of thickness: the holes are punched in their extremities, and they are trimmed on the edge with a smooth file; the springs are then tied up with binding-wire in a loose *open* coil, and heated over a charcoal fire upon a perforated revolving plate; they are hardened in oil, and blazed off.

The spring is now distended in a long metal frame, similar to that used for a saw-blade, and ground and polished with emery, and oil, between lead blocks; by this time its elasticity appears quite lost, and it may be bent in any direction; its elasticity is, however, entirely restored by a subsequent hammering on a very bright anvil, which "*puts the nature into the spring.*" sition will take fire on a flaming body being presented to its surface, but which must be instantly extinguished again by putting on the cover of the vessel."—Manufactures in Metal, Vol. i. p. 33t. Lardner's Cyclopaxlia. *Fee* also p. 311, Ibid.
Gun lock springs are sometimes literally *fried in oil* for a considerable time over a fire in an iron tray; the thick parts are then sure to be sufficiently reduced, and the thin parts do not become the more softened from the continuance of the blazing heat. CHRONOMETER SPRINGS, ADAMS' BOW-SPRINGS, ETC. 251
The colouring is done over a flat plate of iron, or hood, under which a little spirit lamp is kept burning; the spring is continually drawn backwards and forwards, about two or three inches at a time, until it assumes the orange or deep blue tint throughout, according to the taste of the purchaser; by many the colouring is considered to be a matter of ornament, and not essential. The last process is to coil the spring into the spiral form, that it may enter the barrel in which it is to be contained; this is done by a tool, with a small axis and winch handle and does not require heat.

The balance springs of marine chronometers which are in the form of a screw, are wound into the square thread of a screw of the appropriate diameter and coarseness; the two ends of the spring are retained by side-screws, and the whole is carefully enveloped in plat inum-foil, and tightly bound with wire. The mass is next heated in a piece of gun-barrel closed at the one end, and plunged into oil, which hardens the spring almost without discolouring it, owing to the exclusion of the air by the close platinum covering, which is now removed, and the spring is let down to the blue, before removal from the screwed block.

The balance or hair-springs of common watches are frequently left soft; those of the best watches are hardened in the coil upon the plain cylinder, and are then curled into the spiral form between the edge of a blunt knife and the thumb, the same as in curling up a narrow riband of paper, or the filaments of an ostrich feather.

Mr. Dent says that 3,200 balance springs weigh only one ounce; but springs also include the heaviest examples of hardened steel works uncombined with iron: for example, of Mr. Adams' patent bow-springs for all kinds of vehicles, some intended for railway use, measure 3 feet long, and weigh 50 pounds each piece; two of these are used in combination; other single springs are 6 feet long, and weigh 70 pounds.f

In hardeuing them they are heated by being drawn backwards The soft springs are worth *2t. 6d.* each; the hardened and tempered springs, 10z. 6ii. each. This raises the value of the steel, originally less than twopence, to 400/. and 1600/. respectively.—*Mr. Dent's Lectures on Timt-pucet, d-c.*

T The principle of these bo springs will be immediately seen, by conceiving the common archery bow fixed horizontally with its cord upwards; the body of the carriage being attached to the cord sways both perpendicularly and sideways with perfect freedom.
252 SPRINGS. Stodart's REMARKS. and forwards through an ordinary forge fire, built hollow, and they are immersed in a trough of plain water; in tempering them they are heated until the black red is just visible at night; by daylight the heat is denoted by its making a piece of wood sparkle when rubbed on the spring, which is then allowed to cool in the air. The metal is nine-sixteenths of an inch thick, and Mr. Adams considers five-eighths the limit to which steel will harden properly, that is sufficiently alike to serve as a spring; he tests their elasticity far beyond their intended range. SECTION IV.—LESS COMMON EXAMPLES OF HARDENING, AND PRECAUTIONARY MEASURES.
One of the most serious evils in hardening steel, especially in thick blocks, or those which are unequally thick and thin, is their liability to crack, from the sudden transition; and in reference to hardening razors, a case in point, Mr. Stodart mentions it as the observation and practice of one of his workmen,

"that the charcoal fire should be made up with shavings of leather;" and upon being asked what good he supposed the leather could do, this workman replied, "that he could take upon him to say that he never had a razor crack in the hardening since he had used this method, though it was a frequent occurrence before."

"When," says Mr. Stodart, "brittle substances crack in cooling, it always happens from the outside contracting and becoming too small to contain the interior parts. But it is known that hard steel occupies more space than when soft; and it may Great diversity of opinion exists respecting the cause of elasticity in springs; by some it is referred to different states of electricity; by others the elasticity ia considered to reside in the thin, blue, oxidized surface, the removal of which is thought to destroy the elasticity, much in the same manner that the elasticity of a cane is greatly lost by stripping off its siliceous rind. The elasticity of a thick spring is certainly much impaired by grinding off a small quantity of its exterior metal, which is harder than the inner portion; and perhaps thin springs sustain in the polishing a proportional loss, which is to them equally fatal.

Mr. Dent stated at the British Association, 1841, that he found experimentally the bare removal of the blue tint from a pendulum spring, by its immersion in weak acid, caused the chronometer to lose nearly one minute each hour; a second and equal immersion scarcely caused any further loss. He also stated it as a wellknown fact, that such springs get stronger, in a minute degree, during the first two or three years they are in use, from some atmospheric change; when the springs are coated with gold by the electrotype process, no such change is observable, and the covering although perfect may be so thin as not to compensate for the loss of the blue oxidized surface.

easily be inferred that the nearer the steel approaches to the state of iron, the less will be this increase of dimensions. If, then, we suppose a razor, or any other piece of steel, to be heated in an open fire with a current of air passing through it, the external part will, by the loss of carbon, become less steely than before; and when the whole piece comes to be hardened, the inside will be too large for the external part, which will probably crack. But if the piece of steel be wrapped up in the cementing mixture, or if the fire itself contain animal coal, and is put together so as to operate in the manner of that mixture, the external part, instead of being degraded by this heat, will be more carbonated than the internal part, in consequence of which it will be so far from splitting or bursting during its cooling, that it will be acted upon in a contrary direction, tending to render it more dense and solid.

"The cracking which so often occurs on the immersion of steel articles in water, does not appear to arise so much from any decarbonisation of the surface merely, as from the sudden condensation and contraction of a superficial portion of the metal, while the mass inside remains swelled with the heat, and probably expands for a moment, on the outside coming in contact with the water."

The file-makers, to save their works from *clinking* or cracking partly through in hardening, draw the files through yeast, beergrounds, or any sticky material, and then through a mixture of common salt and animal hoof roasted and pounded. This is corroborative of the above, as in the like manner it supplies a little carbon to the outside, and also renders the steel somewhat harder and less disposed to crack; the composition also renders the more important service of protecting the fine points of the teeth from being injured by the fire.

An analogous method is now practised in hardening Jones's patent axletrees, which are of wrought iron, with two pieces of steel welded into the lower side, where they rest upon the wheels and sustain the load. The work is heated in an open forge fire, quite in the ordinary way, and when it is removed, a mixture, principally the prussiate of potash, is laid upon the steel; the axletree is then immediately immersed in water, and additional water is allowed to fall upon it from a cistern. The steel is considered to become very materially harder for the 254

Perkins's Engraved Steel Plates. treatment, and the iron around the same is also partially hardened.

These are, in fact, applications of the case-hardening process, which is usually applied to wrought iron for giving it a steely exterior, as the name very properly implies. Occasionally, steel which hardens but imperfectly, either from an original defect in the material, or from its having become deteriorated by bad treatment, or too frequent passage through the fire, is submitted to the case-hardening process in the ordinary way, by inclosing the objects in iron boxes, as will be explained. This in part restores the carbon which has been lost, and the steel admits of being hardened, but this practice is not to be generally recommended, although it is well employed for the purposes of transfer engraving explained at foot; a method introduced by Mr. Jacob Perkins, and which took its origin in the curious transfer processes of the calico works, wherein however copper is the material principally used.f

Various methods-have been likewise attempted to prevent the distortions to which work is liable in the operation of hardening, These axletrees are used for Jones's Patent wrotight-iron suspension wheels, which have iron naves made of chilled castings; to the prussiate is added one-third of the carbonate of ammonia; the effect of the carbonate is principally considered to be the more minute subdivision of the prussiate over the surface of the steel.

t Mr. Perkins's admirable process of transfer engraving may be thus explained. A soft steel plate was first engraved with the required subject in the most finished style of art, either by hand or mechanically, or the two combined, and the plate was then hardened. A decarbonized steel cylinder was next rolled over the hardened plate by powerfid machinery, until the engraved impression appeared in relief, the hollow lines of the original becoming ridges upon the cylinder. The roller was recon-

verted to the condition of ordinary steel and hardened, after which it served for returning the impression to any number of decarbonized plates, every one of which became absolutely a *counterpart* of the original; and every plate when hardened would yield the enormous number of 150,000 impressions without any perceptible difference between the first and the last.
In the event of any accident occurring to the transfer roller, the original plate still existed, from which another or any required number of rollers could be made, aud from these rollers any number of new plates, all capable of producing as many impressions as above cited.

Specimens of Mr. Perkins's process may be seen in the 38th volume of the Transactions of the Society of Arts; and it is there stated by the inventor, that to decarbonize the plates they were placed in the vertical position in cast-iron boxes not less than f of an inch thick, and surrounded on all sides by a stratum of iron filings not less than J an inch thick; the boxes were then placed in a furnace, and after having been heated, were suffered to cool in the most gradual manner by stopping off PRACTICE IN THE BANK OF ENGLAND.
255 but without any very advantageous results; for instance, it has been recommended to harden small cylindrical wires, by rolling all the air-paasages and covering the boxes with a layer of cinders 6 or 7 inches deep. The re-conversion was similarly accomplished, but with the charcoal from leather sifted fine, and on removal from the boxes at the end of from three to five hours the plates were immediately plunged vertically iuto cold water.

The late Mr. Warren was instrumental in bringing into common use the thin steel plates, similar to those previously used for copper; these were annealed at a higher temperature, in earthenware boxes filled with pounded oyster-shells. —*Sec* Trans. Soc. of Arts, vol. xli. p. 88.

The practice at the Bank of England, introduced by the late Mr. John Oldham, and followed under the superintendence of his son Mr. Thomas Oldham, is to anneal at one time four cast iron boxes, each containing from three to six steel plates, surrounded on all sides with fine charcoal mixed with an equal quantity of chalk and driven in hard.

The reverberatory furnace employed, has a circular cast-iron plate or bed upon which the four boxes are fastened by wedges, and as the plate revolves very slowly and continually by the steam engine employed in working the printing presses and other machinery, the plates are exposed in the most equal manner to the heat, and w hen the proper temperature is attained, all the apertures are carefully closed and luted to extend the cooling over a space of at least forty-eight hours.

The surfaces of the cylinders and plates are thus rendered exceedingly soft, to the depth of about the 32nd of an inch, " so as to become more like lead than anything else," and thus much of their surfaces must be turned or planed off: the device is raised in the transfer-press upon the natural soft steel of the rollers, under a pressure of some tons, and these are hardened without any intentional application of the case-hardening process, as the simple steel is undoubtedly very superior in all respects to that which has been decarbonized and reconverted.

The plates themselves are used in the soft state, as they then admit of reparation by the transfer rollers; and the process is found to be more economical, as the risk of warping is avoided, and they may be easily repaired. The dates and numbers are at present printed as a second process by letter press printing, with the machines invented by the late Mr. Braniah, and which have been engraved and described in ditfei ent books.

In hardening engraved plates, rollers, dies, and similar works, it is of the greatest importance to preserve the surface unimpaired; and as steel is very liable to oxidation at the red heat if exposed to the air for even a few seconds, and which oxidized scale will in some cases nearly remove, or at any rate injure, the subject produced upon ita surface, it is of great importance to conduct the heating and cooling with the most complete exclusion of the air.

Jir. Thomas Oldham introduced a mode of hardening the transfer rollers which appears as near to perfection as possible, and by it, instead of the works acquiring the ordinary black and grey tints, and a minute roughness, like the surface of the finest emery paper, the steel comes out of the water as smooth to the touch as at first, and mottled with all the beautiful tints seen on case-hardened gun-locks. The method is simply as follows:

The work to be hardened is inclosed in a wrought-iron box with a loose cover, a false bottom, and with three cars projecting from its surface about midway; the steel is surrounded on all sides with carbon from leather, driven in hard, and the 256 DR. Wollaston's, AND PERKINS'S METHODS.
them when heated between cold metallic surfaces to retain them perfectly straight. This might probably answer, but unfortunately cylindrical steel wires supply but a very insignificant portion of our wants.

Another mode tried by Dr. "Wollaston was to inclose the piece of steel in a tube filled with Newton's fusible alloy, the whole to be heated to redness and plunged in cold water; the object was released by immersion in boiling water, which melted the alloy, and the piece came out perfectly unaltered in form, and quite hard. This mode is too circuitous for common practice, and the reason why it is to be always successful is not very apparent.

Mr. Perkins resorted to a very simple practice with the view of lessening the distortion of his engraved steel plates, by boiling the water in which they were to be hardened to drive off" the air, and plunging them vertically; and as the plates were required to be tempered to a straw colour, instead of allowing them to remain in the water until entirely cold, he removed them whilst the inside was still hot, and placed them on the top of a clear fire until the tallow with which they were rubbed, smoked; the plate was then returned to the water for a few moments, and so on alternately

until they were quite cold, the surface never being allowed to exceed the tempering heat.
cover and bottom are carefully luted with moist clay. Thus prepared, the case is placed iu the vertical position, in a bridle fixed across a great tub, which i9 then tilled with water almost to touch the false bottom of the case. The latter is now heated in the furnace as quickly as will allow the uniform penetration of the heat.

When sufficiimtly hot, it is removed to its place in the hardening tub, the cover of the iron box is removed, and the neck or gudgeou of the cylinder is grasped *bentath the surface of the carbon,* with a long pair of tongs, upon which a coupler is dropped to secure the grasp. It only remains for the individual to hold the tongs with a glove whilst a smart tap of a hammer is given on their extremity; this knocks out the false bottom of the case, and the cylinder and tonga are instantly immersed in the water; the tongs prevent the cylinder from falling on its side, and thus injuring its delicate but still hot surface. For square plates, a suitable frame is attached by four slight claws, and it is the frame which is seized by the tongs; the latter are sometimes held by a chain, which removes the risk of accident to the individual. In some cases, the work assumes a striated and mackled appearance, evident to the touch as well as the sight, and which is to be attributed to an imperfect manufacture of the steel.

Mr. Oldham informed me that in the Paris Mint, the dies are inclosed in the soot of burnt wood; and that in our own Mint, the dies are hardened by a powerful jet of water. Ho also added, that his workpeople have the impression that steel is reduced to its softest state by inclosure with lime and ox-gall. The reader is referred to Sir J. Robinson's note *V,* in the appendix of vol. ii. page 970.

Gooch's Hardened Steel Railway Tires. 257

From various observations, it appears on the whole to be the best in thick works thus to combine the hardening and tempering processes, instead of allowing the objects to become entirely cold, and then to re-heat them for tempering. To ascertain the time when the plate should be first removed from the water, Mr. Perkins heated a piece of steel to the straw colour, and dipped it into water to learn the sound it made; and when the hardened plate caused the *same* sound, it was considered to be cooled to the right degree, and was immediately withdrawn

I will conclude these numerous examples and remarks by one of a very curious, massive, and perfect kind, in which the hardening is sure to occur without loss of figure, unless the work break under the process. I refer to the locomotive wheels with hardened steel tires, patented by Mr. Daniel Gooch, and which may be viewed as the most ponderous example of hardening as the tires of the eight-foot wheels weigh about 10 cwt., and consist of about one-third steel, and there seems no reason why this diameter might not be greatly exceeded.

The materials for the tires are first swaged separately, and then welded together under the heavy hammer at the steelworks, after which they are bent to the circle, welded, and turned to certain gages. The tire is now heated to redness in a circular furnace; during the time it is getting hot, the iron wheel, previously turned to the right diameter, is bolted down upon a face-plate; the tire expands with the heat, and when at a cherry-red, it is dropped over the wheel, for which it was previously too small, and is also hastily bolted down to the surface plate, the whole load is quickly immersed by a swing crane into a tank of water about five feet deep, and hauled up and down until nearly cold; the steel tires are not afterwards tempered.f Trans. Soc. of Arts, vol. xxxviii. p. 55. t The patentees consider the steel tires to be eight times harder than those of wrought-iron, and to be proportionately durable, and from the very ingenious method employed in their construction they are little exposed to accident. The pokes are forged out of flat bars with T formed heads; these are arranged radially in the founder's mould, whilst the cast-iron center is poured around them: the ends of the T heads are then welded together to constitute the periphery of the wheel or inner tire, and little wedge-form pieces are inserted where there is any deficiency of iron.

The wheel is then chucked on a lathe, bored, and turned on the edge, not cylindrically, but like the meeting of two cones, and about one quarter of an inch higher in the middle than on the two edges. The compound tire is turned to the VOL. I." 258 STEEL AND CAST-IKON. CHILLED IKON CASTINOS.

SECT. V.—HARDENING AND SOFTENING CAST-IEON.

The similitude of chemical constitution between steel, which usually contains about one per cent, of carbon, and cast-iron that has from three to six or seven per cent., naturally leads to the expectation of some correspondence in their characters, and which is found to exist. Thus some kinds of cast-iron will harden almost like steel, but they generally require a higher temperature; and the majority of cast-iron, also like steel, assumes different degrees of hardness, according to the rapidity with which the pieces are allowed to cool.

The casting left undisturbed in the mould, is softer than a similar one exposed to the air soon after it has been poured. Large castings cannot cool very hastily, and are seldom so hard as the small piedfes, some of which are hardened like steel bj the moisture combined with the moulding sand, and cannot be filed until they have been annealed after the manner of steel, which renders them soft and easy to be worked.

Chilled iron-castings, present as difficult a problem as the hardening and tempering of steel; the fact is simply this, that iron castings, made in iron moulds under particular circumstances, become on their outer surfaces perfectly hard, and resist the file almost like hardened steel; the effect is however superficial, as the chilled exterior shows a distinct line of demarcation when the objects are broken.

The late Mr. Ransome, of Ipswich, took out a patent in 1803, for ploughshares cast on this principle: the under sides and points are hard from the chilling process, and these from resisting abrasion more than the softer parts, maintain a comparatively thin edge;—an imitation of the beautiful provision in nature for preserving the acute edges of the front teeth of the rat, squirrel, and other rodentiae, the external enamel of which is always left in advance from the wearing away of the softer bone or ivory beneath, and acts almost in the manner of a carpenter's chisel.

The production of chilled castings is always a matter of some corresponding form, and consequently larger within or under-cut, so that the shrinking secures the tire without the possibility of obliquity or derangement, and no rivets are required. It sometimes happens that the tire breaks in sin-inking when by mismanagement the diameter of the wheel is in excess.
uncertainty, and depends upon the united effect of several causes: the quality of the iron, the thickness of the casting, the temperature of the iron at the time of pouring, and the condition or temperature of the iron mould, which has a greater effect in "striking in" when the mould is *heated* than if quite cold: a very thin stratum of earthy matter will almost entirely ohviate the chilling effect.
Mr. May, a member of the present firm, has furnished me with specimens of chilling, in which the hard portion varies from less than one-sixteenth to more than one-fourth of an inch in thickness: he considers it to be an effect of crystallization, and conjectures that the tendency to chill properly, will be found to depend upon the due atomic proportions of carbon and iron, combined with the circumstances before alluded to.

There is this remarkable difference between cast-iron thus hardened, and steel hardened by plunging whilst hot into water; that whereas the latter is softened again by a dull red-heat, the chilled castings on the contrary are turned out of the moulds as soon as the metal is set, and are allowed to cool in the air; yet although the whole is at a bright red heat, no softening of the chilled part takes place. This material was employed by the late Mr. Peter Keir for punches for red-hot iron; the punches were fixed in cast-iron sockets, from which they only projected sufficiently to perforate the wheel tires in the formation of which they were used, and from retaining their hardness they were more efficient than those punches made of steel.

Chilled castings are also commonly employed for axletree boxes, and naves of wheels, which are finished by grinding only; also for cylinders for rolling metal, for the heavy hammers and anvils or stithies for iron works, the stamp-heads for pounding metallic ores, &c. Cannon balls, as well *ae* ploughshares, are examples of chilled castings; with the destructive engine the chilling is unimportant, and occurs alone from the method essential to giving the balls the required perfection of form and size.

Malleable iron-castings are at the opposite extreme of the scale, and are rendered externally *soft* by the abstraction of A cold mould doea not generally chill so readily as one heated nearly to the ex. tent called " black-hot:" but the reverse conditions occur with some cast-iron.
their carbon, whereby they are nearly reduced to the condition of pure malleable iron, but without the fibre which is due to the hammering and rolling employed at the forge.

The malleable iron-castings are made from the rich Cumberland iron, and are at first as brittle as glass or hardened steel; they are enclosed in iron boxes of suitable size, and surrounded with pounded iron-stone, or some of the metallic oxides, as the scales from the iron forge, or with common lime, and various other absorbents of carbon, used either together or separately. The cases, which are sometimes as large as barrels, are luted, rolled into the ovens or furnaces, and submitted to a good heat for about five days, and are then allowed to cool very gradually within the furnaces.

The time and other circumstances determine the depth of the effect; thin pieces become malleable entirely through; they are then readily bent, and may be slightly forged; cast-iron nails and tacks thus treated admit of being clenched, thicker pieces retain a central portion of cast-iron, but in a softened state, and not brittle as at first, on sawing them through, the skin or coat of soft iron is perfectly distinct from the remainder.

The mode is particularly useful for thin articles that can be more economically and correctly cast, than wrought at the forge, as bridle-bits, snuffers, parts of locks, culinary and other vessels, pokers and tongs, many of which are subsequently case-hardened and polished, as will be explained, but malleable cast-iron should never be used for cutting-tools.

SECT. VI.— CASE-HARDENING WROUGHT AND CAST-IRON.

The property of hardening is not possessed by pure malleable iron; but I have now to explain a rapid and partial process of cementation, by which wrought-iron is first converted exteriorly into steel, and is subsequently hardened to that particular depth; leaving the central parts in their original condition of soft fibrous iron. The process is very consistently called *case-hardening,* and is of great importance in the mechanical arts, as the pieces combine the economy, strength, and internal flexibility of iron, with a thin casing of steel; which although admirable as an armour of defence from wear or deterioration as regards the surface, is unfit for the formation of cutting edges or tools, owing to the entire absence of hammering, subsequent to the cementation with the carbon. Cast-iron obtains in like manner a coating of steel, which surrounds the peculiar shape the metal may have assumed in the iron-foundry and workshop.

The principal agents used for case-hardening are animal matters, as the hoofs, horns, bones, and skins of animals; these are nearly alike in chemical constitution; they are mostly charred and coarsely pounded: some persons also mix a little common salt with some of the above; the works should be surrounded on all sides with a layer from half an inch to one inch thick.

The methods pursued by different individuals do not greatly differ; for example, the gunsmith inserts the iron work of the gun-lock, in a sheet-iron case in the midst of partially charred bone-dust; the lid of the box is tied on with iron wire, and the joint is luted with clay; it is then slowly heated to redness, and retained at that heat from half an hour to an hour, and the contents are immersed in cold water as quickly as possible, to prevent the access of air. The objects sought are a steely exterior, and a clean surface covered with the pretty mottled tints, apparently caused by oxidation from the partial admission of air.

Some of the malleable iron castings, such as snuffers, are casehardened to admit of a better polish; it is usually doue with burnt bone-dust, and at a dull red heat; they remain in the fire about two or three hours, and should be immersed in oil, as it does not render them quite so brittle as when plunged into water. It must be remembered they are sometimes changed throughout their substance into an inferior kind of steel, by a process that should in such instances be called cementation, and not casehardening, consequently they will not endure violence.

The mechanician and engineer, use horns, hoofs, bone-dust, and leather, and allow the period to extend from two to eight hours, most generally four or five; sometimes for its greater penetration, the process is repeated a second time with new carbonaceous materials. Some open the box and immerse the work in water direct from the furnace; others, with the view to preserve a better surface, allow the box to cool without being opened, and harden the pieces with the open fire as a subsequent operation; the carbon once added, the work may be annealed::nd hardened much the same as ordinary steel.

When the case-hardening is required to terminate at any particular part, as a shoulder, the object is left with a band or projection; the work is allowed to cool without being immersed in water, the band is turned off, and the work when hardened in the open fire is only affected so far as the original cemented surface remains. This ingenious method was introduced by Mr. Roberts, of Manchester, who considers the success of the case-hardening process to depend on the gentle application of the heat; and that, by proper management not to overheat the work, it may be made to penetrate three-eighths of an inch in four or five hours.

A new substance for the case-hardening process, but containing the same elements as those more commonly employed, has of late years been added, namely, the prussiate of potash, (a salt consisting of two atoms of carbon and one of nitrogen,) which is made from a variety of animal matters.

It is a new application without any change of principle the time occupied in this steelifying process, is sometimes only minutes instead of hours and days, as for example when iron is heated in the open fire to a dull red, and the prussiate is either sprinkled upon it or rubbed on in the lump, it is returned to the fire for a few minutes and immersed in water; but the process is then exceedingly superficial, and it may if needful be limited to any particular part upon which alone the prussiate is applied. The effect by many is thought to be partial or in spots, as if the salt refused to act uniformly; in the same manner that water only moistens a greasy surface in places.

The prussiate of potash has been used for case-hardening the bearings of wrought-iron shafts, but this seems scarcely worth the doing: it has been also employed with the view of giving an additional and extreme, although superficial hardness to steel, as in Jones's axletrees, Perkins's engraved steel plates, &c.; but 1 have only heard of one individual who has encased work with this salt; it was for case-hardening the iron rollers and side plates of glaziers' vices employed for milling window-lead.

In the general way, the conversion of the iron into steel, by case-hardening, is quite superficial, and does not exceed the sixteenth of an inch; if made to extend to one quarter or three-eighths of an inch in depth, to say the least it would be generally useless, as the object is to obtain durability of surface, with strength of interior, and this would disproportionately encroach on the strong iron within. The steel obtained in this adventitious manner is not equal in strength to that converted and hammered in the usual way, and if sent in so deeply, the provision for wear would far exceed that which is required.

Let us compare the case-hardening process with the usual conversion of steel. The latter requires a period of about seven days, and a very pure carbon, namely wood charcoal, of which a minute portion only is absorbed; and it being a simple body, when the access of air is prevented by the proper security of the troughs, the bulk of the charcoal remains unconsumed, and is reserved for future use, as it has undergone no change. The hasty and partial process of cementation is produced in a period commonly less than as many hours with the animal charcoal, or than as many minutes with the prussiate of potash; but all these are compound bodies, (which contain cyanogen, a body consisting of carbon and nitrogen,) and are never used a second time, but on the contrary the process is often repeated with another dose. It would be therefore, an interesting inquiry for the chemist, as to whether the cyanogen is absorbed after the same manner as carbon in ordinary steel, (and which in Mackintosh's patent process was driven through the crucible in the form of carbonic acid gas, and is stated to be absorbed at the rate of one-thirtieth of an inch in depth, each hour;) or whether the nitrogen assists in any way in hastening the admission of the carbon, *by* some as yet untraced affinity or decomposition.

This hasty supposition will apply less easily to cast-iron, which contains from three to seven times as much carbon as steel, and although not always hardened by simple immersion, is constantly under the influence of the case-hardening process; unless we adopt the supposition, that the carbon in cast-iron which is mixed with the metal in the shape of cinder in the blast furnace, when all is in a fluid state, is in a less refined union than that instilled in a more aeriform condition in the acts of cementation and case-hardening.f It may happen that the carbon is not essential, as the Indian steel or wootz is stated to contain alumine and silex; and manganese is used in Heath's and Vicker'a patents. + It would have been an easy task to have multiplied the examples and remarks upon these curious subjects of hardening and tempering, which have already far, very far, exceeded the intended limits. The reader will find much useful matter upon the same in Lardner's Cyclopedia in the volume on Iron and Steel, and in Gill's CHAPTER XIII. THE METALS AND ALLOYS MOST COMMONLY USED.

The thirteen metals before referred to have now to be considered, namely, Antimony, Bismuth, Copper, Gold, Lead, Mercury, Nickel, Palladium, Platinum, Rhodium, Silver, Tin, and Zinc. Unlike iron and steel, they do not admit of being hardened beyond that degree which may be produced by simple mechanical means, such as hammering, rolling, &c. , neither, (with the exception of platinum,) do they submit to the process of welding.

On the other hand, their fusibility offers an easy means of uniting and combining many of these metals with great readiness, either singly, or in mixtures of two or several kinds, which are called alloys. By the process of founding, any required form may be given to the fusible metals and alloys; their malleability and ductility are also turned to most useful and varied accounts; and by partial fusion neighbouring metallic surfaces may he united, sometimes *per se,* but more generally by the interposition of a still more fusible metal or alloy called solder.

The author intends therefore to commence with a brief notice of the physical characters and principal uses of the thirteen metals before named, and of their more important alloys. Tables of the cohesive force and of the general properties of metals will be next added to avoid the occasional necessity for reference to other works.

These tables will be followed by some remarks on alloys, which as regards their utility in the arts, may be almost considered as so many distinct metals: this will naturally lead to the processes of melting, mixing and casting the metals; a general notice and explanation of many works, taking their origin in the malleable and ductile properties, will then follow; and the consideration of the metals, and of materials from the three kingdoms, will be concluded by a descriptive account of the modes of soldering.

Technological Repository, in each of which the subjects are perhaps treated in a more practical manner than in many other works in which they are touched upon. The Journal of the Royal Institution, vol. ix. p. 819 to 333, contains much curioua information upon alloys of steel with silvor, rhodium, platinum, nickel, &c.

DESCRIPTION OF THE

Bfjgsi'tal eCfjanutrt nnlj SHscs *OF* THE METALS AND ALLOYS COMMONLY EMPLOYED IN TILE MECHANICAL AND USEFUL ARTS.

N.B.—The first paragraph upon each metal is extracted from "A Manual of Chemistry," by William Thomas Brande, Esq., of Her Majesty's Miut, F. B.8., &c. Edition 1841. *Sote.*—The alloys are in general arranged under these metals which constitute respectively their largest proportional parts, but in some few instances uuder these from which they derive their peculiar characters. The greater part of these alloys have been obtained from various worksheps and manufactories; the autherities for these derived from books are for the most part stated. ANTIMONY is of a silvery white colour, brittle and crystalline in its ordinary texture. It fuses at about 800, or at a dull red heat, and is volatile at a white heat. Its specific gravity is 6712. *Satchett, Phil Tram.* 1803. *Brande,* 849.

Antimony expands on cooling; it is scarcely used alone, except in combination with similar bars of other metals for producing thermo-electricity: but antimony, which in the metallic state is frequently called " regulus," is generally combined with a large portion of lead, and sometimes with tin, and other metals. *See* Lead and Tin.

"Antimony and tin, mixed in equal proportions, form a moderately hard, brittle, and very brilliant alloy, capable of receiving an exquisite polish, and not easily tarnished by exposure to the air; it has been occasionally manufactured into speculums for telescopes. Its s. g, according to *Qellert,* is less than the mean of its constituent parts."—*A ikint Dictionary.* BISMUTH is a brittle white metal with a sbght tint of red: its specific gravity is 9-822. *(llatchctt, Phil. Trant.* 1803.) It fuses at 476 *(Crichton)* 507, *(Rudbery),* and always crystallises on cooling. According to Chaudet, pure bismuth is somewhat flexible. A cast bar of the metal *(see Sennit),* onetenth of an inch diameter, supports, according to Muschenbroeck, a weight of forty-eight pounds. Bismuth is volatile at a high heat, and may be distilled in close vessels. It transmits heat more slowly than most other metals, perhaps in consequence of its texture. *(Brande,* 861.) 266 BISMUTH ALLOYS.—CLICHEE PROCESS. COPPER.

BISMUTH—*continued.*

Bismuth is scarcely used alone, but it is employed for imparting fusibility to alloys, thus: 8 bismuth, 5 lead, 3 tin, constitute Newton's fusible alloy, which melts at 212. F.

2 bismuth, 1 lead, 1 tin, Rose's fusible alloy, which melts at 201 F. 5 bismuth, 3 lead, 2 tin, when combined melt at 199. 8 bismuth, 5 lead, 4 tin, 1 type-metal, constitute the fusible alloy used on the Continent for producing the beautiful casts of the French medals, by the *clichic* process. The metals should be repeatedly melted and poured into drops until they are well mixed. Mr. Charles V. Walker substituted antimony for the type-metal, and strongly recommends this latter in preference to the first-named fusible alloy. *Electrotype Manipulation,* Part II. pp. 9—11, where the *clichie* process is described. 1 bismuth and 2 tin make the alloy Mr. Cowper found to be the most suitable for rose-engine and eccentric-turned patterns, to be printed from after the manner of letter-press. He recommends the

thin plates to be cast upon a cold surface of metal or stone, upon which a piece of smooth paper is placed, and then a metal ring; the alloy should neither burr nor crumble; if proper, it turns soft and silky; when too crystalline, more tin should bo added.

All these alloys must be cooled quickly to avoid the separation of the bismuth; they are rendered more fusible by a small addition of mercury.

COPPER, with the exception of titanium, is the only metal which has a red colour: it has much lustre, is very malleable and ductile, and exhales a peculiar smell when warmed or rubbed. It melts at a bright-red or dull-white heat; or, according to Daniell, at a temperature intermediate between the fusing points of silver and gold = 1996 Fahr. Its specific gravity varies from 8 86 to 8-89; the former being the least density of cast copper, the latter the greatest of rolled or hammered copper *(Brande,* 812).

CorpER is used alone for many important purposes, and very extensively for the following; namely, sheathing and bolts for ships, brewing, distilling, and culinary vessels. Some of the fire-boxes for locomotive engines, boilers for marine engines, rollers for calico-printing and paper-making, plates for the use of engravers, &c.

Copper is used in alloying gold and silver, for coin, plate, Sc., and it enters with zinc and nickel into the composition of German silver. Sec p. 279. Copper alloyed with one-tenth of its weight of arsenic is so similar in appearance to silver, as to have been substituted for it.

The alloys of copper, which are very numerous and important, are principally included under the general name, *Bran.* In the more common acceptation, brass means the yellow alloy of copper, with about half its weight of zinc; this is often called by engineers " yellow brass." COPPER—*continued.*

Copper alloyed with about one-ninth its weight of tin, is the metal o£ brass ordnance, which is very generally called gun-metal; similar alloys used for the " *brasset"* or bearings of machinery, are called by engineers, *hard* brass, and also gun-metal; and such alloys when employed for statues and medals are called bronze. The further addition of tin leads to bell metal, and speculum metal, which are named after their respective uses; and when the proportion of copper is exceedingly small, the alloy constitutes one kind of pewter. See page 284.

Copper, when alloyed with nearly half its weight of lead, forms an inferior alloy, resembling gun-metal in colour, but very much softer and cheaper, lead being only about one-fourth the value of tin, and used in much larger proportion. This inferior alloy is called pot-metal, and also cock-metal, because it is used for large vessels and measures, for the large taps or cocks for brewers, dyers, and distillers, and those of smaller kinds for household use.

Generally, the copper is only alloyed with one of the metals, zinc, tin, or lead; occasionally with two, and sometimes with the three in various proportions. In many cases, the new metals are carefully weighed according to the qualities desired in the alloy, but random mixtures more frequently occur, from the ordinary practice of filling the crucible in great part with various pieces of old metal, of unknown proportions, and adding a certain quantity of new metal to bring it up to the colour and hardness required. This is not done solely from motives of economy, but also from an impression which appears to be very generally entertained, that such mixtures are more homogeneous than those composed entirely of new metals, fused together for the first time.

The remarks I have to offer on these copper alloys will be arranged in the tabular form, in four groups; and to make them as practicable as possible, they will be stated in the terms commonly used in the brass foundry. Thus, when the founder is asked the usual proportions of yellow brass, he will say, 6 to 8 oz. of zinc (to every pound of copper being implied). In speaking of gun-metal, he would not say, it had one-ninth, or 11 per cent. of tin, but simply that it was 1 J, 2, or 2j oz. (of tin,) as the case might be; so that the quantity and kind of the alloy, or the *addition* to the pound of copper, is usually alone named; and to associate the various ways of stating these proportions, many are transcribed in the forms in which they are elsewhere designated. See remarks on mixing these alloys, pp. 311—316.

AfLOTs Qf Copper And Zinc Oijlt. (Jfote.—The marginal numbers denote the ounces of zinc added to every pound of copper.)

I to 4 oz. Castings are seldom made of pure copper, as under ordinary circumstances it does not cast soundly: about half an ounce of zinc is usually added, frequently in the shape of *i* oz. of brass to every pound of copper; and by others 4 oz. of brass are added to every two or three pounds of copper. 208 ALLOYS OF COPPEE. AND ZINC; COPPER—*continued.* 1 to 1 i oz. Gilding metal, for common jewellery: it is made by mixing 4 parts of copper with one of calamine brass; or sometimes 1 lb. of copper with 6 oz. of brass. The sheet gilding-metal will be found to match pretty well in colour with the cast gun-metal, which latter does not admit of being rolled; they may be therefore used together when required.

3 oz. Red sheet brass, made at Hegermiihl, or *5* parts copper, 1 zinc, (Pre.) 3 to 4 oz. Bath metal, pinchbeck, Mannheim gold, similor, and alloys bearing various names, and resembling inferior jeweller's gold greatly alloyed with copper, are of about this proportion: some of them contain a little tin; now however, they are scarcely used. 6 oz. Brass, that bears soldering welL 6 oz. Bristol brass is said to be of this proportion. 8 oz. Ordinary brass, the general proportion; less fit for soldering than 6 oz. it being more fusible. 8 oz. Emerson's patent brass (1781, see specification, foot-note, p. 312,) was of this proportion; and so is generally the ingot brass, made by simple fusion of the two metals. 9 oz. This proportion is the one extreme of Muntz's patent sheathing. See 10j. 10J oz. Muntz's metal, or 40 zinc and 60 copper. "Any proportions," says the patentee, " between the extremes, 50 zinc and 50 copper, and 37 zinc, 63 copper, will roll and work at the red heat;" but the first-named proportion, or 40

zinc to 60 copper is preferred.

The metal is cast into ingots, heated to a red-heat, and rolled and worked at that heat into ship's bolts and other fastenings and sheathing. 12 oz. Spelter-solder for copper and iron is sometimes made in this proportion; for brass work, the metals are generally mixed in equal parts. See 16 oz. 12 oz. Pale yellow metal, fit for dipping in acids, is often made in this proportion.

16 oz. Soft spelter-solder, suitable for ordinary brass work, is made of equal parts of copper and zinc. About 14 lb. of each are melted together and poured into an ingot mould with cross ribs, which indents it into little squares of about 2 lb. weight; much of the zinc is lost. These lumps are afterwards heated nearly to redness upon a charcoal fire, and are broken up one at a time with great rapidity on an anvil or in an iron pestle and mortar. The heat is a critical point; if too great, the solder is beaten into a cake or coarse lumps, and becomes tarnished; when the heat is proper, it is nicely granulated, and remains of a bright yellow colour; it is afterwards passed through a sieve. Of course the ultimate proportion is less than 16 oz. of zinc. 16 oz. Equal parts is the one extreme of Muntz's patent sheathing. See lOj. 16 J oz. Hamilton and Parker's patent Mosaic gold, which is dark-coloured when first cast, but on dipping assumes a beautiful golden tint. When cooled and broken, say the patentees, "all yellowness must cease, and the tinge vary from reddish fawn or salmon colour, to a light purple or lilac, and COPPER—*continued.* from that to whiteness." The proportions are stated as from 52 to 58 zinc to 50 of copper, or 16 J to 17 oz. to the pound. *Repertory,* vol. iii. p. 248. 32 oz. or 2 zinc to 1 copper, a bluish-white, brittle alloy, very brilliant, and so crystalline that it may be pounded cold in a pestle and mortar. 128 oz. or 2 ounces of copper to every pound of zinc; a hard crystalline metal, differing but little from zinc, but more tenacious, it has been used for laps or polishing disks. Remarks On The Alloys or Cotter And Zinc.

These metals seem to mix in all proportions.

The addition of zinc continually increases the fusibility, but from the extremely volatile nature of zinc, these alloys cannot be arrived at with very strict regard to proportion. See remarks on these alloys, pages 811 to 816.

The red colour of copper slides into that of yellow brass at about 4 or 5 oz. to the pound, and remains little altered unto about 8 or 10 oz.; after this it becomes whiter, and when 32 oz. of zinc are added to 16 of copper, the mixture has the brilliant silvery colour of speculum metal, but with a bluish tint.

These alloys, from about 8 to 16 oz. to the pound of copper, are extensively used for dipping, as in an enormous variety of furniture work: in all cases the metal is annealed before the application of the scouring or cleaning processes, and of the acids, bronzes and lackers subsequently used.

The alloys with zinc retain their malleability and ductility well, unto about 8 or 10 ounces to the pound; after this the crystalline character slowly begins to prevail. The alloy of 2 zinc and 1 copper, before named, may be crumbled in a mortar when cold.

The ordinary range of good yellow brass, that files and turns well, is from about 4 4 to 9 oz. to the pound. With additional zinc, it is harder and more crystalline; with less, more tenacious and it hangs to the file like copper; the range is wide, and small differences are not perceived.

Allots Op CorPER And Tin Only. *(Note.*—The marginal numbers denote the ounces of *tin* added to every pound of copper.) *Ancient Copper and Tin Alloy).* j oz. Ancient bronze nails flexible, or 20 copper, 1 tin. *Vre.* 'According to Pliny, as quoted by Wilkinlj oz. Soft bronze, or.. 9 to 1 son.

Ancient weapons and tools, by various 2 oz. Medium bronze, or 8 to 1 / analyses, or 8 to 15 per cent. tin; medals from 8 to 12 per cent. tin, with 2 parts zinc 2 J oz. Hard bronze, or. 7 to 1 added to each 100, for improving the bronze colour.— *Vre.* 6 to 8 oz. Ancient mirrors.

270 ALLOYS OF COPPER AND TIN. COPPER—*continued. Modern Copper and Tin Alloys.* 1 oz. Soft gun-metal, that bears drifting, or stretching from a perforation. 14 oz. A little harder alloy, fit for mathematical instruments; or 12 copper and 1 very pure grain tin. 14 oz. Still harder, fit for wheels to be cut with teeth.

U to 2 oz. Brass ordnance, or 8 to 12 per cent. tin; but the general proportion is one ninth part of tin. 2 oz. Hard bearings for machinery.

2 J oz. Very hard bearings for machinery. By Muschenbroeck's Tables it appears that the proportion 1 tin and 6 copper is the most tenacious alloy; it is too brittle for general use, and contains 2 oz. to the pound of copper. See p. 289, For some other alloys used in machinery, *tec* Alloys of Copper, Zinc, Tin, and Lead, p. 272. 3 oz. Soft musical bells. 34 oz. Chinese gongs and cymbals, or 20 per cent. tin. *Ure,* fol. 191.) 4 oz. House bells.,1A oz. Large bells. 5 oz. Largest bells. 7i to 81, oz. Speculum metal. Sometimes one ounce of brass is added to every pound as the means of introducing a trifling quantity of zinc, at other times small proportions of silver are added; the employment of arsenic was strongly advocated by the Rev. John Edwards. Lord Oxniantown, now the Earl of Rosse, says, " tin and copper, the materials employed by Newton in the first reflecting telescope, are preferable to any other with which I am acquainted; the best proportions being four atoms of copper to one of tin (turner's numbers); in fact, 126 4 parts of copper to 58 9 of tin."—ZVaxt. *Royal Sue.,* 1840, p. 504.

The object agreed upon by all experimentalists appears to be the exact saturation of the copper with the tin, and the proportionate quantities *differ very materially, (in this and all other alloys,) according to the respective degrees of purity* of the metals: for the most perfect alloys of this group, Swedish copper, and grain tin, should be used.

Mr. Ross says, "When the alloy is perfect, it should be white, glassy, and flaky. When the copper is in excess, it imparts a red tint easily detected; when the tin is in excess, the fracture is granulated and also less white." ilia practice is to pour the melted tin into the fluid copper when it is at the lowest temperature that a mixture by stirring can be effected, then to pour the mixture into an ingot and to complete the combination by remelting in the most gradual manner, by putting the metal into the furnace as soon almost as the fire is lighted: trial is made of a little piece taken from the pot immediately prior to pouring.

See Dr. Edward's Paper on Speculums, published in the Nautical Almanack, 1787, reprinted in Gill's Technological Repository, vol. i. p. 240, and 264. *See* likewise Lord Oxmantown's Paper, Trans. Royal Society, 1810, pp. 503—527. COPPER—*continued.* 32 oz. of tin to one pound of copper, make the alloy called by the pewterers "*temper,*" which is added in small quantities to tin, for some kinds of pewter, called "*tin and temper,*" in which the copper is frequently much less than 1 per cent. See Pewter, page 284, and mixing the same, page 310. Remarks On The Alloys Of Copper And Tin Onlt.

These metals seem to mix in all proportions.

The addition of tin continually increases the fusibility, although when it is added cold it is apt to make the copper pasty, or even to set it in a solid lump in the crucible.

The red colour of the copper is not greatly impaired in those proportions used by the engineer, namely, up to about 2& ounces to the pound; it becomes greyish white at 6, the limit suitable for bells, and quite white at about 8, the speculum metal, after this, the alloy becomes of a bluish cast.

The tin alloy is scarcely malleable at 2 ounces, and soon becomes very hard, brittle, and sonorous; and when it has ceased to serve for producing sound, it is employed for reflecting light.

The tough tenacious character of copper under the tools rapidly gives way; alloys of 1J cut easily, 24 assume about the maximum hardness without being crystalline; after this they yield to the file by crumbling in fragments rather than by ordinary abrasion in shreds, until the tin very greatly predominates, as in the pewters, when the alloys become the more flexible, soft, malleable, and ductile, the less copper they contain.

Alloys Of Copper And Lead Only. *(Note.*—The marginal numbers denote the ounces of lead added to every pound of copper.) 2 oz. A red-coloured and ductile alloy.

4 oz. Less red and ductile; neither of these is so much used as the following, as the object is to employ as much lead as possible. 6 oz. Ordinary pot-metal, called dry pot-metal, as this quantity of lead will be taken up without separating on cooling; this is brittle when warmed. 7 oz. This alloy is rather short, or disposed to break. 8 oz. Inferior pot-metal, called wet pot-metal, as the lead partly oozes out in cooling, especially when the new metals are mixed; it is therefore always usual to fill the crucible in part with old metal, and to add new for the remainder. This alloy is very brittle when slightly wanned. More lead can scarcely be used as it separates on cooling.

Remarks On Thk Alloys Of Copper And Lead Only.

These metals mix in all proportions until the lead amounts to nearly half; after this they separate in cooling.

The addition of lead greatly increases the fusibility.

The red colour of the copper is soon deadened by the lead; at about *i* ounces to the pound the work has a bluish leaden hue when first turned, but changes in an hour or so to that of a dull gun-metal character.

272 ALLOTS OP COPPER, Z1JJC, TIN, LEAD, ETC. COPPER—*continued.*

When the lead does not exceed about 4 oz. the mixture is tolerably malleable, but with more lead it soon becomes very brittle and rotten: the alloy is greatly inferior to gun-metal, and is principally used on account of the cheapness of the mixture, and the facility with which it is turned and filed.

Allots Of Copper, Zinc, Tin, And Lead, 4c.

This group refers principally to gun-metal alloys, to which more or less zinc is added by many engineers; the quantity of tin in every pound of the alloy, which is expressed by the marginal numbers principally determines the hardness.

Keller's statues at Versailles are found as the mean of four analyses, to ;of—

Copper... 91-40 or about *Hi* ounces.
Zinc.... 5-53 „ 1 „
Tin.... 170 „ Oi „
Lead.... 1-37 „ 0J „

In 100' parts or the 16 ounces.

14 to 2 oz. tin to 1 lb. copper used for bronze medals, or 8 to 15 per cent. tin, with the addition of 2 parts in each 100 of zinc, to improve the colour. *(I're.)* The modern so-called bronze medals of our Mint are of pure copper, and afterwards bronzed superficially.

14 oz. tin 4 zinc to 16 oz. copper. Pumps and works requiring great tenacity. 14 oz. tin 2 oz. brass 16 „ „ „ . „ i For wheels to be cut into teeth.

J „ 2 „ 16 „ 1 2 „ 14 „ 16 „ For turning work.

2 „ 14 „ 16 „ For nuts of coarse threads, and bearings.

The engineer who uses these five alloys recommends melting the copper alone; the small quantity of brass is then melted in another crucible, and the tin in a ladle; the two latter are added to the copper when it has been removed from the furnace; the whole are stirred together and poured into the moulds without being run into ingots. The real quantity of tin to every pound of copper is about one-eighth oz. less than the numbers stated, owing to the addition of the brass, which increases the proportion of copper.

1J oz. tin, 1J oz. zinc, to 1 lb. copper. This alloy, which is a tough, yellow, brassy gun-metal, is used for general purposes by a celebrated engineer; it is made by mixing 14 lb. tin, 14 lb. zinc, and 10 lb. of copper; the alloy is first run into ingots. 24 oz. tin, 4 oz. of zinc, to 1 lb. copper, used for bearings to sustain great weights. 2i oz. tin, 24 oz. zinc, to 1 lb. copper, were mixed by the late Sir F. Chantrey, and a razor was made from the alloy; it proved nearly as hard as tempered steel, and exceedingly destructive to new files, and none others would touch it. (*Wilkinson.)* 1 oz. tin, 2 oz. zioc, 16 oz. brass. Best hard white metal for buttons. 4 oz. tin, 14 oz. zinc, 16 oz. brass. Common ditto. *(Phillips't Dictionary.)* COPPER— *continued.* 10 lb. tin, 6 lb. copper, 4 lb. brass constitute white solder. The copper and brass are first melted together, the tin is added, and the whole stirred ami poured through birch twigs into water to granulate it; it is afterwards dried and pulverised cold in an iron pestle and mortar. This white solder was introduced as a substitute for silver solder in making gilt buttons. Another button solder consists of 10 parts copper, 8 of brass, and 12 of spelter or zinc.

Remarks On Alloys or Copper, Zinc, Tin, And Lead, &c.

Ordinary Yellow Brass, (copper and zinc,) is rendered very sensibly harder, so as not to require to be hammered, by a small addition of tiu, say or 4 oz. to the lb. On the other hand, by the addition of i to J oz. of lead, it becomes more malleable and 'casta more sharply. Brass becomes a little whiter for the tin, and redder for the lead. The addition of nickel to copper and zinc constitutes the so-called German silver. *See* Nickel, page 279.

Gun Metal, (copper and tin,) very commonly receives a small addition of zinc; this makes the alloy mix better, and to lean to the character of brass by increasing the malleability without materially reducing the hardness. The standard measures for the Exchequer were made of a tough alloy of this kind. The zino, which is sometimes added in the form of brass, also improves the colour of the alloy, both in the recent and bronzed states. Lead in small quantity improves the ductility of gun-metal, but at the expense of *its* hardness and colour; it is seldom added. Nickel has been proposed as an addition to gun-metal by Mr. Donkin, and antimony by Dr. Ure.

Pot Metal (copper and lead) is improved by the addition of tin, and the three metals will mix in almost any proportions: when the tin predominates, the alloy so much the more nearly approaches the condition of gunmetal. Zinc may be added to pot-metal in very small quantity, but when the zinc becomes a considerable amount, the copper takes up the zinc, forming a kind of brass, and leaves the lead at liberty, and which in great measure separates in cooling. Zinc and lead are also very indisposed to mix alone, although a little arsenic assists their union by " killing" the lead, as in shot metal. Antimony also facilitates the combination of pot-metal; 7 lead, 1 antimony, and 16 copper, mix perfectly well the first fusion, and the alloy was decidedly harder than 4 lead and 16 copper; and apparently a better metal. "Lead and antimony, though in small quantity, have a remarkable effect in diminishing the elasticity and sonorousness of the copper alloys." GOLD is of a deep and peculiar yellow colour. It melts at a bright red heat equivalent, according to Daniell, to 2016 of Fahrenheit's scale, and when in fusion appears of a brilliant greenish colour. Its specific gravity is 19"3. It is so malleable that it may be extended into leaves which do not exceed the one two-hundred and eighty-two thousandth of an inch in thickness, or a single grain may be extended over 56 square inches of surface. This VOL. L T GOLD—*continued.* extensibility of the metal is well illustrated by gilt buttons, 144 of wbioh are gilt by 5 grains of gold, and less than even half that quantity is adequate to give them a very thin coating. It is also so ductile that a grain may be drawn out into 500 feet of wire. The pure acids have no action upon gold. *(Brande,* 972.)

Gold in the pure or fine state is not employed in bulk for many purposes in the arts, as it is then too soft to be durable. The gold foil used by dentists for stopping decayed teeth is perhaps as nearly pure as the metal can be obtained: it contains about 6 grains of alloy in the pound troy, or the one-thousandth part. Every superficial inch of this gold foil or leaf weighs j of a grain, and is 42 times as thick as the leaf used for gilding.

The wire for gold lace prepared by the refiners for gold lace manufacturers, requires equally fine gold, as when alloyed it does not so well retain its brilliancy. The gold in the proportion of about 100 grains to the pound troy of silver, or of 140 grains for double-gilt wire, is beaten into sheets as thin as paper; it is then burnished upon a stout red-hot silver bar, the surface of which has been scraped perfectly clean. When extended by drawing, the gold still bearing the same relation as to quantity, namely, the 57th part of the weight, becomes of only one-third the thickness of ordinary gold-leaf used for gilding. In water-gilding, fine gold is amalgamated with mercury, and washed over the gilding metal, (copper and tin,) the mercury attaches itself to the metal, and when evaporated by heat, it leaves the gold behind in the dead or frosted state: it is brightened with the burnisher. (See Technological Repository, vol. ii., p. 361—182S.) By the electrotype process a still thinner covering of pure gold may be deposited on silver, steel, and other metals. Mr. Dent has introduced this method of protecting the steel pendulum springs of marine chronometers and other time-pieces from rust.—*See* Note, p. 252.

Fine gold is also used for soldering chemical vessels made of platinum.

Gold Allots.

Gold-leaf for gilding contains from 3 to 12 grains of alloy to the Ob., but generally 6 grains. The gold used by respectable dentists, for plates, is nearly pure, but necessarily contains about 6 grains of copper in the ox. troy, or one 80th part; others use gold containing upwards of one-third of alloy, the copper is then very injurious.

With *copper,* gold forms a ductile alloy of a deeper colour, harder and more fusible than pure gold: this alloy, in the proportion of 11 of gold to 1 of copper, constitutes *standard gold;* its density is 17'157, being a little below the mean, so that the metals slightly expand on combining. One troy pound of this alloy is coined into 46t sovereigns, or 20 troy pounds into 934 sovereigns and a half. The pound was formerly coined into 44 guineas and a half. The standard gold of France consists of 9 parts of gold and 1 of copper. *(Brande,* 979.)

G O LD—*eon tin ued.*

For *Gold Plate* the French have three different standards: 92 parts gold, 8 copper; also 84 gold, 16 copper; and 75 gold, 25 copper.

In England, the purity of gold is expressed by the terms 22, 18, 16, carats, &c. The pound troy is supposed to be divided into 24 parts, and the gold, if it could be obtained perfectly pure, might be called 24 carats fine,

The " Old Standard Gold," or that of our present currency, is called fine, there being 22 parts of pure gold to 2 of copper.

The "New Standard," for watch-cases, &c. is 18 carats of fine gold, and 6 of alloy. No gold of inferior quality tol8 carats, or the " New Standard," can receive the Hall mark; and gold of lower quality is generally desoribed by its commercial value, as 60 or 40 shilling gold, &o.

The alloy may be entirely silver, which will give a green colour, or entirely copper for a red colour, but the copper and silver are more usually mixed in the one alloy according to the taste and judgment of the jeweller.

The following alloys of gold are transcribed from the memoranda of the proportions employed by a practical jeweller of considerabloexperience.

Firit Group. Different kinds of gold that are finished by polishing, burnishing,
&c. without necessarily requiring to be coloured:—. The gold of 22 carats fine, or the "Old Standard," is so little used on accouut of its expense and greater softness that it has been purposely omitted. 18 carats, or New Standard gold, of yellow tint: 15 dwt. 0 grs. gold. 2 dwt. 18 grs. silver.

2 dwt. 6 grs. copper. 20 dwt. 0 grs. 18 carats, or New Standard gold, of red tint:

15 dwt. 0 grs. gold.

1 dwt. 18 grs. silver.

3 dwt. 6 grs. copper. 20 dwt. 0 grs. 16 carats, or Spring gold: this, when drawn or rolled very hard, makes springs little inferior to those of steel. 1 oz. 16 dwt. gold, or 112 6 dwt. silver. — 4 12 dwt. copper.— "12 2 oz. 14 dwt. 2-8 When it is not otherwise expressed, it will be understood all these alloys are made with fine gold, fine silver, and fine copper, obtained direct from the refiners. And to ensure the standard gold passing the test of the Hall, three or four grains additional of gold are usual y added to every ounce. 60t. gold of yellow tint, or the fine gold of the jewellers; 16 carats nearly.

1 oz. 0 dwt. gold.

7 dwt. silver.

5 dwt. copper.

1 oz. 12 dwt. 40s. gold, or the old-fashioned jewellers' gold, about 11 carats fine: no longer used: 1 oz. 0 dwt. gold.

12 dwt. silver.

12 dwt. copper.

2 oz. 4 dwt. 276 COLOORED GOLDS, AND GOLD SOLDERS. GOLD—*continued. Second Group.* Coloured golds: these all require to be submitted to the process of wet-colouring, which will be explained: they are used in much smiller quantities, and require to be very exactly proportioned.

Full red gold:, Grey gold: (Platinum is also called 5 dwt. gold. grey gld b7 jewellers.) 5 dwt. copper. 3 dwt. 15 grs. gold.

— 1 dwt. 9 grs. silver.

Dr. Hermstadt's imitation of gold, which is stated not only to resemble gold in colour, but also in specific gravity and ductility, consists of 16 parts of platinum, 7 parts of copper, and 1 of zinc, put in a cruoible, covered with charcoal powder, and melted into a mass.—*Hanoverian Magazine.*

Gold alloyed with platinum is also rather elastic, but the platinum whitens the alloy more rapidly than silver.

By others, 4 grains of bnu'i are added to the solder; it then futes beautifully and is of good colour. Zinc is sometimes added to other gold solders to increase their fusibility, the zinc, (or brass when used.) sheuld be added at the last momeut, to lessen the volatilization of the cine. LEAD appears to have been known in the earliest stages of the world. Its colour is bluish white; it has much brilliancy, is remarkably flexible and soft, and leaves a black streak on paper: when handled it exhales a peculiar odour. It melts at about 612, and by the united action of heat and air, is readily converted into an oxide. Its specific gravity, when pure, is 11"445; but the lead of commerce seldom exceeds 11-35. *(Brandt,* 833.)

Lead is used in a state of comparative purity for roofs, cisterns, pipes, vessels for sulphuric acid, &c. Ships were sheathed with lead and with wood, from before the Christian era to 1450, after which wood was more commonly employed, and in 1790 to 1800 copper sheathing became general; of late years, lead with a little antimony has likewise been used, also Muntz's sheathing, an alloy of copper and zinc and galvanised sheet iron. The most important alloys of lead are those employed for printers' type, namely, about 3 lead, 1 antimony, for the smallest, hardest, and most brittle types.

4 lead, 1 antimony, for small, hard, brittle types. 5 lead, 1 antimony, for types of medium size. 6 lead, 1 antimony, for large types. 7 lead, 1 antimony, for the largest and softest types.

The small types generally contain from 4 to 6 per cent of tin, and sometimes also 1 to 2 per cent. of copper; but as old metal is always used with the new, the proportions are not exactly known. —See pp. 293, 310, and 323.

Stereotype-plates contain about 4 to 8 parts of lead to 1 of antimony.

Baron Wetterstedt's patent sheathing for ships, consists of lead with from 2 to 8 per cent. of antimony; about 3 per cent, is the usual quantity. The alloy Ib rolled into sheets.

Similar alloys, and those of lead and tin in various preparations, are much used for emery wheels and grinding-tools of various forms by the lapidary, engineer and others. The latter also employs these readily-fused alloys for temporary bearings, guides, screw-nuts, &c.—See foot-note, p. 293.

Organ pipes consist of lead alloyed with about half its quantity of tin to harden it. The mottled or crystalline appearance so much admired shows an abundance of tin.

Shot metal is said to consist of 401b. of arsenic to one ton of lead.

In casting sheet-lead, the metal was poured from a swing-trough upon a long and nearly horizontal table covered with a thin layer of coarse damp and, previously levelled with a metal rule or strike. The thickness of the fluid metal was determined by running the strike along the table before the lead cooled, the excess being thus swept into a spill-trough at the lower end of the table; but the sheet lead now more commonly used, is cast in a thick slab, and reduced between laminating rollers; it is known as "milled-lead."

The metal for organ-pipes is prepared by allowing the metal to escape through the slit in the trough, as it is slid along a horizontal table, so as to leave a trail of metal behind it; the thickness of the metal is regulated by the width of the slit through which it runs, and the rapidity of the traverse; 278 LEAD PIPES. SHOT. MEKCUKY. NICKEL.
LEAD—*continued.* a piece of cloth or ticken is stretched upon the casting table. The metal is planed to thickness, bent up, and soldered into the pipes.
Lead pipes are cast as hollow cylinders, and drawn out upon triblets; they are also cast of indefinite length without drawing. A patent was taken out for casting a sheath of tin within the lead, but it has been abandoned.

Lead shot are cast by letting the metal run through a narrow slit, into a species of colander at the top of a lofty tower; the metal escapes in drops, which for the most part assume the spherical form before they reach the tank of water into which they fall at the foot of the tower, and this prevents their being bruised. The more lofty the tower, the larger the shot that can be produced; the good and bad shot are separated by throwing small quantities at a time upon a smooth board nearly horizontal, which is slightly wriggled, the true or round shot run to the bottom, the imperfect ones stop by the way, and are thrown aside to be remelted, the shot are afterwards riddled or sifted for size, and churned in a barrel with black lead.

Mr. Joseph Manton took out a patent for amalgamating the surface of leaden shot with mercury. One pound of mercury was added to every cwt. of shot; they were churned together in a revolving barrel nearly full of water, until the shot assumed a silvery coat. These shot were stated to foul the barrel of the gun in a less degree than others, and also to be less injurious to the game after it had been killed.
MERCURY is a brilliant white metal, having much of the colour of silver, whence the terms *hydrargyrum, argentum vivum,* and *quicksilver.* It has been known from very remote ages. It is liquid at all common temperatures; solid and malleable at 40—F., and contracts considerably at the moment ot congelation. It boils and becomes vapour at about 670. Its specific gravity at 60 is 13 5. In the solid state its density exceeds 14. The specific gravity of mercurial vapour is 6'976. (ddmas, *Ann. de Ch. et Ph.* xxxiii. *Brande,* 928.)
Mercury is used in the fluid state for a variety of philosophical instruments, and for pressure gages for steam-engines, etc. It is sometimes, although rarely, employed for rendering alloys more fusible; it is used with tin-foil for silvering looking-glasses, and it has been employed as a substitute for water in hardening steel. Mercury forms amalgams with bismuth, copper, gold, lead, palladium, silver, tin, and zinc.

Mercury is commonly used for the extraction of gold and silver from their ores by amalgamation, and also in water gilding. *Set* Qolo. NICKEL is a white brilliant metal, which acts upon the magnetic needle, and is itself capable of becoming a magnet. Its magnetism is more feeble than that of iron, and vanishes at a heat somewhat below redness, 630, (fakiDat.) It is ductile and malleable. Its specific gravity varies from 8-27 to 8-40 when fused, and after hammering, from 8'69 to 9-00. It is not oxidized by exposure to air at common temperatures, but when heated in the air it NICKEL— *continued.* acquires various tints like steel; at a red-heat it becomes coated by a grey oxide. *(Brandt,* 802.)

Nickkl is scarcely used in the simple state; Mr. Brande mentions, however, that he has seen a Bavarian coin that had been struck in it; but it is principally used together with copper and zinc, in alloys that are rendered the harder and whiter the more nickel they contain; they are known under the names of albata, British plate, electrum, German silver, pakfong, teutanag, &c.: the proportions differ much according to price; thus the

Commonest are 3 to 4 parts Nickel, 20 copper, and 16 zinc.

Best. are 5 to 6 parts Nickel, 20 copper, and 8 to 10 zinc.

About two-thirds of this metal is used for articles resembling plated goods, and some of which are also plated, *tee* silver; the remainder is employed for harness, furniture, drawing and mathematical instruments, spectacles, the tongues for accordions, and numerous other small works.

The *while copper* of the Chinese, which is the same as the German silver of 'the present day, is composed, accordingfto the analysis of Dr. Fyfe, of 31-6 parts of nickel, 40-4 of copper, 25"4 of zinc, and 2 6 of iron 17-48 63-39 13-0. *Frick't Imitative Silver.*

The white copper manufactured at Sutil in the duchy of Saxe Hildburghausen, is said by Keferstein, to consist of copper 88"000, nickel 8'753, sulphur with a little antimony 0-750, silex, clay, and iron, 1"75. The iron is considered to be accidentally introduced into these several alloys, along with the nickel, and a minute quantity is not prejudicial.

Iron and steel have been alloyed with nickel; the former, (the same as the meteoric iron which always contains nickel,) is little disposed to rust: whereas the alloy of steel with nickel, is worse in that respect than steel not alloyed.
PALLADIUM is of a dull-white colour, malleable and ductile. Its specific gravity is about 11-3, or 11-86 when laminated. It fuses at a temperature above that required for the fusion of gold. *(Brande,* 998.)
"Palladium is a soft metal, but its alloys are all harder than the pure metal. "With silver it forms a very tough malleable alloy, fit for the graduations of mathematical instruments, and for dental surgery, for which it is much used by the French; with silver and copper, palladium makes a very springy alloy, used

for the points of pencil-cases, inoculating lancets, tooth-picks, or any purpose where elasticity and the property of not tarnishing are required; thus alloyed it takes a high polish. Pure palladium is not fusible at ordinary temperatures, but at a high temperature it agglutinates so as to be afterwards malleable and ductile."—*W. Cock.*

This useful metal was discovered by Dr. Wollaston in 1303, and it has recently been found in some abundance in the gold ores of the Minas Geraes district; the process now employed for its separation was discovered by Mr. P. N. Johnson. Palladium is calculated thoroughly to fulfil many of the purposes to which platinum and gold are applied in the useful arts, and 280 PALLADIUM. MANUFACTURE OF PLATINUM.

PALLADIUM— *continued.* from its low specific gravity, it may be obtained at about half the price of an equal *bulk* of platinum, anoVat one-eighth that of gold; and it equally resists the action of mineral acids and sulphuretted hydrogen.—*London Journal of Science for* 1840.

Palladium was used in the construction of the ballances for the United States' mint. *Sec* also Mallett's paUadiumizing process, page 302.

PLATINUM is a white metal extremely difficult of fusion, and unaltered by the joint action of heat and air. It varies in density from 21 to 21 "5, according to the degree of mechanical compression which it has sustained; it is extremely ductile, but cannot be beaten into such thin leaves as gold and silver. *(Brande,* 4th Ed. p. 822.)

The particles of the generality of the metals, when separated from the foreign matters with which they are combined, are joined into solid masses by simple fusion; but platinum being nearly infusible when pure, requires a very different treatment, which was introduced by Dr. WoUaaton, and is now conducted in the following manner by Messrs. Johnson and Cock, of London, the celebrated metallurgists.

The platinum is first dissolved chemically, and it is then thrown down in the state of a precipitate; next it is partly agglutinated in the crucible into a spongy mass, and is then compressed whilst cold in a rectangular mould by means of a powerful fly-press or other means, which in operating upon 500 ounces, converts the platinum into a dense block about S inches by 4, and 2 J inches thick. This block is heated in a smith's forge, with two tuyeres meeting at an angle, at which spot the platinum is placed amidst the charcoal fire; when it has reached the welding point, or almost a blue heat, it receives one blow under a heavy *drop,* or a vertical hammer somewhat like a pile-driving engine; it then requires to be re-heated, and it thus receives a fresh blow about every 20 minutes, and in *a* week or ten days, it is sufficiently welded or consolidated on all sides to admit of being forged into bars, and converted into sheets, rods, or wires by the ordinary means.

The motive for operating upon so great a quantity is for making the large pans for concentrating sulphuric acid, in only two or three pieces, which are soldered together with fine gold. In France, 2,000 ounces are sometimes welded into one mass, so that the vessels may be absolutely entire; a practice which is considered in this country to be unnecessarily costly. For small quantities the treatment is the same, but in place of the drop, the ordinary flatter and sledge hammer are used.

Platinum is exceedingly tough and tenacious, and "hangs to the file worse than copper," on which account, when it is used for the graduated limbs of mathematical instruments, the divisions should be cut with a diamond point, and which is the best instrument for fine graduations of all kinds, and for ruling grounds, or the lined surfaces for etching. *See* p. 180.

Platinum is employed in Russia for coin. This valuable metal is also used for the touch-holes of fowling-pieces, and in various chemical and PLATINUM ALLOYS. RHODIUM, IRIDIUM, OSMIUM. 281 PLATINUM—*continued.* philosophical apparatus, in which resistance to fusion or to the acids is essential.

The alloys of platinum are scarcely used in the arts; that with a small quantity of copper is employed in Paris for dental surgery. For alloys of platinum and steel, *see* Quarterly Journal of the Royal Inst. vol. ix. p. 328. The alloy of equal parts of steel and platinum is therein highly spoken of as a mirror.

"Dr. Von Eckart's alloy, contains platinum 2 40, silver 3'53, and copper 1171. It'is highly elastic, of the same specific gravity as silver, and not subject to tarnish, it can be drawn to the finest wire from J of an inch diameter without annealing, and does not lose its elasticity by annealing. It is highly sonorous, and bears hammering red-hot, rolling and polishing."

Mr. Boss added to silver, one-fourth of its weight of platinum, and he considers that it took up one-tenth its weight. The alloy became much harder than silver, capable of resisting the tarnishing influences of sulphur and hydrogen, and was fit for graduations.

An alloy of platinum with ten parts of arsenic is fusible at a heat a little above redness, and may therefore be cast in moulds. On exposing the alloy to a gradually-increasing temperature in open vessels, the arsenic is oxidized and expelled, and the platinum recovers its purity and inf usibility.—*Turner's Chemistry.*

Tin also so greatly increases the fusibility of platinum, that it is hazardous to solder the latter metal with tin-solder, although gold is so used.

Platinum, as well as gold, silver and copper, are deposited by the. electrotype process; and silver plates thus platinized are employed in Smee'a Galvanic Battery.

RHODIUM is a white metal very difficult of fusion: its specific gravity is about 11: it is extremely hard. When pure the acids do not dissolve it. *(Brandc,* 1001.) Rhodium was discovered in 1803 by Dr. Wollaston, and has been long employed for the nibs of pens, which have been also made of ruby, mounted on shafts of spring gold; these kinds have had to endure for the last 7 or 8 years the rivalry of " Hawkins's everlasting Pen," of which latter the author from many months' constant use can speak most favourably. "The everlasting pen," says the inventor, "is made of gold tipped with a natural alloy, which is as much harder than rhodium as steel is harder than lead; will endure longer than the ruby; yields ink as freely as the quill, is as easily wiped, and if left unwiped is Not Corroded." See also Mec. Mag. 1840, p. 554. Mr. Hawkins employs the natural alloy of iridium and osmium, two scarce metals discovered by Tennant amongst the grains of platinum; the alloy is not malleable, and is so hard as to require to be worked'with diamond powder. The metals rhodium, iridium, and osmium, are not otherwise employed in the arts than for pens; although steel has been alloyed with rhodium. See also the Quarterly Journal Royal Inst., voL ix.

SILVER is of a more pure white than any other metal; it has considerable brilliancy, and takes a high polish. Its specific gravity varies between 1(H, which is the density of cast silver, and 10'5 to 10-6, which is the density of rolled or stamped silver. It is so malleable and ductile, that it may be extended into leaves not exceeding a ten-thousandth of an inch in thickness, and drawn into wire much finer than a human hair. Silver melts at a bright-red heat, estimated by Mr. Daniell at 1873 of Fahrenheit's scale, and when in fusion appears extremely brilliant. (*Brand?,* 953.)

Silver is but little used in the pure unalloyed state on account of its extreme softness, but it is generally alloyed with copper in about the same proportion as in our coin, and none of inferior value can receive the " Hall mark." Diamonds are set in fine silver, and in silver containing 3 to 12 gre. of copper in the ounce, the work is soldered with pure tin.

The sheet metal for plated works is prepared by fitting together very truly, a short stout bar of copper, and a thinner plate of silver; when scraped perfectly clean they are tied strongly together with binding wire, and united by partial fusion without the aid of solder. The plated metal is then rolled out, and the silver always remains perfectly united and of the same proportional thickness as at first. Additional silver may be burnished on hot, when the surfaces are scraped clean as explained under gold: this is done either to repair a defect, or to make any part thicker for engraving upon, and the uniformity of surface is restored with the hammer. In addition to its use for articles of luxury, the important service of copper plated with silver for the parabolic reflectors of lighthouses must not be overlooked; these are worked to the curve with great perfection by the hammer alone.

Plated spoons, forks, harness, and many other articles are made of iron, copper, brass, and German silver either cast or stamped into shape; the objects are then filed and scraped perfectly clean; and fine silver often little thicker than paper is attached with the aid of tin solder and heat: the silver is rubbed close upon every part with the burnisher.

The electrotype process is also used under Elkington and Co.'s patent for plating several of the metals with silver, which it does in the most uniform and perfect manner; the silver added is charged by weight at about three times the price of the metal; the German silver or albata is generally used for the interior substance, as when the silver is partially worn through, the white alloy is not so readily detected as iron or copper.

Siltir Allots.

Mr. Brande says, " The alloy with *copper* constitutes plate and coin; by the addition of a small proportion of copper to silver, the metal is rendered harder and more sonorous, while its colour is scarcely impaired. Even with equal weights of the two metals, the compound is white: the maximum of hardness is obtained when the copper amounts to one-fifth of the silver. The *siandard silver* of this country consists of lift pure silver and £ copper, or 1110 silver and 0 90 copper. A pound troy, therefore, is composed of 11 oz. 2 dwt. pure silver, and 18 dwt. of copper. Its density is SILVER—*continued.* 10-3; its calculated density is 10'5; so that the metals dilate a little on combining. The French silver coin is constituted of 9 silver and 1 copper." *(Brande,* 968.)—The French *billon* coin, is 1 silver and 4 copper. *(Kelly.)*

"For *silver plate,* the French proportions are, 94 parts silver, 4 copper; and for trinkets, 8 parts silver, 2 copper."

Silver solders are made in the following proportions:—

Hardest silver solder, 4 parts fine silver, and 1 part copper; this is difficult to fuse, but is occasionally employed for figures.

Hard silver solder, 3 parts sterling silver, and 1 part brass wire, which is added when the silver is melted, to avoid wasting the zinc.

Soft silver solder for general use, 2 parts fine silver, and 1 part brass wire. By some few, J part of arsenic is added, to render the solder more fusible and white, but it becomes less malleable; the arsenic must be introduced at the last moment, with care to avoid its fumes.

Silver is also soldered with tin solder (2 tin, 1 lead), and with pure tin.

Silver and mercury are used in the plastic metallic stopping for teeth, see Appendix to VoL II. Note V. page 970.

TIN has a silvery white colour with a slight tint of yellow; it is malleable, though sparingly ductile. Common tinfoil, which is obtained by beating out the metal, is not more than 1-1000th of an inch in thickness, and what is termed *white Dutch metal* is in much thinner leaves. Its specific gravity fluctuates from 7'28 to 7 '6, the lightest being the purest metal. When bent it occasions a peculiar crackling noise, arising from the destruction of cohesion amongst its particles.

When a bar of tin is rapidly bent back'wards and forwards, several times successively, it becomes so hot that it cannot be held in the hand. When rubbed it exhales a peculiar odour. It melts at 442, and, by exposure to heat and air, is gradually converted into a protoxide. *(Brande,* 779.)

Pure tin is commonly used for dyer's kettles; it is also sometimes employed for the bearings of locomotive carriages and other machinery. This metal is beaten into very large sheets, some of which measure 200 by 100 inches, and are of about the thickness of an ordinary card;

the small sized foil is stated not to exceed one-thousandth of an inch in thickness. The metal is first laminated between rollers, and then spread one sheet at a time upon a large iron surface or anvil, by the direct blows of hammers with very long handles; great skill is required to avoid beating the sheets into holes. The large sheets of tin-foil are only used for silvering looking' glasses by amalgamation with mercury. *See* Mr. Farrow's apparatus, Trans. Soc. of Arts, vol. 49, p. 146. Tin-foil is also used for electrical purposes. The amalgam used for electrical machines, is 7 tin, 3 zinc, and 2 mercury.

Tin is drawn into wire, which is soft and capable of being bent and unbent many times without breaking; it is moderately tenacious and completely inelastic. Tin tube is extensively used for gas fittings and many other purposes; it has been recently introduced in an ingenious manner for the 284 TIN PLATE. KINDS OP PEWTER.

TIN—*continued.* formation of very cheap vessels, for containing artists' and common colours, besides numerous other solid substances and fluids, required to be hermetically sealed, with the power of abstracting small quantities. Rand's Patents.

Tin plate as an abbreviation of tinned iron plate; the plates of charcoal iron are scoured bright, pickled, and immersed in a bath of melted tin covered with oil, or with a mixture of oil and common resin; they come out thoroughly coated. Tinned iron wire is similarly prepared; there are several niceties in the manipulation of each of these processes which cannot be noticed in this place.

Tin is one of the most cleanly and sanatory of metals, and is largely consumed as a coating for culinary vessels, although the quantity taken up in the tinning is exceedingly small, and which was noticed by Pliny.

Tin imparts hardness, whiteness, and fusibility to many alloys, and is the basis of different solders, pewters, Britannia metal, and other important alloys, all of which have a low power of conducting heat.

Pewteb is principally tin; mostly lead is the only addition, at other times copper, but antimony, zinc, &c., are used with the above, as will be separately adverted to. The exact proportions are unknown even to those engaged in the manufacture of pewter, as it is found to be the better mixed when it contains a considerable portion of old metal to which new metal is added by trial.

In order to regulate the quality of pewter wares, the Pewterers' Company published in 1772 "A Table of the Assays of Metal, and of the Weights and Dimensions of the several sorts of Pewter Wares," and they threatened with expulsion from their guild, any who departed from the regulations given in this now scarce and disregarded pamphlet.

The assay is made by casting a small button of the metal to be tried in a brass mould, which is so proportioned that the button if pure tin weighs exactly 132 grains; all the metals added to the tin being heavier than the latter, the buttons or.assays are the heavier the less tin they contain, and at page 14 of the pamphlet the following scale is given:—

Assay of pure tin 182 grains.

Ditto of fine or plate metal 14 grains heavier than tin or.. 183 „

Ditto of trifling metal 3$... 185J „

Ditto of ley metal 16 —... 198 „ and it may be added although an unauthorised addition, that equal parts of tin and lead are about fifty grains heavier than tin or 232 grains.

Some pewters are now made nearly as common as the last proportion; when cast they are black, shining and soft; when turned, dull and bluish. Other pewters only contain J or J of lead; these when cast are white, without gloss and hard; such are pronounced very good metal, and are but little darker than tin. The French legislature sanctions the employment of 18 per cent, of lead with 82 of tin as quite harmless in vessels for wine and vinegar.

The finest pewter, frequently called " tin and temper," consists mostly of tin with a very little copper, which makes it hard and somewhat sonorous, but the pewter becomes brown-coloured when the copper is in excess. The copper TIN—*continued.* is melted, and twice its weight of tin is added to it, and from about J to 7 lb. of this alloy or the " temper," are added to every block of tin weighing from 360 to 390 lb.

Antimony is said to harden tin and to preserve a more silvery colour, but is little used in pewter. Zinc is employed to cleanse the metal rather than as an ingredient; some stir the fluid pewter with a thin strip, half zinc and half tin; others allow a small lump of zinc to float on the surface of the fluid metal whilst they are casting, to lessen the oxidation.

Britannia metal, or white metal, is said to consist of 34 cwt. of block tin, 28 lb. antimony, 8 lb. copper, and 8 lb. brass; it is cast into ingots and rolled into very thin sheets. This manufacture was introduced in about the year 1770, by Jessop and Hancock.—*(Lardncr'i Cyclopaedia. Set* page 311.)

Tin solders are very much used in the arts, and according to Dr. Turner 1 tin 3 lead, the coarse plumber's solder, melts at about 500 F.

2 — 1 — the ordinary or fine tin solder 860 F.

See also Bismuth, and the chapter on soldering.

ZINC is a bluish-white metal, with considerable lustre, rather hard, of a specific gravity of about 6'8 in its usual state, but when drawn into wire, or rolled into plates, its density is augmented to 7 or 72. In its ordinary state at common temperatures, it is tough, and with difficulty broken by blows of the hammer. It becomes very brittle when its temperature approaches that of fusion, which is about 773; but at a temperature a little above 212, and between that and 300, it becomes ductile and malleable, and may be rolled into thin leaves, and drawn into moderately fine wire, which, however, possesses but little tenacity. When a mass of zinc, which has been fused, is slowly cooled, its fracture exhibits a lamellar and prismatic crystalline texture. *Brande,* 770.)

Zinc, which is commercially known as "Spelter," although it is always brittle when cast, has of late years taken its place amongst the malleable metals; the early stages of its manufacture into sheet, foil and wire are stated to be conducted at a temperature somewhat above that of boiling water; and it may be afterwards bent and hammered cold, but it returns to its original crystalline texture when remelted. It has been applied to many of the purposes of iron, tinned-iron, and copper; it is less subject to oxidation from the effects of the atmosphere than the iron, and much cheaper, although less tenacious, ductile, or durable than the copper. The sheet metals when bent lengthways of the sheet, (or like a roll of cloth,) are less disposed to crack than if bent sideways, in this respect zinc and sheet iron are the worst: the risk is lessened when they are warmed.

The pipes of the great organ in the Town Hall, Birmingham, and of York Cathedral, are made principally of sheet zinc.

Zinc is applied as a coating to preserve iron from rust. *See* the accounts of Mallett's patent zincing process, page 301, and those of Craufurd's galvanized iron, and of Morewood's galvanized tinned-iron; *tee* Notes X and Y, page 971-2, Appendix to Vol. II.

286 ZINC. BIDDER? WARE, ETC. ZINC—*continued.*

Zinc mixed with one-twentieth its weight of speculum metal, may be melted in an iron ladle, and made to serve for soine of the purposes of brass such as common chucks. The alloy is sufficient to modify the crystalline character, but reserves the toughness of the zinc; it will not however bear hammering either hot or cold. *A. Host.*—Four atoms of zinc and one of tin, or 133-2 and 57 9, make a hard, malleable, and less crystalline alloy.

Biddery ware, manufactured at Biddery, a large city, 60 miles N.W. of Hyderabad in the East Indies, and also at Benares, is said by Dr. Heyne to consist of copper 16 oz., lead 4 oz., and tin 2 oz., melted together; and to every 3 oz. of this alloy, 16 oz. of spelter or zinc are added. The metal is used as an inferior substitute for silver, and resembles some soris of pewter.

The foregoing alloys are mostly derived from actual practice, and although it has been abundantly shown that alloys are most perfect, when mixed according to atomic proportions, or by multiples of their chemical equivalents, yet this excellent method is little adopted, owing to various interferences.

For example, it is in most cases necessary from an economic view, to mix some of the old alloys, (the proportions of which are uncertain,) along with the new metals. In most cases also unless the fusion and refusion of the alloys are conducted with considerably more care than ordinary practice ever attains, or really demands, the loss by oxidation completely invalidates any nice attempts at proportion; and which proportions can be alone exactly arrived at, when the combined metals are nearly or quite pure.

For the convenience however of those who may desire to pursue the scientific course, the chemical equivalents of the metals upon the hydrogen scale now most usually adopted, are appended to the list of metals, p. 290.

Thus, for mixtures of any metals, say tin and zinc; instead of taking arbitrary quantities, one atom of tin, or 57'9 parts by weight, should be combined with 1, 2, 3, *4,* or 5 atoms of zinc, or any multiple of 32 3 parts, and so with all other metals. See Speculum Metal, page 270; Zincing Process, page 301; Article Brass, Suppt. Ency. Brit., &c.

Remarks On Tredoold'b Tables Of Cohesion.

The following tables are a part of these upon the strength of materials, colloctod by Trodgold; thoy arc transcribed literatim from Vol. 50 of the Philosophical Magazine, p. 4'21—3. (1817.)

Tbcy will be found of considerable service in all inquiries which regard the employment of the metals and alloys in sustaining loads. The first column of figures denotes the comparative strength of the metals, glass boing considered as unity; thus steel of razor temper is nearly 16 times as strong as glass of equal size; and by the second column, it is sodu that a bar of steel one inch square is pulled asunder by a load of 150,000 lb.

It has been considered unnecessary for this place to extend the tables by any of the snbsequent labours of Telford, Rennie, Barlow, Brown, Bramah, and others; but the render interested in the earlier inquiries into the strength of materials, will find them enumerated in the introduction to Ronnie's experiments. Trans. Roy. Society, 1818, and Phil. Mag. Vol. 53, page 161.

TABLES OF THE COHESIVE FORCE OF SOLID BODIES. Table I.—Metals. (A) and *(I)* mark the highest and lowest result which Muschenbroek obtained from each kind of iron. t Kirwmn, Elem. Miner. ii. 155. This 1t the mean result of thirty-three experiments. t Calculated from experiments on the transverse strength, by arts. 14 and 16. (Yielded to the file without difficulty. 1 TABLES OF THE COHESIVE FORCE OF SOLID BODIES. f In the operation of casting, the surfate of the iron always becomes much harder, and is more tenacious than the internal parts: hence the strength of a small specimen is always greater than that of a lnrgo one. These of M. Ramus, bowever, are unexceptiouable lo thu respect, as the area of the section at the place of fracture was above nine squ.ro incies. JV". *U.* When the specific *gravity* is not referred to a separate autherity, it is to be considered that of the specimen of which the cohesive force is given. t Kirwan'fl Miner. voL ii Themson's Chemistry, vol. i.

J TABLES OF THE COHESIVE FORCE OF SOLID BODIES. Table II.—Allots. TABULAR VIEW OF SOME OF THE PROPERTIES OF METALS, *(Extracted principally from Branddi Manual of Chemistry,* 1841.)

Fusibility.' f Mercury

Potassium.

Sodium.

Tin.

Bismuth.

Lead.

Tellurium, rather less fusible than

Arsenic, undetermined. lead.

Zinc.... 773 deg.

Antimony, a little below a red heat.

Cadmium, about.. 442 deg.

Silver.... 1873 deg. Copper... 1996 „ Gold.... 2016 „ Cobalt, rather less fusible than iron. Iron, cast.... 27S6 deg.

requiring the highest heat of a smith's forge. Nickel, nearly the same as Cobalt. Palladium,!) Molybdenum

Almost infusible and not to be procured in buttons by the heat of a smith's forge, but fusible! before the oxyhy drogen blowpipe.

Iron, malleable Manganese

Tungsten Chromium Titanium Cerium Osmium Iridium Rhodium Platinum Columbium

Alloys having a Specific Gravity inferior to the mean of their components. Gold and Silver.

— Iron.

— Lead.

— Copper.

— Iridium.

—, Nickel.

Silver and Copper. 1

Copper and Lead.

Iron and Bismuth.

— Antimony.

— Lead. Tin and Lead.

— Palladium.

— Antimony. Nickel and Arsenic. Zinc and Antimony.

Extracted from Dr. Turner's Chemistry, Seventh Edition, 1841. TABUL AR VIEW OF METALS—*continued.*

Titanium

Manganese.

Platinum Palladium. Copper Gold Silver Tellurium. Bismuth. Cadmium. Tin.

Chromium. Rhodium. Nickel Cobalt Iron. Antimony. Zinc. Lead.

Potassium.

Sodium

Mercury

Hardness. Harder than Steel.

Scratched by Calcspar.

Scratch glass.

y Scratched by glass.

Scratched by the nail.

Soft as wax (at 60 deg.) Liquid.

BniTTLENESS. The following Metals are brittle, and most of them may even be reduced to powder. Antimony. Manganese. Arsenic. Molybdenum. Bismuth. Rhodium. Cerium. Tellurium. Chromium. Titanium. Cobalt. Tungsten. Columbium. Uranium.

CHAPTER XIV. REMARKS ON THE CHARACTERS OF THE METALS AND ALLOYS.

SECT. I.— HARDNESS, FRACTURE, AND COLOUR OF ALLOTS.

The object of the present chapter, is to explain in a general way some of the peculiarities and differences amongst alloys, in the manner of a supplement to the list; prior to entering in the next chapter on the means of melting the metals, without which process alloys cannot be made: yet notwithstanding that the list contains the greater number of the alloys in ordinary use, and many others, it is merely a small fraction of those which might be made, for, says Dr. Turner, "It is probable that each metal is capable of uniting in one or more proportions with every other metal, and on this supposition the number of alloys would be exceedingly numerous."

It is also stated by the same distinguished authority, that "Metals appear to unite with one another in every proportion, precisely in the same manner as sulphuric acid and water. Thus there is no limit to the number of alloys of gold and copper." The same might be said of many other metals, and when the alloys compounded of three, four, or more metals, are taken into account, the conceivable number of alloys becomes almost unlimited. "It is certain, however, that metals have a tendency to combine in definite proportion; for several atomic compounds of this kind occur native." "It is indeed possible that the variety of proportions in alloys is rather apparent than real, arising from the mixture of a few definite compounds with each other, or with uncombined metal; an opinion not only suggested by the mode in which alloys are prepared, but in some measure supported by observation." f

It appears to be scarcely possible to give any sufficiently general rules, by which the properties of alloys may be safely Dr. Turner's Chemistry, Seventh Edition, 1841, p. 558.

t Ibid., p. 559.

iuferred from those of their constituents; for although, in many cases, the-working qualities and appearance of an alloy may be nearly a mean proportional between the nature and qualities of the metals composing it; yet in other and frequent instances the deviations are excessive, as will be seen by several of the examples referred.to.

Thus, when lead, a soft and malleable metal, is combined with antimony, which is hard, brittle, and crystalline, in the proportions of from twelve to fifty parts of lead to one of antimony; a flexible alloy is obtained, resembling lead, but somewhat harder, and which is rolled into sheets for sheathing ships. Six parts of lead and one of antimony, are used for the large soft printers' types, which will bend slightly but are considerably harder than the foregoing; and three parts of lead and one of antimony are employed for the smallest types, that are very hard and brittle, and will not bend at all; antimony being the more expensive metal, is used in the smallest quantity that will suffice. The difference in specific gravity between lead and antimony constantly interferes, and unless the type metal is frequently stirred, the lead, from being the heavier metal, sinks to the bottom and the antimony is disproportionally used from the surface.f

In the above examples, the differences arising from the proportions appear intelligible enough, as when the soft lead prevails, the mixture is much like the lead; and as the hard, brittle antimony is increased, the alloy becomes hardened, and more In this alloy the antimony fulfils another service besides that of imparting hardness: antimony somewhat expands on cooling, whereas lead contracts very much, and the antimony, therefore, within certain limits, compensates for this contraction, and causes the alloy to retain the full size of the moulds.

Sometimes from motives of economy the neighbouring parts of machinery are not wrought accurately to correspond one with the other, but lead is poured in to fill up the intermediate space, and to make contact. As around the brass nuts in the heads of some screw presses, in the guides or followers for the same, and some other parts of either temporary or permanent machinery. Antimony is quite essential in all these cases to prevent the contraction the lead alone would sustain, and which would defeat the intended object, as the metal would otherwise become smaller than the space to be filled.

t A little tin is commonly introduced into types, and likewise copper in minute quantity; iron and bismuth are also spoken of; the last is said to be employed on account of its well-known property of expanding in cooling, so as to cause the types to swell in the mould, and copy the face of the letter more perfectly, but although I find bismuth to have been thus used, it appears to be neither common nor essential in printing types. 294 DISSIMILARITY OF THE COPPER ALLOYS. brittle: with the proportion of four to one, the fracture is neither reluctant like that of lead, nor foliated like antimony, but assumes very nearly the grain and colour of some kinds of steel and castiron. In like manner, when tin and lead are alloyed, the former metal imparts to the mixture some of its hardness, whiteness, and fusibility, in proportion to its quantity, as seen in the various qualities of pewter, in which however copper, and sometimes zinc or antimony, arc found.

The same agreement is not always met with: as nine parts of copper, which is red, and one part of tin, which is white, each *very* malleable and ductile metals, make the tough, rigid metal used in brass ordnance, from which it obtains its modern name of gun-metal, but which neither admits of rolling nor drawing into wire; the same alloy is described by Pliny as the soft *bronze* of his day. The continual addition of the tin, the *softer metal,* produces a gradual increase of hardness in the mixture; with about onesixth of tin the alloy assumes its maximum hardness consistent with its application to mechanical uses: with one-fourth to onethird tin it becomes highly elastic and sonorous, and its brittleness rather than its hardness is greatly increased.

When the copper becomes two, and the tin one part, the alloy is so hard as not to admit of being cut with steel tools, but crumbles under their action; when struck with a hammer or even suddenly warmed, it flies in pieces like glass, and clearly shows a structure highly crystalline, instead of malleable. The alloy has no trace of the red colour of the copper, but it is quite white, susceptible of an exquisite polish, and being little disposed to tarnish, it is most perfectly adapted to the reflecting speculums of telescopes and other instruments, for which purpose it is alone used.

Copper, when combined *in the same proportions* with a different metal, also light-coloured and fusible, namely two parts of copper, with one of zinc, (which latter metal is of a bluish-white, and *crystalline,* whereas tin is very ductile,) makes an alloy of entirely opposite character to the speculum metal; namely, the soft yellow brass, which becomes by hammering very elastic and ductile, and is very easily cut and filed.

Again, the same proportions, namely, two parts of copper, and one of lead, make a common inferior metal, called pot-metal or cock-metal, from its employment in those respective articles. This alloy is much softer than brass, and hardly possesses MALLEABILITY AND DUCTILITY OF ALLOYS.

295 malleability; when for example, the beer-tap is driven into the cask, immediately after it has been scalded, the blow occasionally breaks it in pieces, from its reduced cohesion.

Another proof of the inferior attachment of the copper and lead, exists in the fact that if the moulds are opened before the castings are almost cold enough to be handled, the lead will ooze out, and appear on the surface in globules. This also occurs to a less extent in gun-metal, which should not on that account be too rapidly exposed to the air; or the tin *strikes to the surface,* as it is called, and makes it particularly hard at those parts, from the proportional increase of the tin. In casting large masses of gun-metal, it frequently happens that little hard lumps, consisting of nearly half tin, work up to the surface of the runners or pouring places, during the time the metal is cooling.

In brass, this separation scarcely happens, and these moulds may be opened whilst the castings are red-hot, without such occurrence; from which it appears that the copper and zinc are in more perfect chemical union, than the alloys of copper with tin, and with lead.

SECT. II. MALLEABILITY AND DUCTILITY OF ALLOYS.

The malleability and ductility of alloys are in a great measure referable to the degrees in which the metals of which they are respectively composed, possess these characters.

Lead and tin are malleable, flexible, ductile, and inelastic, whilst cold, but when their temperatures much exceed about half way towards their melting heats, they are exceedingly brittle and tender, owing to their reduced cohesion.

The alloys of lead and tin partake of the general nature of these two metals; they are flexible when cold, even with certain additions of the brittle metals, antimony and bismuth, or of the fluid metal mercury; but they crumble with a small elevation of temperature, as these alloys melt at a lower degree than either of their components, to which circumstance we are indebted for the tin solders.

Zinc, wheu cast in thin cakes, is somewhat brittle when cold, but its toughness is so far increased when it is raised to about 300 F. that its manufacture into sheets by means of rollers is then admissible; it becomes the malleable zinc, and retains the malleable and ductile character, in a moderate degree, even when 296 MALLEABILITY AND DUCTILITY OF ALLOYS.

cold, but in bending rather thick plates it is advisable to warm them to avoid fracture; when zinc is remelted, it resumes its original crystalline condition.

Zinc and lead will not combine without the assistance of arsenic, unless the lead is in very small quantity; the arsenic makes this and other alloys very brittle, and it is besides dangerous to use. Zinc and tin make, as may be supposed, somewhat hard and brittle alloys, but none of the zinc alloys, except that with copper to constitute brass, are much used.

Gold, silver, and copper, which are greatly superior in strength to the fusible metals above named, may be forged either when red-hot or cold, as soon as they have been purified from their earthy matters, and fused into ingots; and the alloys of gold, silver, and copper, are also malleable, either red-hot or cold.f

Fine, or pure gold and silver, are but little used alone; the alloy is in many cases introduced less with the view of depreciating their value, than of adding to their hardness, tenacity, and ductility; the processes which the most severely test these qualities, namely, drawing the finest wires, and beating gold and silver leaf, are not performed with the pure metals, but gold is alloyed with copper for the red tint, with silver for the green, and with both for intermediate shades. Silver is alloyed with copper only, and when the quantity is small its colour suffers but slightly from the addition, although all its working qualities are greatly improved, pure silver being little used.

The alloys of similar metals having been considered, it only remains to observe that when dissimilar metals are combined, as those of the two opposite groups; namely, the fusible lead, tin, or zinc, with the less fusible copper, gold, and silver: the malleability of the alloys when cold, is less than that of the superior metal; and when heated barely to redness, they fly in pieces under the hammer; and therefore, brass, gun-metal, &c., when red-hot, must be treated with precaution and tenderness. Muntz's patent metal, which is a species of brass and is rolled red-hot, appears rather a contradiction to this; but in all probability this alloy, like the ingots of cast-steel, requires at first a very nice attention to the force applied. It will be also remembered the action of rollers is more regular than that of the hammer; and soon gives It is considered that most of the sheet zinc contains a very little lead.

T Gold alloyed with copper alone is not very malleable when hot.

rise to the fibrous character, which so far as it exists in metals, is the very element of strength, when it is uniformly distributed throughout their substance. This will be seen by the inspection of the relative degrees of cohesion possessed by the same metal when in the conditions of the casting, sheet, or wire, shown by the table, page 288, and to which quality, or the tenacity of alloys, we shall now devote a few lines.

SECT. III. STRENGTH OR COHESION OF ALLOYS.

The strength or cohesion of the alloys, is in general greatly superior to that of any of the metals of which they are composed. For example, on comparing some of the numbers of the table, on pages 288 and 289, it will be seen that the relative weights, which tear asunder a bar one inch square of the several substances, stand as follows; all the numbers being selected from Muschenbrok's valuable investigations, so that it may be presumed, the same metals, and also the same means of trial, were used in every case.

The inspection of these numbers is highly conclusive, and it shows that the engineer agrees with theory and experiment, in selecting the proportion 6 to 1 as the strongest alloy; and that the philosopher, in choosing the most reflective mixture, employs the weakest but one; its strength being only one-third to onesixth that of the tin, or one-twentieth that of the copper, which latter constitutes two-thirds its amount.

It is much to be regretted that the valuable labours of Muschenbrok have not been followed up by other experiments upon the alloys in more general use. One curious circumstance will be observed however, in those which are given, namely, that in the following alloys, which are the strongest of their respective groups, the tin is always four times the quantity of the other metal; and they all confirm the circumstance, of the alloys having 298 Roberts's Alloy Balance. mostly a greater degree of cohesion than the stronger of their component metals.

Fig. 137 represents a very ingenious instrument, denominated an alloy-balance, by its inventor, Mr. Roberts, of the present firm of Roberts, Fothergill and Dobinson, of Manchester. It is intended for weighing those metals the proportions of which are stated decimally, its principle, which is so simple as hardly to require explanation, depends upon the law that weights in equilibrium are inversely as their distances from the point of support.

For weighing out any precise number of pounds or ounces, in the common way, the arms of the ordinary scale-beam are made as nearly equal as possible; so that the weights, and the articles to be weighed, may be made to change places, in proof of the equality of the instrument. But to weigh out an alloy, say of 17 parts tin and 83 copper, unless the quantities were either 17 and S3 lb. or ounces, would require a little calculation.

This, in truth, is an exception; it barely equals in strength the alloys with 8 and 2'parts of tin to 1 of zincbut is superior to that of equal parts: it corroborates the greatjiucrease of strength in alloys generally. TABLE OP PROPORTIONS OF ALLOYS. 299

This is obviated, if the point of suspension *a,* of the alloy balance, which is hung from any fixed support *b,* is adjusted until the arms are respectively as 17 to 83; and for this purpose the half of the beam is divided into fifty equal parts numbered from the one end; and, prior to use, it only remains to adjust the weight, *w,* so as to place the empty balance in equilibrium. A quantity of copper, rudely estimated, having been suspended from the short arm of the balance, the proportionate quantity of the tin will be denoted with critical accuracy, when, by its gradual addition, the beam is exactly restored to the horizontal line; should the alloy consist of three or more parts, the process of weighing is somewhat more complex.

The annexed table was calculated by the author, for converting the proportions of alloys stated decimally, into avoirdupois weight. It applies with equal facility to alloys containing two or many components, so as to bring them readily within the power of ordinary scales.

TABLE FOR CONVERTING DECIMAL PROPORTIONS *Application of the Table.* The Chinese Packfong, similar to. our German silver, according to Dr. Fyfe's analysis, p. 279, is said to consist of—40.4 parts of Copper 1 6 oz. 4 drams, nearly. 25.4 — Zinc ., 4 — 0 — full. 31. 6-Nickel j" equivalent to 6 _ j _ nearly 2.6 — Iron J (.0 — 8 — full. 1 CO. 0 Parts 16 oz. 0 — Avoirdupois. 300 FUSIBILITY OF ALLOYS.

All nice attempts at proportion, are however entirely futile, uulcss the metals are perfectly pure; for example, it is a matter of common observation that for speculums, a variable quantity of from seven and a half to eight and a half ounces of tin is required for the exact saturation of every pound of copper, and upon which saturation the efficiency of the compound depends; bells of exactly similar quality sometimes thus require the dose of tin to vary from three and a half to five ounces, to the pound of copper, according to the qualities of the metals.

The variations in the purity of the metals obtained from different localities, are abundantly demonstrated by the disagreement in the cohesive strengths of the two in question, more particularly the tin, as seen on page 297, and which can be only ascribed to their respective amounts of impurity. Any other supposition than the presence of foreign matter, would necessarily go to disprove the fact of the metals being simple bodies, and therefore strictly alike when absolutely pure, wheresoever they may have been obtained.

SECT. IV.—FUSIBILITY OF ALLOYS.

In concluding this slight view of some of the general characters of alloys, it remains to consider the influence of heat, both as an agent in their formation, and as regards the degree in which it is required for their after-fusion; the lowest available temperature being the most desirable in every such case.

"Metals do not combine with each other," says Dr. Turner, "in their solid state, owing to the influence of chemical affinity being counteracted by the force of cohesion. It is necessary to liquefy at least one of them, in which case they always unite, provided their mutual attraction is energetic. Thus, brass is formed when pieces of copper are put into melted zinc; and gold unites with mercury at common temperatures by mere contact."

The agency of mercury in bringing about *triple* combinations of the metals, both with and without heat, is also very curious and extensive. Thus, in *water-gilding,* the silver, copper, or gilding metal, when chemically clean, are rubbed over with an amalgam of gold containing about eight parts of mercury; this Turner's Chemistry, p. 559.

COLD TINNING Mallett's ZINCING PROCESS. 301 immediately attaches itself, and it is only necessary to evaporate the mercury, which requires a very moderate heat, and the gold is left behind. *Water-silvering* is similarly accomplished.

Cast-iron, wrought-iron, and steel, as well as copper and many other metals, may be tinned in a similar manner. An amalgam of tin and mercury is made so as to be soft and just friable; the metal to be tinned is thoroughly cleaned either by filing or turning, or if only tarnished by exposure, it is cleaned with a piece of emery-paper or otherwise, without oil, and then rubbed with a thick cloth moistened with a few drops of muriatic acid. A little of the amalgam then rubbed on with the same rag, thoroughly coats the cleaned parts of the metal by a process which is described as *cold-tinning;* other pieces of metal may be attached to the tinned parts by the ordinary process of tin-soldering.

In making the tinned-iron plates, the scoured and cleaned iron plates are immersed in a bath of pure melted tin, covered with pure tallow; the tin then unites with every part of the surfaces; and in the ordinary practice of tinning culinary vessels of copper, pure tin is also used. The two metals, however, must then be raised to the melting heat of tin; but the presence of a little mercury enables the process to be executed at the atmospheric temperature, as above explained.

In Mr. Mallett's recently patented "processes for the protection of iron from oxidation and corrosion, and for the prevention of the fouling of ships," one proceeding consists in covering the iron with zinc.

The ribs or plates for iron ships are immersed in a "cleansing bath " of equal parts of sulphuric or muriatic acid and water, used warm; the works are then hammered, and scrubbed with emery or sand, to detach the scales and to thoroughly clean them; they are then immersed in a "preparing bath" of equal parts of saturated solutions of muriate of zinc and sal-ammoniac, from whieh the works are transferred to a fluid " metallic bath," consisting of 202 parts of mercury and 1292 parts of zinc, both by weight; to every ton weight of which alloy, is added about one pound of either potassium or sodium, (the metallic bases of potass and soda,) the latter being preferred. As soon as the cleaned iron works have attained the melting heat of the triple alloy, they are removed, having become thoroughly coated with zinc.

Being in the proportion of one atom of mercury to forty atoms of zinc. 302 ALL METALS CAPABLE OF FORMING ALLOYS.

"The affinity of this alloy for iron is, however, so intense, and the peculiar circumstances of surface as induced upon the iron presented to it by the preparing bath are such, that care is requisite lest by too long an immersion the plates are not partially or wholly dissolved. Indeed where the articles to be covered are small, or their parts minute, such as wire, nails or small chain, it is necessary before immersing them to permit the triple alloy to dissolve or combine with some wrought-iron, in order that its affinity for iron may be partially satisfied and thus diminished. At the proper fusing temperature of this alloy which is about 680 Fahr., it will dissolve a plate of wroughtiron of an eighth of an inch thick in a few seconds. "

"*The palladiumizing process.*—The articles to be protected are to be first cleansed in the same way as in the case of zincing; namely, by means of the double salts of zinc and ammonia, or of manganese and ammonia; and then to be thinly coated over with palladium, applied in a state of amalgam with mercury." f

In the opinion of eminent chemists and metallurgists, *all* the metals, even the most refractory, which nearly or quite refuse to melt in the crucible when alone, will gradually run down when surrounded by some of the more fusible metals in the. fluid state; in a manner similar to the solution of the metals in mercury, as in the amalgams, or the solutions of solid salts in water. The surfaces of the superior metals are, as it were, dissolved, washed down, or reduced to the state of alloys, layer by layer, until the entire mass is liquefied.

Thus rfickel, although it barely fuses alone, enters into the composition of German silver by aid of the copper, and whilst it gives whiteness and hardness, it also renders the mixture less fusible. Platinum combines very readily with zinc, arsenic, and also with tin and other metals; so much so, that it is dangerous to melt either of those metals in a platinum spoon; or to solder platinum with common tin solder, which fuses.at a very low temperature: although platinum is constantly soldered with fiue gold, the melting point of which is very high in the scale.

The reader is directed to the important and much-used methods of zincing iron, described in notes X and Y, pages 971—*2,* of the Appendix to vol. II. f Mechanics' Magazine, 1842, vol. 36, p. 41, where the details of this zincing process, and of the preparation of the varnishes and paints used for additional protection, are minutely transcribed from the specification of the patent. MK-TALUO UNIONS IN WHICH ALLOYS ARE FORMED. 308

Again, the circumstances that some of the fusible bismuth alloys melt below the temperature of boiling water, or at less than half the melting heat of tin, their most fusible ingredient, show that the points of fusion of alloys, are equally as difficult of explanation or generalization as many other of the anomalous circumstances concerning them.

This much however may be safely advanced, that the alloys, without exception, are more easily fused than the superior metal of which they are composed; and extending the same view to the *relative* quantities of the components, it may be observed that the hard solders for the various metals and alloys, are in general made of the selfsame material which they are intended to join, but with small additions of the more fusible metals. The solder should be, as nearly as practicable, equal to the metal on which it is employed, in hardness, colour, and every property except fusibility; in which it must excel just to an extent that, when ordinary care is used, will avoid the risk of melting at the same time, both the object to be soldered and likewise the softer alloy or solder by which it is intended to unite its parts.

It would appear as if every example of soldering in which a more fusible alloy is interposed, were also one of superficial alloying. Thus, when two pieces of iron are united by copper, used as a solder, it seems to be a natural conclusion that each surface of the iron becomes alloyed with the copper; and that the two alloyed surfaces are held together from their particles having been fused in contact, and run into one film. It is much the same when brass or spelter solder is used, except that triple alloys are then formed at the surfaces of the iron, and so with most other instances of soldering.

And in cases where metallic surfaces are coated by other metals, the latter being at the time in a state of fusion, as in tinned-iron plates and silvered copper; may it not also be conceived, that between the two exterior surfaces wihch are doubtless the simple metals, a thin film of an alloy compounded of the two does in reality exist? And in those cases in which the coating is laid on by the aid of mercury, and without heat, the circumstances are very similar, as the fluidity of mercury is identical with the ordinary state of fusion of other metals, although the latter require higher temperatures than that of pur atmosphere.

"When portions of the same metal are united by partial fusion, 304 METALLIC UNIONS IN WHICH ALLOYS ARE NOT FORMED.

and without solder, as in the process described as *burning together,* and more recently known as the " *autogenous* " mode of soldering, no alloy is formed, as the metals simply fuse together at their surfaces.

Neither can it be supposed that any formation of alloy can occur, where the one metal is attached to the other by the act of burnishing on with heat, as in making gilt wire, but without a temperature sufficient to fuse either of the metals. The union in this case is probably mechanical, and caused by the respective particles or crystals of the one metal being forced into the pores of the other, and becoming attached by a species of entanglement, similar to that which may be conceived to exist throughout solid bodies. This process, almost more than any other in common use, requires that the metals should be perfectly or chemically clean; for which purpose they are scraped quite bright before they are burnished together, so that the junction may be next approaching to that of solids generally.

And, lastly, when metals are deposited upon other metals by chemical or electrical means, the addition frequently appears to be a detached sheath, and which is easily removed; indeed, unless the metal to be coated is chemically clean, and that various attendant circumstances are favourable, the sound and absolute union of the two does not always happen, even when carefully aimed at.f

It is time, however, that we proceed to the description of the methods of forming the ordinary alloys, the subject of the succeeding chapter.

Delbruck's patent autogenous soldering process, which will be described in the chapter on soldering, is here alluded to.

T The table of the metals and alloys, pp. 2ss—286, contains many illustration of these two pages; but it should be observed, the energies of the respective unions depend very considerably on the chemical affinities of the metals concerned.

The Parts I. and II. of Electrotype Manipulation, by Mr. Charles V. Walker, 1841, may be likewise consulted by those interested in electro-plating, gilding, and etching.

CHAPTER XV.

MELTING AND MIXING THE METALS.

SECT. I. — THE VARIOUS FURNACES, ETC., FOR MELTING THE METALS.

The subject upon which we have now to enter consists of two principal divisions, namely, the melting and combining of the metals, and the formation of the moulds into which the fluid metals are to be poured. In the foundry the two processes are generally carried on together, so that by the time the mould is completed, the metal may be ready to be poured into it; but as in conveying these several particulars the one process must have precedence, I propose to commence with the means ordinarily employed in melting and mixing the metals, in order to associate more closely all that concerns the alloys. In accordance with ordinary practice, the formation of the moulds will be described whilst the metals may be supposed to be in course of fusion; the concluding remarks will be on pouring, or filling the moulds, the act strictly speaking of casting, and which completes the work.

The fusible metals, or those not requiring the red-heat, are melted when in small quantities in the ordinary plumber's ladle over the fire; otherwise larger cast-iron ladles or pans are used, beneath which a fire is lighted; for very large quantities and various manufacturing purposes, such as casting sheet-lead, and lead-pipe, and also for type-founding, the metals are melted in iron pans set in brickwork, with a fire-place and ash-pit beneath, much the same as an ordinary laundry copper, and the metals are removed from the pans with ladles.

The pewterers and some others call the melting pan *a pit,* although it is erected entirely *ahbve* the floor; and as their meltings are made up in great part of old metal which is sometimes wet or damp, they have iron doors to enclose the mouth of the pan, in case any of the metal should be splashed about from the moisture reaching the fluid metal.

VOL. i.-x 306 BRASS FOUNDERS', Oil AIR FURNACE.

Antimony, copper, gold, silver, and their alloys, are for the most part melted in crucibles within furnaces similar to the kind used by the brass founders, which is represented at a, fig. 138; the entire figure represents the imaginary section of a brass foundry, with the moulding trough, *b,* for the sand on the side opposite the furnace, the pouring or spill trough, *c,* in the center, and the core oven, *d,* which is usually built in the wall close against one of the flues, but these matters will be described hereafter.

The brass furnace is usually built within a cast-iron cylinder, about 20 to 24 inches diameter and 30 to 40 inches high, which is erected over an ash-pit, arrived at through a loose grating on a level with the floor of the foundry. The mouth of the furnace stands about 8 or 10 inches above the floor, and its central aperture is closed with a plate now usually of iron, although still called a *tile;* the inside of the furnace is contracted to about 10 inches diameter by fire-bricks set in Stourbridge clay, except a small aperture at the back about 4 or 5 inches square, leading into the chimney.

There are generally three or four such furnaces standing in a row, and separate flues proceed from all into the great chimney or stack, the height of which varies from about twenty to forty feet and upwards, the more lofty it is the greater the draught; every furnace has also a damper to regulate its individual fire.

It is quite essential for constant work to have several furnaces, in order that one or two may be in use, whilst the others lie idle to allow of their being repaired, as they rapidly burn away, and

KEFINEIts' FUItNACE, ETC.

307 when the space around the crucible exceeds about 2 or 3 inches, the fuel is consumed unnecessarily quick; the furnace is then contracted to its original size with a dressing of road drift and water applied like mortar, the fire is lighted immediately, and urged vigorously to glaze the lining.

It is also convenient to have several furnaces for another reason, as when a single casting requires more than the usual charge of one furnace, namely, about 40 to 60 lbs., two or more fires can be used. When the quantity of brass to be melted exceeds the charge of three or four ordinary air furnaces, the common blast furnace for iron is sometimes used as a temporary expedient, the practice however is bad, as it causes great oxidation and waste. The greatest quantities of metal, as for large bells,statues and ordnance, amounting sometimes to several tons, are commonly melted in reverberatory furnaces; these are, of course, never required by the amateur, and those who may desire to know their construction will find them described in various works on manufactures and chemistry.

The furnaces used by the gold and silver refiners are in many respects similar to the brass furnace *a,* fig. 13S, but they are built as a stunted wall along one or more sides of the refinery, and entirely above the floor of the same. The several apertures for the fuel and crucible are from 9 to 16 inches square, or else cylindrical, and 12 to 20 inches deep; the front edge of the wall is horizontal, and stands about 30 inches from the ground, but from the mouth of the furnace backwards it is inclined at an angle of about 20 to 40 degrees, so that the tiles, or the iron covers of the furnaces, lie at that angle. A narrow ledge cast in the solid with the iron plates covering the upper surface of the wall, retains the tiles in their position.

The small kinds of air furnaces are of easy construction, but as a temporary expedient almost any close fire may be used, including some of the German stoves and hot-air stoves, that is for melting brass, which is more fusible than copper; although it is much the most convenient that the fire be open at the top, so that the contents of the crucible may be seen without the necessity for its removal from the fire. Such stoves however radiate Road drift, or the scrapings of the ordinary turnpike roads, principally silex and alumina, is often used for the entire lining of the furnace. The refuse sand from the glass grinders, which contains flint glass, is also used for repairing them.

308 PORTABLE FURNACES, ETC.; MAKING VP. heat in a somewhat inconvenient manner, and to a much greater extent than the various portable furnaces,

most of which are lined with fire-brick or clay; the lining concentrates the heat and economises the fuel. Many of these portable furnaces answer not only for copper but also for iron, when they have a good draft; it may happen however, that the chemical furnaces are equally as inaccessible to the amateur, as those expressly constructed for the metals.

Country blacksmiths, who are frequently called upon to practise mauy trades, sometimes melt from ten to fifteen pounds of brass in the ordinary forge fire, but there is considerable risk of cracking the earthen crucible at the point exposed to the blast; a wrought iron pot is sometimes resorted to, but this is not very enduring, as the brass will soon cause it to burn into holes and leak. The forge, p. 203, may be thus used for a small quantity of brass, but it answers better when provided with the furnace prepared for it, as the blast then enters at the bottom and acts on all parts of the fuel alike. Some years back the blast was almost as commonly used for brass as it now is for iron; in which case the stream of air from the bellows entered beneath the fire-bars in a close chamber, which also served as the ash-pit, just as if the front wall of the furnace *a,* fig. 138, were continued down to the ground, with only a small hole for the blast. See Appendix, Note Z, page 973, Vol. II.

SECT. II.—OBSERVATIONS ON THE MANAGEMENT OF THE FURNACE, AND ON MIXING ALLOYS.

The fuel for the brass furnace is always hard coke, which is prepared in ovens and broken into lumps about the size of hens' eggs: in lighting the fire, a bundle of shavings, chips of cork, or any similar combustible, is first thrown in and ignited, and then some coke or charcoal is added. It is also usual to put the pot in the fire at an early stage, and with its mouth downwards: by this means the thin edge which admits the most easily of expansion gets hot first, and the heat plays within the crucible, so as to warm it gradually, it is not reversed until the whole is red hot; putting it in bottom downwards would be almost certain to cause it to crack.

The pot is now bedded upon the fuel, and the brass-founder, whilst making up the fire, puts an iron cover with a long central THE FIRE; FLUXES, CRUCIBLES, ETC.

309 handle over the mouth of the pot, to prevent the small cokes which are now thrown on from entering the same. Next, the charge of metal is put in the crucible, and three or four large pieces of coke are placed across the mouth of the pot; the tile is put on the furnace, the damper is then adjusted to heat the crucible quickly, and the whole is left to itself until the metal is run down.

The gold and silver refiners and jewellers manage their furnaces much in the same way, except that they support the crucible upon a hollow earthen stand placed on the fire-bars to catch any leakage, and also put an earthen cover over its mouth. They generally use coke, although charcoal is a purer fuel, and is laid *upon* the fluid metals to prevent oxidation.

The generality of the metals are far more disposed to oxidation when in the melted condition than when solid, it is therefore usual, whilst they are in the crucible, to protect their surfaces from the air with some flux, to lessen their disposition to oxidize.

In the iron furnace, the slag from the lime floats on the metal and fulfils this end; many brass-founders always throw broken glass, charcoal dust, sandiver, or sal-enixon, into the meltiug-pot by others these precautionary measures are altogether neglected. The black and white fluxes, borax and saltpetre, are also used for the precious metals, and oil or resin for the more fusible, as lead or tin, but excess of heat should be at all times avoided.

The generality of the fusible metals may be mixed in all proportions. Those in which the melting points are tolerably similar may be easily combined, such as lead with tin, or gold,with silver, or copper, these appear to call for no instructions Descriptions of the methods of constructing the white melting pot s used for lrasa, iron and steel, will be found in the Transactions of the Society of Arts: uamely, Mr. Marshall's method, vol. 41, p. 52; Mr. Anstey's, vol. 43, p. 32, and Sir. Charles Sidney Smith's, vol. 47, p. 54. The above, and the so called blue pots, or black-lead pots, are not burned until they are put into the fire for Ubo; but the Hessian pots, the English brown or clay pots, the Cornish and the Wedgwood crucibles, are all burned before use. The account of some comparative trials on all of these kinds will be found attached to Mr. Anstey's paper.

It may be further observed, that the pots for brass are too porous for gold and silver, as they *such up* too much of the same: the black-lead pots are closer and better for the precious metals, and they withstand change of temperature best of any kind; they are however the most expensive, but cannot be safely used with iluxes. The Hessian crucibles resist the fluxes, and serve with care for several consecutive meltings; the English clay pots, which resemble the Hessian, are safe for one or sometimes for more meltings, and their cost is trifling. The pots for gold and silver are occasionally coated or luted externally with clay as a prote-tiou.

310 MIXING PEWTER, TYPE METAL, ETC. beyond moderation in the beat employed, but the difficulty of making definite and uniform alloys increases, when the melting points of the metals, or their qualities or quantities, are widely dissimilar.

In mixing alloys with new metals, it is usual to melt the less fusible first, and subsequently to add the more fusible; the mixture is then stirred well together, and common opinion seems to be in favour of running the metal into an ingot mould, as the second fusion is considered more thoroughly to incorporate the mixture. Sometimes, with the same view, the alloy is granulated, by pouring it from the crucible into water, either from a considerable height through a colander, or over a bundle of birch twigs, which subdivide it into small pieces; others condemn such practices, and greatly prefer the first fusion, in order to avoid oxidation, and departure from the intended proportions.

But in many, and perhaps in most

cases, it is the practice to fill the melting pan, or the crucible, in part with old alloy, consisting of fragments of spoiled or worn-out work; and to which is added, partly by calculation but principally by trial, a certain quantity of new metals. This is not always done from motives of economy alone, but from the opinion that such mixturcs cast and work better than those made entirely of new metals.

When small quantities of a metal of difficult fusion, are added to large proportions of others which are much more fusible, the whole quantities are not mixed at once. Thus, in pewter, it would be scarcely possible to throw into the melted tin the half per cent, or the one per cent. of melted copper with any certainty of the two combining properly, and it is therefore usual to melt the copper in a crucible, and to add to it two or three times its weight of melted tin: this as it were, dilutes the copper, and makes the alloy known as *temper,* which may be fused in a ladle, and added in small quantities to the fluid pewter or to the tin, as the case may be, until on trying the mixture by the assay (see page 2S4) its proportions are considered suitable.

The metal for printer's type is often mixed nearly in the same manner; the copper is first melted alone in a crucible, the antimony is melted in another crucible, and is poured into the copper, sometimes a little lead is also added. The hard alloy and the tin are then introduced to the mass of typc-mctal or lead, also in great measure by trial, as old metal mostly enters into the mixture.

MIXING BRITANNIA METAL, BRASS, ETC. 311

In Lardner's Cyclopedia, Manufactures in Metal, Vol. III., p. 103, it states:

"The composition of Britannia metal is as follows:—3£ cwt. of best block tin; 28 lbs. of martial regulus of antimony; 8 lbs. of copper, and 8 lbs. of brass. The amalgamation of these metals is effected by melting the tin, and raising it just to a red heat in a stout cast-iron pot or trough, and then pouring into it, first the regulus, and afterwards the copper and brass, from the crucibles in which they have been respectively melted, the caster meanwhile stirring the mass about during this operation, in order that the mixture may be complete."

It would appear, however, much more likely and consistent that a similar mode is adopted in making this alloy, as in pewter and type-metal; namely, that the copper and brass arejmelted together in one crucible, the antimony then added from another crucible, and perhaps also a little tin; this would dilute the hard metals, and make a fusible compound, to be added to the remainder of the tin when raised a very little beyond its fusing point, so as to maintain fluidity when the whole were mixed and stirred together, previously to being poured into ingots. By this treatment the tin would be much less exposed to waste.

When a very oxidizable or volatile metal, as zinc, is mixed with another metal the fusing point of which is greatly higher, as with copper for making the important alloy brass, whatever weight of each may be put into the crucible, it is scarcely possible to speak with anything like certainty of the proportions of the alloy produced, from the rapid and nearly uncontrollable manner in which the waste of the zinc occurs.

Various means have been devised at different times for combining these two metajs. Thus the author of the article "Brass," Supplement to Encyclopedia Britannica, says:

"Although the most direct way of forming the different kinds of brass is by immediately combining the metals together, one of them, which is most properly called brass, was manufactured long before zinc, one of its component parts, was known in its metallic form. The ore of the latter metal was cemented with sheets of copper, charcoal being present. The zinc was formed and united with the copper, without becoming visible in a distinct form. The same method is still practised for making brass." VAK10US MODES OK SIAKIKG UKASS.

Under Emerson's patent, the more nearly direct fusion of the two metals was accomplished, as will he seen by the extract.

The author of the "Britanuica" also states:—"The best way of uniting zinc with copper, in the first instance, will be to introduce the copper in thin slips into the melted zinc, till the alloy requires a tolerable heat to fuse it, and then to unite it with the melted copper."

Some persons thrust the whole of the copper, in thin plates, into the melted zinc, which rapidly dissolves them; and the mass is kept in a pasty condition until within a few miuutes of the time of pouring, when they augment the heat to the degree required for the casting process.

But the plan which is the most expeditious, and now most usually adopted, is to thrust the broken lumps of ziuc beneath the surface of the melted copper with the tongs, which mode will be more particularly described; but howsoever conducted, a considerable waste of the zinc will inevitably occur.

It is also certain that every successive fusion wastes, in some degree, the more oxidizable metal, so that the original proportion is more and more departed from, especially with the least excess of heat; and when the metals are not well covered with flux. The loose oxide frequently mixes with the metal, this in brass gives rise to the white-coloured stains, and the little cavities filled with.

" I'atcnt granted to James Emerson, dated July 13, 1781.

"I take spelter in ingots and melt them down in an iron boiler; I then run the melted spelter through a ladle with holes in it, fixed over a tub of cold water, by which means the spelter is granulated or sholed, and is then fit for making bia--s on my plan. 1 theu mix about 54 lb. of copper shot, about 10 lb. of calcined calamine ground fine, and about one bushel of ground charcoal together; I then put into a casting pot a handful of the mixture, and upon it I put about 3 lb. of the sholed spelter; I then fill up the pot with the said mixture of cupper shot, calcined calamine and ground charcoal. In the same manner I fill eight other pots, so that 541b. of copper shot, 27 lb. of sholed spelter, about 101b. of calciued calamine, aud about 1 bushel of ground charcoal, make a charge for one furnace, containing 9 pots, for miking brass on my plan. My chief reason for using the small quantity of calamine in the process is more for confiuing the spelter by its weight, than for the sake of any increase arising from it, and 1 have frequently omitted the calamine in the process.

"The pots being so filled are respectively put into the furnace, and about 12 hours completes the process, aud from this charge I have on the average 32 lb. of pure fine brass, fit for making ingots or casting plates for making brass battery ware, or brass latten; and my brass, made as aforesaid, is of a superior quality to any brass made from copper and calamine."—Repertory of Arts, 1795, vol. 5, p. 21.
the white oxide of zinc; and in gunmetnl the stains and streaks are blacker, and the oxide of tin, (or putty powder,) being much harder than the former, is sadly destructive to the tools. The vitreous fluxes collect these oxides, and are therefore serviceable; but when in excess, they are liable to run into the mould when the metal is poured. The chemist generally uses covers to the crucibles, to lessen the access of air, and therefore the oxidation; but the brass-founder frequently leaves the metal entirely uncovered: no considerable waste occurs until the metal is entirely fused, and rather hotter than is required for pouring, which is indicated by the zinc beginning to burn at the surface with a blue flame.

In collecting the several alloys given at pages 265 to 286, especially those of copper, I found great difficulty in reconciling many of the statements derived from books; and therefore, to place the matter upon a surer basis, and also with some other views, I determined to mix a series of the copper alloys, in quantities of from one to two pounds each, pursuing, as nearly as possible, the common course of foundry-work, to make the results practical and useful.

My first intention was to weigh the metals into the crucible, and to find, by the weight of the product, the amount of loss in every case, as well as the quality of the alloy. Commencing with this view with copper and zinc, the several attempts entirely failed; owing to the extremely volatile nature of the latter metal, especially when exposed to the high temperature of melted copper. The difficulty was greatly increased, owing to the very large extent of surface exposed to the air, compared with that which occurs when greater quantities are dealt with, and the increased rapidity with which the whole was cooled.

The zinc was added to the melted copper in various ways; namely, in solid lumps, in thin sheet hammered into balls, poured in when melted in an iron ladle; and all these, both whilst the crucible was in the fire and after its removal from the same. The surface of the copper M as in some cases covered with glass The loss which occurs in melting brass filings is a proof that the granulation of the metals is not always desirable; and unless the brass filings are well *drawn,* by a group of magnets, to free them from particles of iron and steel, the latter often spoil the castings, as they become so exceedingly hard as to resist the file or turning tool, and can be only removed by the hammer and cold-chisel.

THE AUTHOR'S EXPERIMENTS OX or charcoal, and in others uncovered, but all to no purpose; as from one-eighth to one-half the zinc was consumed with most vexatious brilliancy, according to the modes of treatment: and these methods were therefore abandoned as hopeless.
I was the more diverted from the above attempts, by the wellknown fact that the greatest loss always occurs in the first mixing of the two metals, and which the founder is in general anxious to avoid: thus, when a very small quantity of zinc is required, as for the so-called copper castings, about 4 oz. of brass are added to every 2 or 3 lb. of copper. And in ordinary work, a pot of brass weighing 40 lb., is made up of 10, 20, or 30 lb. of old brass, and two-thirds of the remainder of copper, these are first melted: a short time before pouring, the one-third of the new metals, or the zinc is plunged in, when the temperature of the mass is such that it just avoids sticking to the iron rod with which it is stirred.

In mixing the copper and zinc for my experiments on brass, an entirely different course was therefore determined upon, namely, to melt the metals on a. much larger scale, and in the usual proportion, that is, 24 lb. of copper to 12 lb. of zinc, to learn the first loss of zinc when conducted with ordinary care. Then to remclt a quantity of the alloy over and over again, taking a trial bar every time, in order to ascertain the average loss of zinc in every fusion. From the residue of the original mixture, to make the alloys containing less zinc, by a proportional addition of copper; and those alloys containing more zinc, by a similar addition of zinc. And lastly, to have the whole of the bars assayed, to determine the absolute proportions of copper and ziuc contained in all, and from these analyses to select my series of specimens, as nearly in agreement as I could with the proportions in common use. This method answered every expectation.

Twenty-four pounds of copper, namely clean ship's bolts, were first melted alone to ascertain the loss sustained by passing through the fire, which was found to be barely oz. on the whole. A similar weight of the same copper was weighed out, and also 12 lb. of the best Hamburg zinc, in cakes about f inch thick, which were broken into pieces.

The copper was first melted, and when the whole was nearly run down, the coke was removed to expose the top of the pot, Twenty-four assays were made from as many bars, from 1J to 2 lbs. weight; besides which, several failures were laid aside.

ALLOYS OF COPPER AND ZINC. 315
which was watched until the *boiling* of the copper, arising probably from the escape of bubbles of air locked up at the lower part of the semi-fluid mass, ceased, and the copper assumed a bright red, but sluggish appearance; the zinc was then added.
Precaution is necessary in introducing the first quantity of zinc, not to *set* the copper, which is liable to occur if a

large quantity of cold metal is thrown in, simply from the abstraction of heat; and it is also necessary to warm the zinc that it may be perfectly dry, as the least moisture would drive the metal out of the pot with dangerous violence. A small lump of the zinc, therefore, was taken in the tongs, held beside the pot for a few moments, and then put in with the tongs with an action between a stir and a plunge, regardless of the flare, and of the low crackling noise, just as if butter had been thrown in; the zinc was absorbed, and the surface of the pot was clear from its fumes almost immediately. The remainder of the zinc was then directly added, in about eight pieces, one at a time, much in the same manner, but the danger of setting the copper nearly ceases when a small quantity of the spelter is introduced. After every addition the pot was free from flame in a few moments, a handful of broken glass was then thrown in, the tile replaced, and the whole allowed to stand for about fifteen minutes to raise the metal to the proper heat for pouring, which is denoted by the *commencement* of the blue fumes of the zinc.

The pot was then taken from the fire, well stirred for one minute and poured; the weight of the brass yielded was 34 lb. 12i oz., showing a loss of 1 lb. 3J oz., or one-tenth of the zinc, or the one-thirtieth part of the whole quantity. This experiment was repeated, and the loss was then 1 lb. 3 oz., the difference being only an ounce. By analysis, the mean of the two brasses was 31J per cent. zinc; or instead of being 8 oz. to the pound, it was only 7 oz.

Twelve pounds of each of these experimental mixtures were remelted six times, a bar weighing about one pound and a half, being taken every time; the two series of trials were conducted in different foundries, by different men, and quite in the ordinary course of work; but the loss per cent. of zinc was in the six experiments exactly alike in each series, that is, each bar, after the sixth melting, contained 22 J per cent. or 4$ oz. to the pound of copper. The second fusion in each case sustained the greatest 31G ALLOYS OF COPPER WITH ZINC, TIN, AND LEAD.
loss (say nearly two-fold); and in the others, taking all the accidental circumstances into account, the loss might be pronounced nearly alike every fusion.

In making the alloys with more zinc, the calculated weight of the first alloy was melted, and the amount of zinc was warmed and plunged in with the tongs, whilst the pot was in the fire, the whole was stirred and quickly poured: the losses in weight were rather large, but this is common when the zinc is in great quantity. To make the alloys containing less zinc than the alloy, the calculated weight of copper was first made red-hot and the respective portion of the brass alloy was then put in the pot, by which means the two ran down nearly together: it being found that the copper, if entirely melted before the brass was added, incurred a risk of being set at the bottom pf the pot; and remelting the mass, wasted the zinc. These alloys came out much nearer to their intended weights.

In making the tin aud copper alloys, very little difficulty was experienced. The copper was put into the pot together with a little charcoal, which was added to assist the fusion and also to cause the alloy to run clean out; as in pouring gun-metal a small quantity is nsually left on the lip of the crucible, which would have been an interference in these experiments. When the copper had ceased boiling, and was at a bright red heat, it was taken from the fire, aud the tin previously melted in a ladle, was thrown in, every mixture was well stirred and poured immediately.

In the fourteen alloys thus formed, each weighing about a pound and a half, namely, , 1, 1 , &c., up to 8 oz. of tin to the pound of copper, (missing the 6£ and 7,) no material loss was sustained in nine instances, and in the other five it never exceeded oz. and that quantity was probably lost rather in fragments than by oxidation.

Alloys of 2,4, 6, and 8 ounces of lead to the pound of copper, were made exactly under the same circumstances as the last.
Lord Oxmantown required the proportion of 2.75 copper and 1 sine, to be very carefully preserved, as that alloy was found to expand equally with the speculum metal to which it had been soldered. After many trials, Lord Oxinantown found that by employing a furnace deeper than usual, and by covering the metal with a layer of charcoal powder two inches thick, the loss every time was the smallest, and almost exactly the 180th each casting. To renew the charcoal dust it was folded up in paper and thrown in. *See* Trans. Royal Society, 1840, p. 507. CHAPTER XVI. CASTING AND FOUNDING.

——

SECT. I.—GENERAL REMARKS. METALLIC MOULDS.

We are indebted to the fusibility of the metals, for the power of giving them with great facility and perfection, any required form, by pouring them whilst in the fluid state into moulds of various kinds, of which the castings become in general the exact counterparts. This property is of immeasurable value.

Some few objects are cast in open moulds, so that the upper surface of the fluid metal assumes the horizontal position the same as other liquids, as in casting ingots, flat plates, and some few other objects; but in general the metals are cast in close moulds, so that it becomes necessary to provide one or more apertures or *ingates* for pouring in the metal, and for allowing the escape of the air which previously filled the moulds.

When these moulds are made of metal, they must be sufficiently hot not to chill or solidify the fluid metal before it has time to adapt itself thoroughly to every part of the mould; and when the moulds are made of earthy matters, although moisture is essential to their formation, little or none should remain at the time they are filled.

The earthen moulds must be also sufficiently pervious to air, that any vapour or gases which may be formed, either at the moment of pouring in the metal or during its solidification, may have free vent to escape; otherwise, if these gases are rapidly formed, there is great danger of the metal being driven out of

the mould with a violent explosion, or when more slowly formed and locked ip without sufficient freedom for escape, the casting will be said to be *blown,* as some of the bubbles of air will displace the fluid metal and render it spongy or porous. It not unfrequently happens that castings which appear externally good and sound, are full of hidden defects, because the surface being 318 PRINCIPLES OF MOULDING.
first cooled, the bubbles of air will attempt to break their way through the central and still soft parts of the casting.

Fig. 139. « C 6

The explanatory diagram, fig. 139, is intended to elucidate some of the circumstances concerning the construction of moulds, which in the greater number of cases are made only in two parts, but in other cases are divided into several. The figure to be moulded is supposed to be a rod of elliptical section, the mould for which might be divided into two parts through the line A, B, because no part of the figure projects beyond the lines *a, b,* drawn from the margin of the model at right angles to the line of division, and in which direction the half of the mould would be removed or *lifted;* the model could be afterwards drawn out from the second half of the mould in a similar manner.

The mould could be also parted upon the line C, D, because in that direction likewise, no part of the model extends beyond the lines *c, d,* which show the direction in which the mould would be then lifted.

The mould, however complex, could be also parted either upon A B or upon C D, provided no part of the model outstepped the rectangle formed by the dotted lines *b, c,* or was undercut.

But, considering the figure 139, to be turned bottom upwards, and with the line E, P, horizontal, the removal of the entire half of the mould upon the lines *e, f,* would be impossible, because in raising the mould perpendicularly to E, F, that portion of the mould situated within the one perpendicular *e,* would catch against the overhanging part of the oval towards A. Were the mould of metal, and therefore rigid, it would be entirely locked fast, or it would not *"deliver;"* were the mould of sand, and therefore yielding, it would break and leave behind PRINCIPLES OF MOULDING.
319 that part between A and E which caused the obstruction. Consequently, in such a case, the mould would be made with a small loose part between A and E, so that when the principal portion, from A to F, had been lifted perpendicularly or in the direction of the line *e,* the small undercut piece, A to E, might be withdrawn *sideways,* on which account it would be designated by the iron founder a *drawback,* by the brass founder a *false core.*
All the patterns in the mould, fig. 140, could be extracted from each half of the mould, because none of them encroach beyond the perpendicular line, or that in which the mould is lifted; *a* and *b,* could be laid in exactly upon the diagonal, or upon one flat side, or partly embedded; and in like manner/, *g, h,* might be sunk more or less into the mould, their sides being perpendicular; but the patterns in fig. 141 being undercut, the division of the mould into two parts *only* would be impracticable, and false cores of subdivisions would be required in the manner represented, the construction of which will be hereafter detailed.

Extending these same views to a more complex object, such as a bust, it will be conceived that the mould must be divided into so many pieces, that none of them will be required to embrace any overhanging part of the figure. For instance, were it attempted to mould a human head, so that the parting might pass through the central line of the face and down the back; the two halves could not be separated if they were made each in a single piece; as the inner angles of the eyes, the spaces behind the ears and the curls of the hair, would obstruct it, and the head could be only thus moulded by making false cores or loose pieces at these particular places, in the manner illustrated by the former figures. These would require to be accurately adapted to the surrounding parts, by pins or contrivances to 320 METAL MODLDS FOIt VEWlEIt WORKS.
ensure their re-taking their true positions. These remarks, however, are only advanced by way of general illustration, as figure casting is the most refined part of the art of moulding.

Metal moulds are employed for many works in the easilyfused metals, which are required to be produced in large quantities, and with great similitude and economy: the examination of which moulds will serve to demonstrate many of the points of construction and proceeding. Thus the common bullet-mould is made like a pair of pliers, the jaws of which are conjointly pierced with a hole or passage leading into a spherical cavity; the aperture is equally divided between the two halves of the mould, so that in fact the division is truly upon the diametrical line both of the sphere and the runner, or the largest part of each, otherwise the pliers could not be opened to remove the bullet when cast. Iron shot for great guns, are likewise cast in iron moulds, by which they also possess great accuracy of form and size.

Figs. 142. 143.

Figs. 142 to 145 represent the moulds for casting pewter inkstands: these moulds are a little more complex, and are each made in four parts; the black portions represent the sections of the inkstands to be cast. The moulds each consist of a top piece or *cap t,* a bottom or *core b,* and two sides or *cottles, s s;* in fig. 145, the one side is removed, in order to expose the casting, and the top piece / is supposed to be sawn through to make the whole more distinct.
METAL MOULDS FOR PEWTEIt WOItKS.
321
It will be seen, the top and bottom parts have each a rebate like the lid of a snuff-box, which embrace the external edges of the two side pieces *s s,* and the latter divide as in the bullet mould, exactly upon the diametrical line of the inkstand, which in a circular object is of course the largest part; the positions of the parts are therefore strictly maintained.

When the mould has been put together, laid upon its side, and filled through,r, the ingate, or, as it is technically called, the *tedge,* it is allowed to stand about a minute or two, and then the top *t*, is knocked off by one or two light blows of a pewter mallet; the mould is then held in the hand, and the bottom part or core is knocked out of the casting by the edge; lastly, the two sides are pulled asunder by their handles, and the casting is removed from the one in which it happens to stick fast; but it requires cautious handling not to break it. The face of the mould is slightly *coated* with red oclire and white of egg, to prevent the casting adhering to the same, and to give the works a better face: the first few castings are generally spoiled, until in fact the mould becomes properly warmed.

Most of the works made in the very useful material, pewter, are cast in gunmetal moulds, which require much skill in their construction; thus a pewter tankard, with a hinged cover and spout, consists of six species, every one of which requires a differet mould, thus, 1. The body has a mould in four part?, like that for the inkstand, but it is filled in the erect position through two ingates, which are made through the top piece *t*, of the mould: 2. The bottom requires a mould in two parts, and is poured at the edge: 3. The cover is cast in the same manner; and thus far the moulds are all made in the lathe, in which useful machine these castings are also finished before being soldered to-Fig. 146. 147. which divides in four parts, as shown in fig. 146, and much resembles, except in external form, the remaining mould: namely, 6. For the handle, which mould, like the last, consists of four pieces fitted together with various ears and projections; they are represented iu their relative positions in fig. 149, with the exception of the piece *a*, fig. 150, which is detached and shown bottom upwards. Fig. 148 shows the pewter handle separately with the three knuckles for joining on the cover; and on reference to fig 149, of the five parts through which the pin *p*, is thrust, the two external pieces belong respectively to the sides *c*, and *d*, of the mould, the others are parts of the casting, and the two hollows are formed by the two solid knuckles fixed to the detached piece of the mould *a*, fig. 150. At the time of pouring, the pin *p*, serves to connect the three parts, *a*, *c*, *d*, together, and also to form the whole in the casting, for the pin of the joint. Fig. 151 shows the section of the mould upon the dotted line *s:* by this it will be seen the handle is cast hollow, as almost immediately the mould has been filled through *t*, all but the thin external shell is poured out again, and the weight is reduced to less than half. To extract the handle, the pin *p*, is first twisted out; then the joint piece a, is removed; next the backpiece *b*; and lastly the two sides *c*, *d*, are pulled asunder.

Tin or pewter bearings for locomotive carriages, have been cast in appropriate metal moulds; and such materials are very METAL MOULDS FOE. PRINTING TYPES.

323 useful to the mechanist for many temporary purposes, such as collars, bearings, screws and nuts, either for difficult positions, or where no screw tap is at hand and the resistance is moderate; in such cases the parts of the machine constitute one portion of the mould, the apertures being closed with moist loam: the processes are most successful when the parts can be made warm and the clay is nearly dry. See Note W, page 970, Vol. II.

The most important, exact, and interesting example of casting in metallic moulds is that of type-founding, the description of which, as well as drawings of the mould, have been repeatedly given; some of the peculiarities only of this art, will be therefore noticed. Each complete set of types consists of five alphabets, A, A, a, *A*, *a*, besides many other characters, in all about two hundred, and which are required to be most strictly alike in every respect except in *device* and *width*; the width is the greatest for the W and M, and the least for the i and!. Every required measure of the types, (represented on an enlarged scale in fig. 152,) is determined by the *mould alone,* and not by any after correction.

If the moulds for the rectangular shafts of the types were made ns in figs. 153 or 151, the usual, forms of square moulds, they would not admit of alteration in width, as shifting *a*, fig. 153, would produce no change, and fig. 154 would thereby produce the form *b*. The mould which is used, is made in two L formed parts as in fig. 155; whence it follows that shifting the part *a*, to 324. TYPES, CLICHEE, l'KOCESS. the right or left increases or decreases the width of the type, without interfering with its thickness, or as it is technically called, its *body,* (b, fig. 152,) the width, w, is adjusted by a piece called the *register,* fixed at the bottom of the mould.

The device is changed by placing across the bottom of the mould one of the two hundred little pieces of copper, fig. 156, called *matrices,* into which the face of the letter is impressed by very beautifully formed punches. The length of the letter is determined by a contraction at the upper part of the mould, as shown at *c*, fig. 157, which represents the type as it leaves the mould; the metal is poured with *a jerk,* to make a sharp impression of the matrix: the mould, which is held in the left hand, and the ladle in the right, being jerked simultaneously upwards, at the moment of filling the mould, and without which the face of the type would be rounded and quite imperfect. The *breaks c,* or the runners of the types, are first broken off, and after a slight correction of the sides, the hollows or channels in the feet are planed out of a whole column of them, fixed between bars of wood, without touching the square shoulders which determine the lengths of the types, and are left as originally cast.

In some types with a large face and much detail, such as the illustrations given on the last page, the motion of the hand is barely sufficient to give the momentum required to throw the metal into the matrix, and produce a clean sharp impression. A machine is then used, which may be compared to a small forcingpump, by which the mould is filled with the fluid metal; but from the greater difficulty of allowing the air to escape, such types are in general considerably more unsound in the shaft or body; so that an equal bulk of them only weigh about threefourths as much as types cast in the ordinary way by hand, and which for general purposes is preferable and more economical Some other variations are resorted to in type-founding; sometimes the mould is filled at twice, at other times the faces uf the type are *dabbed,* (the *clichee* process,) many of the large types and ornaments are stereotyped, and either soldered to metal bodies, or fixed by nails to those of wood. The music type and ornamental borders and dashes, display much very curious power of combination.

The *die/tie* process is rather stamping than casting. The melted alloy, (see page 26G,) is placed in a *paper* tray, and stirred with a card until it assumes the pasty condition. The *vietal* die, or mould, is then "dabbed" upon the soft metal, aj in sealing a letter, but with a little more of sluggish force.

STEREOTYPE FOUNDING. 325 SECT. II.—PLASTER OF PARIS MOULDS AND SAND MOULDS.

Other examples of metallic moulds might be given, but there are far more frequent cases in which one single casting is alone required; or else the number is so small, or the pieces themselves are so large or peculiar, that the construction of metal moulds would be found almost or quite impracticable, even without reference to an equally fatal barrier, the expense.

In making these single copies in the metals of considerable fusibility, plaster of Paris is sometimes employed: thus, after the printer has arranged the loose types into a page, and the requisite corrections have been made, a stereotype, or *solid type,* is taken of the whole as a thin sheet of metal, which serves to be printed from almost as well as the original letters; and its small cost enables the printer to retain it for future use, after the types themselves have served perhaps for a hundred similar regenerations, and are ultimately worn out.

The stereotype founder takes a copy of the entire mass of type in plaster of Paris; this is dried in an oven, and placed *face downwards* within a cast iron mould, like a covered box, open at the four top corners. The mould and plaster-cast are heated to the fusing temperature of the type-metal, and gradually lowered into a pan or bath of the same by means of a crane; the hot fluid metal runs in at the corners of the mould, and raises the inverted plaster, which latter would rise entirely to the surface but for the restraint of the cover of the mould.

Type-metal is about eleven times a3 heavy as water: and if the mould be immersed four inches below the surface, it is subjected to a pressure equal to that of a column of water forty-four inches high, or of above two pounds upon every square inch.

The necessity of this arrangement is shown when a few ounces of type-metal are poured from a ladle on the face of the plaster; the metal looks like a dump, almost without any mark of the letters, whereas the stereotype-cast is nearly as sharp as the original type. The immersion fulfils the same end as the jerk of the hand-caster, or of the pump occasionally employed; and the long continuance of the mould in the fluid metal allows ample time for the air to escape in bubbles to the surface; after which the mould is raised and cooled in a vessel of water, and the plaster is mostly destroyed in its removal.

326 SAND MOULDS, FLASKS, OUTLINE OF MOULDING.

Plaster of Paris, although it may be, and frequently is used for the fusible metals, such as lead, tin, and pewter, cannot be employed alone for iron, copper, brass, and many other metals, the intense melting heats of which would calcine the material, and cause it to crumble; even the soft metals should not be very hot, or they will make the plaster of Paris blister off iu flakes or dust. We must therefore seek a substitute better capable of enduring the heat, and likewise susceptible of receiving definite forms; for which purpose damp sand, with a small natural or subsequent admixture of clay or loam, is found to be perfectly adapted.

The moulding sand cannot however be used without external support, and which is given by shallow iron frames without tops or bottoms, called flasks, represented in figs. 160 and 161. The bottom part, 4, 5, is supposed to have been rammed full of sand, and to stand upon a flat board, 6. The model of the plain flat bar which is to be cast, is now laid on the surface of the sand, that of the round bar is embedded half way in the same, and the mould is dusted with dry parting sand.

The top part of the flask, 2, 3, is shown still empty, and in the act of being attached to 4, 5, by its pins, which enter corresponding holes in the latter, easily, but without shake: 2, 3, is also rammed full of sand, and covered with a top board, 1, not represented, to avoid confusion. The mould is now opened, the models are removed, and channels are scooped out from the ends of the cavities left by the models, to the hollows or pouring holes at the end of the flask; the parts are all replaced in the order 1 to 6, represented in fig. 160, and the whole are fixed together by screw clamps, so as to assume the condition of fig. 161.

The flask is now placed almost perpendicularly beside the pouring-trough, and the metal is poured into it from the crucible as shown in fig. 138, p. 306; but the flask if small is put on the surface of the pouring or spill-trough, and propped up with a short bar.

This brief sketch of the entire process of moulding and casting in sand moulds, will be now followed by some remarks in greater detail: first, on the patterns of the objects to be cast: secondly, on the conditions required in the sand; and thirdly, the process of moulding simple and solid bodies. The, section then following will be devoted to moulding cored works, and figures, after which a few lines will be given upon the subject of filling the moulds.

SECT. III. PATTERNS, MOULDS, AND MOULDING SIMPLE OBJECTS.

The perfection of castings depends much on the skill of the pattern-maker, who should thoroughly understand the practice of the moulder, or he is liable to make the patterns in such a manner that they cannot be used, or at any rate be well used.

Straight-grained deal, pine, and mahogany, are the best woods for making patterns, as they stand the best; screws should be used in preference to nails, as alterations are then more easily made in the models, and glue joints, such as dovetails, tenons, and dowels, arc also good as regards the after use of the saw and plane for corrections and alterations.

Foundry patterns should be always made a little taper in the parts which enter most deeply into the sand, in order to assist their removal from the same, when their purposes will not be materially interfered with by such tapering. The patternmaker, therefore, works most of the thicknesses, and the sides or edges, both internal and external, a little out of parallel or square, perhaps as much as about one-sixteenth to one-eighth of an inch in the foot, sometimes much more.

When foundry patterns are exactly parallel, the friction of the sand against their sides is so great when they penetrate deeply, that it requires considerable force to extract them; and which violence tears down the sand, unless the patterns are much knocked about in the mould, to enlarge the space around them. This rough usage frequently injures the patterns, and causes the castings to become irregularly larger than intended, and also defective in point of shape, from the mischief sustained by the moulds; all which evils are much lessened when the patterns are made consistently taper and very smooth.

It must be distinctly and constantly borne in mind, that although patterns require all the methods, care, and skill, of good joinery or cabinet-making, they must not, like such works, be made quite square and parallel, for the reasons stated. Sharp, internal angles should in general be also avoided, as they leave a sharp edge or arris in the sand, which is liable to be broken down in the removal of the pattern; or to be washed down on the entry of the metal into the mould. Either the angle of the model should be filled w ith wood, wax or putty, or the sharp edges of the sand should be chamfered off with the knife or trowel. Sharp internal angles are very injudicious in respect also to the strength of castings, as they seem to denote where they w ill be likely to break; and more resemble carpentry than good metallic construction.

Before the patterns reach the founder's hands, all the glue that may have been used in their construction should be carefully scraped off, or it will adhere to and pull down the sand. The best way is to paint or varnish wooden patterns, so as to prevent them from absorbing moisture, as they will then bang to the sand much less, and will retain their forms much better. Whether painted or not, they deliver more freely from the mould when they are well brushed with black lead, like a stove.

In patterns made in the lathe, exactly the same conditions are required; the parts which enter deeply into the sand should be neither exactly cylindrical nor plane surfaces, but either a little coned, or rounding, as the case may be; and the internal angles should not be turned exactly to their ultimate form, but rather filled in, or rounded, to save the breaking down of the sharp edges of the mould.

Foundry patterns are also made in metal; these are very excellent, as they are permanent; and when very small are less apt to be blown away by the bellows used for removing the loose sand and dust from the moulds. To preserve iron patterns from rusting, and to make them deliver more easily, they should be allowed to get slightly rusty, by lying one night on the damp sand; next, they should be warmed sufficiently to melt becs'-wax, which is then rubbed all over them, and in great part removed, and then polished with a hard brush when cold. Wax is also used by the founder for stopping up any little holes in the wooden patterns; whiting is likewise employed, as a quicker but less careful expedient; and very rough patterns arc seared with a hot iron. The good workman however leaves no necessity for these corrections, and the perfection of the pattern is well repaid by the superior character of the castings.

Metal patterns frequently require to have holes tapped into them, for receiving screwed wires, by way of handles for lifting them out of the sand; and in like manner, large wooden patterns should have screwed metal plates let into them for the same purpose, or the founder is compelled to drive pointed wires into them, to serve as handles, which is an injurious practice.

The flasks or casting-boxes for containing the sand, are made of various sizes; each side is about 2 to 3 inches deep; they are poured at the edge when placed nearly vertical, but for large brass works the practice of the iron-founder is generally followed, who mostly pours his work horizontally, through a hole in the top, as will be explained. The pins of the flask should fit easily but without shake, or the two halves will shift about and cause a disagreement or slip in the casting. The tools used in making the moulds are few and simple; namely, a sieve, shovel, rammer, strike, mallet, a knife, and two or three loosening wires and little trowels, which it is unnecessary to describe.

The principal materials for making foundry moulds are very fine sand and loam; they are found mixed in various proportions, so that the respective quantities proper for different uses cannot be well defined; but it is always judicious to employ the least quantity of loam that will suffice. These materials are seldom used in their new or recent states for brass castiugs, although more so for iron, and the moulds made of fresh sand are always dried as will be explained.

The ordinary moulds are made of the old damp sand, and they are generally poured immediately or whilst they are *green;* The brass founders' sand and loam used in the metropolis are principally obtained from Hampstead, the iron-founders' sand from Lewisham likewise.

330 MATERIALS FOR FOUNDRY MOULDS.

sometimes they are more or less dried upon the face. The old working sand is considerably less adhesive than the new, and of a dark brown colour; this arises from the brick dust, flour and charcoal dust, used in moulding, becoming mixed with the general stock, which therefore requires occasional additions of new sand or loam, so that when slightly moist and pressed firmly in the hand, it may form a moderately hard compact lump.

Red brick-dust is generally used to make the partings of the mould, or to prevent the damp sand in the separate parts of the flask from adhering together.

The face of the mould which receives the metal, is generally dusted with meal-dust, or waste-flour; but in large works, powdered chalk, and also wood or tan ashes, are used, from being cheaper. The moulds for the finest brass castings are faced either with charcoal, loamstone, rottenstone, or a mixture of the same; the moulds are frequently inverted and dried over a dull fire of cork shavings, or when dried they are smoked over pitch or black resin lighted in an iron ladle.

The cores or loose internal parts of the moulds for forming holes and recesses, are made of various proportions of new sand, loam and horse-dung, as will be explained in the section on cored works. They all require to be thoroughly dried, and those containing horse-dung must be well burned at a red heat; this consumes the straw and makes them porous and of a brick-red.

In making the various moulds, it becomes necessary to pursue a medium course between the conditions best suited to the formation of the mould, and those best suited to filling them with the red-hot metal, without risk of failure or accident. Thus, within certain limits, the more loam and moisture the sand contains, and the more closely it is rammed, the better will be the impression of the model; but at the same time, the moist arid impervious condition of the mould would then incur the greater risk of accident, both from the moisture and from the non-escape of the air; therefore the policy, on the score of safety, is to use the sand as dry as practicable, so as to avoid the delay of after-drying, and also to keep the mould porous.

The founder, therefore, compromises the matter by using a The gold and silver casters frequently use a lighted link for facing their sandmoulds, and some of the type-founders' metallic moulds are smoked over a lamp; all these modes deposit a fine layer of soot upon the moulds.

INSTRUCTIONS FOR MOULDING ORDINARY WORKS. 331 little *facing* sand containing rather more loam, for the face of the green moulds for general work; and in those cases where much loam is used, the moulds are *thoroughly* dried by heat, which is not generally necessary with ordinary sand moulds.

The power of conducting heat is considerably less in red-hot iron than in copper and brass, and therefore the moulds for the latter require to be in a drier condition than those which may be used for iron; but in either case the presence of superfluous moisture is always attended with some danger to the individual as well as to the work.

Another point has also to be considered: as castings contract considerably in cooling, in moulding large and slight works the face of the mould must not be too strongly rammed, nor too much dried, or its strength may exceed that of the red-hot metal, whilst in the act of shrinking. The result would be, that in contracting, the casting would be rent or torn asunder from the restraint of the mould; whereas it should have the preponderance of strength, so as to pull down the face of the sand instead of being itself destroyed. But the exact condition both of the mould and of the melted metal, must be determined by the nature of the object to be cast; matters which can be only referred to with the development of the practice of the foundry, and upon which we shall now commence.

The sand having been prepared, and the appropriate flask and boards selected, the moulder first examines every pattern separately, to determine the most appropriate way of inserting it in the flask, as explained by fig. 140, p. 319; also to see that patterns, such as/ and *h,* therein shown, are smallest at the parts entering the most deeply into the sand, in order that they may *deliver* well. It should also be noticed whether they are perfectly smooth, and that there is no glue hanging about them, which would cause them to adhere and to pull down the moist sand.

The bottom flask, 4, 5, p. 326, is placed on a board not less than an inch or two longer and wider than itself, with the face 4, The above is the reason generally assigned for the fact, that the iron-founders may and do use their moulds with safety when sensibly more moist than is admissible for brass and copper castings. It is confirmatory of the fact that the more dense the mould, the drier it must be: as the sand used by iron-founders is also coarser and therefore more porous than that employed by brass-founders.

downwards, and it is filled from the side 5. A small portion of the strong facing-sand is rubbed through a fine sieve; the remainder is thrown in from the trough with the shovel, and the moulder drives the whole moderately hard into the flask, either with a mallet, the handle of the spade, or other rammer; or else he jumps up by aid of the rope suspended from the ceiling, and treads the sand in with his feet. The surface is then struck off level with a straight metal bar or scraper, a little loose sand is sprinkled on the surface, upon which another board is placed, and rubbed down close.

The two boards and the flask contained between them, are then all three turned over together; this requires them to be brought to the front of the moulding-trough, so that the individual may rest his chest against them, and his forearms upon the edges of the top board; he then grasps the three together at the back part with his outstretched hands, and thus retained in contact, the whole are quickly turned over upon the front edge of the moulding-trough, and then slid back upon the transverse bearers or *blocks,* to the usual position.

The top board is afterwards taken off, the clean surface of moist sand then exposed, is well dusted over with red brickdust crushed fine and contained in a linen bag: the mouth of the bag is held in the right hand, and the bottom corner in the left, and both hands are shaken up and down together, to scatter the dry powder uniformly over the flask; a part of the loose powder is removed with the hand-bellows, and the bottom half of the mould is then ready for receiving the patterns.

The models are next arranged upon the face of the sand at 4, so as to leave space enough to prevent the parts breaking one into the other, and also for the passages by which the metal is to be introduced, and the air allowed to escape. When there are only two or three pieces to be cast, a separate runner is often made to each of them from one of the holes in the end of the flask; when several small patterns are to be moulded, they are arranged on both sides a central runner, or *ridge,* from which small passages lead into every section of the mould. The whole mass when poured has been compared to a great fern leaf with its leaflets, and is usually called a *spray.*

Those patterns which are cylindrical or thick, are partly sunk in the sand, by scraping out hollow recesses with the bowl of an old copper spoon, and knocking the model into the sand with a mallet; afterwards the general surface is repaired to agreement with the *diametrical line* of the model, or its largest section, as the case may be, by means of a knife or a little piece of sheet steel, something like the worn-out blade of a dessert-knife bent up a little at the end, or else with very small trowels.

After the sand is made good to the edges of the patterns the brick-dust is again shaken over them, so that the patterns may receive a slight share as well as the general surface of the sand. The upper part of the flask, 2, 3, is then fitted to the lower, or 4, 5, by the pins, and this half likewise is made up"; first a little strong sand is sifted in, it is then filled up from the trough, rammed down, and struck off as before, the dry powder serving to prevent the two halves from sticking together.

In order to open the mould for the extraction of the models, aboard is placed on the top of flask, 2, 3, and struck smartly at different parts with the mallet, the tool is then laid aside, and the upper part of the flask and its board are lifted up *very gently and quite level,* after which it is inverted on its board, and now each of the inner faces of the mould is exposed. Should it happen that any considerable portion of the mould, say a part as large as a shilling, is broken down in one piece, the *cavity* is moistened with the end of the knife, the mould is again carefully closed, and lightly struck before the removal of the patterns; it is probable on the second lifting such piece will be picked up.

The breaks are carefully repaired before the extraction of the patterns, to effect which they are driven slightly sideways with blows of the mallet given on a short wire or punch, so as to loosen them by enlarging the space around them; the patterns are then lifted out very carefully with the finger-nails, or sometimes a pointed wire is driven a little way into the pattern to serve as st handle to lift it by: this process requires some delicacy not to tear away the sand, which accident must be carefully repaired, sometimes by replacing the loose pieces, at other times with a little new sand picked out of any unused part of the mould.

Should the flask only contain one or two objects, the ingate or runner is now scooped out of the sand, so as to lead from the object to the pouring hole, and when several objects are contained.

A steel wire, pointed and hardened, is convenient as a *picker out,* and when fixed in the pattern and struck sideways it serves as a *loosening har* likewise. a large central channel, and lesser passages sideways, are made as before mentioned. The entrance round about the pouringhole is smoothed and compressed with the thumb that it may not break down when the metal is poured, and all the loose sand is carefully blown out of the mould, both parts of which may be placed edgeways for the more convenient application of the bellows if necessary.

The succeeding processes are to dust the faces of both halves of the mould with meal dust or waste flour, as explained with regard to the brick dust, and to replace the mould and boards: the whole of them are then carried to the spill-trough, upon the edge of which they are rested whilst the one board is placed exactly level with the end of the flask, but the board on the side from which the crucible will be poured, is placed about two inches below, as in fig. 161, p. 326, and the hand-screws are fixed on as shown. The mould is now held mouth downwards, that any sand loosened in the screwing down may be allowed to fall out, and the flask, according to its size, is supported either on the ground or on the surface of the trough by aid of a little bar resting against the clamp: it is now quite ready to be filled, the particulars of which process will be described when the remarks on moulding are concluded.

In works that require the first side or 3, *i,* to be cut away for embedding the models, it is usual when the second part or 2, 3, has been made, to destroy the first or *false* side, (which is only hastily made,) and to repeat it in a more careful manner by inverting the lower flask upon 2, 3, proceeding in all other respects as before, by which means a much more accurate and sound mould "is produced.

When many copies of the same patterns are required, an *odd side* is prepared, that is, a flask is chosen to which there are two bottom sides, 4,5. One of these latter is very carefully arranged with all the patterns, but which are only embedded *barely half way,* so that when 2, 3, is filled and both are turned over, the whole of the patterns are left in the new side; a second side, 4, 5, is moulded to serve for receiving the metal, as the mould is destroyed every time the metal is poured in. By this plan the trouble ot re-arranging the patterns for every separate mould is saved, as they are merely replaced in the odd side, and the routine of forming the two working sides is repeated.

SECT. IV. MOULDING CORED WORKS.

If the objects to be cast require to be so moulded that when they leave the sand they may contain one or several holes, they are said to be cored, and in such cases a variety of methods are practised for introducing internal moulds or cores, which shall intercept the flow of the metal, and prevent it from forming one solid mass at those respective parts. For example, the pins inserted in the pewterers' moulds, figs. 146 and 149, pages 321 and 322, for producing the holes in the joints, are essentially cores. Various other methods are pursued, the three most usual of which are represented in figs. 162, 163, and 104; the upper figures show the exact sections of the three models or casting patterns; the lower figure represents the two halves of the mould, which are respectively shaded with perpendicular and horizontal lines, the cores are shaded obliquely; and the white open spaces show the hollows to be occupied by the metal when it is poured in.

First. Many works are said to deliver their owu cores; of such kind is fig. 162, in which the cavity extends through the model, and exactly represents that which is required in the casting; the hole is either made quite parallel, or a little larger one side than the other, and gradually taper between the two. In some cases,-when the hole is sufficiently taper, it delivers its own core as a continuation of the general mass of sand filling the one side of the flask; but in many or most cases, the space in the model is rammed full of strong sand at first, and it is then moulded as if to produce a plain solid casting. Before the mould is finally closed for pouring, the sand core is pushed carefully out of the pattern, and inserted in the mould; to denote its precise position, one side of the core is scored with one or two deep marks in the first instance, which cause similar ridges or guides in the mould.

Secondly. When the hole extends only partway through,the hole of the pattern, fig. 103, is fitted with a solid plug, sawn and filed out of soft unburnt brick, principally sand, (or the common Flanders brick,) the core is made long enough to project about as much as its own diameter, and the work is moulded as if to be cast with a solid pin instead of a hole. The last step is to extract the filed core, and to insert it into the hollow formed by itself in the flask.

Thirdly. The patterns for iron work and some others are mostly made with *prints* instead of holes, as in fig.l64,that is, the patternmaker places square or round pieces on one or both sides of the pattern, where the square or round holes are respectively required; and the founder has moulds for forming cores of corresponding diameters or sections, and in lengths of about two to twelve inches; short pieces of which are cut off as may be required.

For example, some core-boxes are made like fig. 105, for cylindrical cores; these divide through the axis, and are kept in position by pins; at the time when they are rammed they are fixed together by wood or iron staples, embracing three sides of the mould, or else by screw clamps. For straight cores, say *one* inch wide, twelve inches long, and half-inch thick, the pieces of wood, fig, 166, are also one inch thick, with an opening between them of twelve inches long and half-inch wide. This core-box is laid on a flat board; it is also held together with clamps, but without pins in the core-box, as the projection at the one end gives the position, it is rammed flush with both sides, aud the two parts can

Fig. 165. 1C6.

be then separated obliquely. If it is preferred to make the cores to the precise lengths instead of cutting them off, this core-hox admits of contraction in length, in the manner of the type mould, fig. 155, p. 323, and by placing thin slips between the two halves it may be temporarily increased in width, but not in thickness. Fig. 167, is a similar core-box for a casting with circular mortises; this requires either pins or projections at each end, as it cannot be opened obliquely. Core-boxes are sometimes made of plaster of Paris, wood is much better, and metal the best of all.

Many works require core-boxes to be made expressly for them; thus the dotted line in fig. 165 shows an enlargement in the center for coring a hole of that particular section. Figs. 168 and 169, represent the two halves of a brass or lead core-box suitable to the stopcock, fig. 170; and fig. 171 shows the core itself after its removal from the part 169, in which it is also figured. In 170, the model from which the objeet is moulded, the shaded parts represent the projections, or *coreprints,* which *imprint* within the mould the places where the extremities of the core, fig. 171, are supported when placed therein.

The various kinds of core-boxes are rammed full of new sand, sometimes with extra loam, the long cores are strengthened by wires; they are carefully removed from the boxes and thoroughly dried before use, in the oven prepared for the purpose.

Fig. 172 represents several examples of coring: in this view the works are represented of their ultimate forms, that is, with the holes in them; in fig. 173, the models are arranged in the flasks, with the runners all prepared, the prints of the cores being in every case shaded for distinction. Thus *a* is the stopcock, of which explanation has been already given; *b,* has a straight and a circular mortise; this pattern *delivers its own core,* in the manner referred to in fig. 162, as the model is made with mortises like the finished work; c, only requires a perpendicular square core*; d,* a round core parallel with the face of the flask, and in this manner all tubes and sockets are cast, whether of uniform or irregular bore, *see* fig. 165; *e,* has two rectangular Others prefer sand, horse-dung, and a very little loam, for making cores; these are dried, and then well burned, for which purpose they are put into an empty crucible within the fire, the last thing at night, and allowed to remain until the morning. This consumes the small particles of straw, and renders them more porous, in consequence of which the works become sounder from the free escape of air, the necessity of which was adverted to in the earlier part of this subject, anil cannot be too much insisted upon.

Vol. I. cores crossing each other at right angles; and /, is the cap of a double-acting pump, the core for which is shown in section by the white part of fig. 174, the shaded portions being the metal: the great aperture leads to the piston, the two smaller are for valves opening inwards and outwards; this of course requires a metal core-box capable of division in two parts, and made exactly to the particular form.

In addition to the cores used for making holes and mortises, much ingenious contrivance is displayed in the cores employed for other works of every-day occurrence, the undercut parts of which would retain them in the sand but for the employment of these and analogous contrivances. It will be now readily understood that if, in the fig. 141, p. 319, the parts shaded obliquely were separate, there would be no difficulty in removing first the upper half of the flask, then the false cores, after which the patterns would be quite free. By such a method, however, the circular edge of a sheave would require at least three such pieces, but fig. 175 shows a different way of accomplishing the same thing, when the pattern is made in two parts in the manner represented.

The term *false core* is employed by the brass founder to express the same thing as the *drawback* of the iron founder. The former calls every loose piece of the mould not intended for holes, a false core.

The entire model is first knocked into the side A, the sand is cut away to the inner margin of the pattern which terminates upon the dotted line *a,* and the side A, of the mould is then well dusted; a layer of sand is now thrown on, and rammed tolerably firm to form an annular core, which is made exactly level with the inner margin *b* of the pattern, and the core is well dusted; lastly, the side B is put on and rammed as usual. To extract the model, the side B is first lifted, the half pattern, *b, b,* (which is shaded,) is removed, and the ingate is cut in the side B, to the edge of the pulley, the mould is well dusted with flour and replaced.

The entire mould is now turned over, A is first removed, then the remaining half pattern *a, a,* which must be touched very tenderly or it will break down the core; and the runner, (which divides in two branches around the core,) is also scooped out in the side A, which is dusted with flour and replaced, ready for pouring. Common patterns not requiring cores are frequently divided into two parts in the above manner, so that when the mould is opened the pattern may divide and remain half in each side; this lessens the risk of breaking down the mould and the attendant trouble of afterwards repairing it.

SECT. V.— REVERSING AND FIGURE CASTING.

Supposing that an ornament, represented in section in fig. 176, has been modelled in relief, either in clay or wax upon a flat board, from which a thin casting in brass is wanted without the tablet, the process is called reversing, and is to be accomplished in any of three ways.

First, an empty flask is placed upon the board, 176, and rammed full of sand; it assumes the appearance of 177; the second part of the flask is attached to 177, and filled to make the part 178, which is called the *back-mould;* some clay is then 340 MODES OF REVERSING, 177.

rolled out to the intended thickness of the casting, with a cylindrical roller running on two slips of wood or on two wires, and a narrow hand of this clay is placed on 178, around the figure, that it may separate 177 and 178, exactly to the required distance, ready for receiving the metal.

By the second mode, 177 is first made, then 178, and from thelatterl79 is moulded, which is a counterpart of 177. A thin sheet of clay is then pressed all over 179, into every cavity, and cut off flush with the plane surface of the mould, by which it assumes the appearance denoted by the double line in 179. After this 17S is destroyed, and made over again in 179, but so much smaller than before as the thickness of the clay *lining;* when the new back-mould, 178, is placed in contact with 177, it leaves the required space for the intended casting. This mode is only preferable to the first, when many parts of the work are nearly perpendicular; in which case, if the first mode be adopted, a portion of the back-mould, *177,* must be pared away at the perpendicular parts, and if incautiously performed there will be a risk of irregularity of thickness, or even of holes in the casting.

The third mode, is to take a casting of 170 in plaster of Paris; when this is thoroughly dry it is oiled, and poured full of a cement of wax, grease, and red-ochre, which is poured out again when partially set, leaving a thin crust behind (as in the pewter handle). A second, a third, or more layers of wax are thus added until the whole is sufficiently thick, when the wax shell is extracted, and then moulded from in the ordinary manner: the first brass casting is finished and chased to serve as the permanent pattern. The management of the wax requires practice.

In constructing such moulds additional care is given to every part of the work; for example, the sand is sifted much finer, the parting is made with fine charcoal dust, and the facing with charcoal and rottenstone mixed together in about equal parts, the mixture being of a slaty colour; sometimes the loamstone, which is found in the pits where clay for making tiles is dug, is used instead of rottenstone. The moulds are well dried in au oven, or over the mouth of the furnace, and the faces are afterwards smoked over a dull fire of cork shavings; this deposits a very fine layer of soot over the face of the mould, which greatly assists the running of the metal; when this additional care is taken the works are known as *fine-castings.*

In casting figures, such as ousts, animals, and ornaments consisting of branches and foliage, considerably more skill is required; the originals are generally solid, but the moulds necessarily divide into very many parts. (See note A A, p. 974, Appendix, Vol. II.) Most persons will have had the opportunity of judging of the complexity of these moulds, from similar works in plaster of Paris, which are frequently purchased by artists and the virtuosi before the seams of the mould are removed.

A glance at these plaster-casts, at the complex and undercut form of many of these ornamental works, and at the explanatory diagram on page 318, will convey some notion of the method to be pursued as well as of the trouble attending them. It is shown for example, by the diagram just referred to, that all figured works approaching to the circular or elliptical section, require that the mould should be divided into at least three parts, except under most favourable circumstances. In the human figure and quadrupeds, the four limbs and the trunk require at least three parts each, and often many more; it will be easily conceived therefore that such moulded works require considerable skill and patience.

Piece after piece of the mould is successively produced, just as in making the core, fig. 175, p. 339, every piece embraciug only so much of the figure, as in no part to require any core to overhang the line in which it is withdrawn. The side of the mould in which the figure is partly embedded is first dusted with charcoal, and then the first core is very carefully rammed into the nook, and pared down to the new line of division; the green or wet sand core is then dusted, and the second core is made, and afterwards dusted, when the moulder proceeds with the third core and so on; every one being carefully adapted to its neighbour and withdrawn, to see that all is right, before the. succeeding core is proceeded with. The relative positions of the cores amongst themselves are readily recognised and maintained by the irregularity of their forms, as in a child's dissected map, or by making a notch or two here and there, which are faithfully copied in the succeeding piece. It is frequently necessary to thrust two or more broken needles through the green cores into the neighbouring parts to connect them together, in imitation of the pins in the flasks.

All the parts of the mould are dried in the oven, and the facings are smoked over a cork fire as before explained; the perfection of the casting is augmented by pouring whilst the mould is still slightly *warm,* as otherwise on cooling it has an increased affinity for damp; but the mould when *hot* is more or less filled with aqueous vapour which is equally prejudicial.

When a figure, such as a bust, is required to be cast hollow from a solid model, it is first moulded exactly as above. The core is now produced as follows: at the foot of the bust a large space, nearly equal in length and bulk to the bust, is cut away in the sand, to serve for fixing the core in the mould, or for the *balance,* as it is called, as the core cannot be propped up at both ends. The entire hollow, that is for the bust and the balance, is filled with a composition of about one part of plaster of Paris and two of sand or fine brick-dust, mixed with a little water and poured in fluid, a few wires being placed amidst the same for additional support.

The mould is now taken to pieces to extract the core, which is then dried, thoroughly burned, and allowed to cool slowly (which the founder calls *annealing,* from a similar method being employed iu annealing or softening the metals and glass): the core is then returned to the mould, to see that it has not become distorted. If needful the fitting around the balance is made good to suit the reduced magnitude of the core, which latter is then so far pared away as to leave room for the thickness of metal; this is frequently regulated by boring holes at many parts of the core with a stop-drill, having a collar to prevent its penetrating beyond the determined depth; the surface of the core is now pared down to the bottoms of the holes, as uniformly as possible. When the mould has been faced, dried and smoked, the whole is put together for pouring, for which purpose the figure is inverted and filled from the pedestal.

Equestrian and other figures are sometimes cast in two, three, or more pieces and joined together by solder, screws, or wires; but in all such works the aim of the founder is to leave little or nothing for the finisher or chaser to do.

Some objects which are either exceedingly complex in their form, or soft and flexible in their substance, and which do not therefore admit of being moulded in sand, in the ordinary manner of figure casting, may be moulded for *a sinyle copy,* provided the originals consist of substances which may be either readily melted or burned into ashes.

A cavity is made in the sand of the moulding-trough, a little larger and longer than the object, or else a wooden box of appropriate size is procured, in the midst of which the wax model may be placed; to the end of the model is added a piece to represent the runner, which will be required for introducing the metal. The composition of one-third plaster of Paris and two-thirds brick-dust, mixed with water, the same as for the core of the bust, is then poured in, entirely to surround the model. The mould is first slowly dried, it is then inverted and made warm to allow the wax to run out, after which it is annealed, or burned to redness, and lastly, when cooled, it is buried in sand and filled with metal. The method necessarily throws the chance of success upon a single trial as the model is destroyed.

Should the face of the casting be required to be particularly smooth, a small quantity of brick-dust is washed, (in the manner practised with emery, and to be explained,) and mixed with very fine plaster; a coat of this is brushed over the model, which excludes air-bubbles, the model is quickly placed in its cavity, and the coarser mixture is poured in as before.

The above method exactly corroborates a mode long since described as being suitable to casting copies of small animals or insects, parts of vegetables and similar objects; these are to be fixed in the center of a small box, by means of a few threads attached to any convenient parts, one or two wires being added to make air-holes, and ingates for the metal. A small quantity of river silt or mud, which had been carefully washed, was first thrown in and spread around the object by swinging the box about; and when partly dry, successive but coarser coats were TliU mode was practised in casting the feather of the equestrian statue of George III. erected in Pall Mall, London.

thrown in, so as ultimately to fill up the box. When it had become thoroughly dry, the wires were first removed from the earthy mould; it was then burned to reduce the object to ashes, and when every particle of the model had been blown out, it was ready to be filled with metal. SECT. VI.—FILLING THE MOULDS. Having traced the formation of various kinds of moulds for brass work, we must now return to the furnace to see if the metal is in condition to be poured, which is indicated by the slight wasting of the zinc from its surface with a lambent flame. When this condition is observed, the large cokes are first removed from the mouth of the pot, and a long pair of crucible tongs are thrust down beside the same to embrace it securely, after which a coupler is dropped upon the handles of the tongs: the pot is now lifted out with both hands and carried to the skimming-place, where the loose dross is skimmed off with au iron rod, and the pot is rested upon the spill-trough, against or upon which the flasks are arranged.

The temperature at which the metal is poured must be proportioned to the magnitude of the works; thus large, straggling, and thin castings, require the metal to be very hot, otherwise it will be chilled from coming in contact with the extended surface of sand before having entirely filled the mould: thick massive castings if filled with such hot metal would be *sand-burned,* as the long continuance of the heat would destroy the face of the mould before the metal would be solidified.

The line of policy seems therefore to be, to pour the metals at that period when they shall be sufficiently fluid to fill the moulds perfectly and produce distinct and sharp impressions, but that the metal shall become externally congealed as soon as possible afterwards.

For slight moulds the carbonaceous facings, whether mealdust, charcoal, or soot, are good, as these substances are bad conductors of heat, and rather aid than otherwise by their ignition: it is also proper to air these moulds for thin works, or slightly warm them before a grate containing a coke fire. But Many beautiful brass castings of vegetable bodies, as ears of wheat, and thick fleshy plants, such as cockscombs, &c. , have been frequently exhibited during the last few years, and appear to have been thus produced, or with the previous composition.

TEMPERATURE OF THE METAL. 345 in massive works these precautions are less required, and the facing of common brick-dust, which is incombustible and more binding, succeeds better.

The founder therefore fills the moulds having the slightest works first, and gradually proceeds to the heaviest; if needful he will wait a little to cool the metal, or will effect the same purpose by stirring it with one of the ridges or waste runners, which thereby becomes partially melted. He judges of the temperature of the melted brass, principally by the eye, as when out of the furnace and very hot, the surface emits a brilliant bluish-white flame, and gives off clouds of the white oxide of zinc, a considerable portion of which floats in the air like snow, the light decreases with the temperature, and but little zinc is then fumed away.

Gun-metal, and pot-metal, do not flare away in the manner of brass, the tin and lead being far less volatile than zinc; neither should they be poured so hot or fluid as yellow brass, or they will become sand-burned in a greater degree, or rather the tin and lead will strike to the surface as noticed at page 295. Gun-metal and the much used alloys of copper, tin, and zinc, described at page 272, are sometimes mixed at the time of pouring; the alloy of lead and copper is never so treated, but always contains old metal, and copper is seldom cast alone, but a trifling portion of zinc is added to it, (see page 267,) otherwise the work becomes nearly full of little air-bubbles throughout its surface.f

Some persons judge of the heat proper for pouring, by applying the skimmer to the surface of the metal; which when very hot has a motion like that of boiling water; this dies away and becomes more languid as the metal cools. Many works are spoiled from being poured too hot, and the management of the heat is much more difficult when the quantity of metal is small. The mixture and temperature of the metal being found to be i When the founder is in doubt as to the quality of the metal, from its containing old metal of unknown character or that he desires to be very exact, ho will either pour a sample from the pot into an ingot mould, or extract a little with a long rod terminating in a spoon heated to redness. The lump is cooled and tried with the file, saw, hammer, or drill, to learn its quality.

t The engraved cylinders for calico-printing are required to be of pure copper, and their unsoundness when cast in the usual way was found to be so serious an evil that it gave rise, in 1819, to Hollingrake's patent for casting the metals under pressure. proper, it is poured in the manner represented in fig. 138, p. 306: the tongs are gradually lowered from the shoulder down the left arm, and the right hand is employed in keeping hack the dross from the lip of the melting pot. A crucible containing the general quantity of 40 or 50 lb. of metal, can be very conveniently managed by one individual, but for larger quantities, sometimes amounting to one hundred weight, an assistant aids in supporting the crucible, by catching hold of the shoulder of the tongs with a *grunter,* an iron rod bent like a hook.

Whilst the mould is being filled, there is a rushing or hissing sound from the flow of the metal and the escape of the air, the effect is less violent when there are two or more passages as in heavy pieces, and then the jet can be kept entirely full, which is desirable. Immediately after the mould is filled, there are' generally small but harmless explosions of the gases, which escape through the seams of the mould, they ignite from the runners, and burn quietly: but when the metal *blows,* from the after-escape of any confined air, it makes a gurgling, bubbling noise, like the boiling of water but much louder, and it will sometimes throw the fluid metal out of the runner in three or four separate spirts: this effect which mostly spoils the castings, is much the most likely to occur with cored works, and *with* such as are rammed injudiciously hard, without being, like the moidds for fine castings, subsequently well dried.

The moulds are generally opened before the castings are cold, and the founder's duty is ended when he has sawn off the ingates or ridges, and filed away the ragged edges where the metal has entered the seams of the mould; small works are additionally cleaned in a *rumble,* or revolving cask, where they soon scrub each other clean.

Nearly all *small* brass works are poured *vertically,* and the runners must be proportioned to the size of the castings that they may serve to fill the mould quickly, and supply at the top a mass of still fluid metal, to serve as a *head* or pressure for compressing that which is beneath, to increase the density and soundness of the casting. Most *large* works in brass, and the greater part of those in iron; are moulded and poured *horizontally,* and the process being exactly alike for both metals, we must refer the reader to the following chapter.

CHAPTER XVII.. CASTING AND FOUNDING CONCLUDED. SECT. I. IRON-FOUNDERS' FLASKS, AND SAND MOULDS.

The process of moulding works in sand is essentially the same both for brass and iron castings; but the very great magnitude of many of the latter gives rise to several differences in the methods: it will suffice however to advert to the more important points in which the two practices differ, or to those which have not been already noticed; I shall therefore commence with a few remarks upon the flasks and the sand.

In the greater number of cases the iron-founder moulds and casts his works horizontally, with the flasks lying upon the ground; frequently the top part only is lifted; and in the largest,works the lower part of the flask is altogether omitted, such pieces being moulded in the sand constituting the floor of the foundry; in these cases the position of the upper flask is denoted by driving a few iron stakes into the earth, in contact with the internal angles of the *lugs,* or projecting ears of the flasks.

The sand would drop out of such large flasks, if only supported around the margin; they are consequently made with cross-bars or wooden *stays* a few inches asunder, which, unless the entire flask is made of wood, are fixed by little fillets cast in the solid with the sides of the iron flasks. A great number of hooks in tbe form of the letter S, but less crooked at the ends, are driven into the bars, and both the bars and hooks are wetted with thick clay water, so that the sand becomes entangled amidst them, and is sustained when the flask is lifted. Some flasks require the force of either two or several men, who raise them up by iron pins or handles projecting from the sides of the flask; they are then placed upon one edge, and allowed to rest against any convenient support whilst they are repaired, or they are sustained by a prop.

The very heavy flasks are lifted with the crane, by means of a transverse beam and two long hangers, called *clutches,* which take hold of two gudgeons in the centers of the ends of the flask; it can be then turned round in the slings, just the same as a dressing-glass, to enable it to be repaired.

The modern iron-founder's flasks are entirely of iron, and do not require the wooden stays, as they are made full of cross ribs nearly as deep as the flask itself, and which divide its entire surface into compartments four or five inches wide, and one to two feet long. On the sides of every compartment are little fillets, sloping opposite ways, so as to lock in the small bodies of sand very effectually. When these top flasks are placed upon middle flasks without ribs, as in moulding thick objects, the two parts are *cottered* or keyed together, by transverse wedges fixed in the steady pins of the flask; lifters or *gaggers* are then placed amidst the sand, these are light T shaped pieces of iron, wetted and placed head downwards, the tails of which are largest at top, so as to hold themselves in the sand, the same as the key-stone of an arch is supported. The gaggers are placed at various parts to combine the sand in the two flasks, and they fulfil the same end as the iron hooks and nails driven into the wooden stays of the old-fashioned flasks.

The bottom flask or drag, has sometimes plain flat cross ribs two inches wide, (like a flat bottom with square holes,) that it may be turned over without a bottom board; and unless the flasks have swivels for the crane, they have two cast-iron pins at each end, and one or more large wrought-iron handles at each side, by which they may be lifted and turned over by a proportionate number of men.

The sand of the iron-founder is coarser and less adhesive than that used by the brass-founder; much of the former kind, used about London, is procured at Lewisham. The parting sand is the burned sand which is scraped off the castings; it loses its sharp, crystalline character from being exposed to the red heat. The facing-sand is sometimes only about equal parts of coal-dust and charcoal-dust, ground very fine; at other times, either old or new sand is added, and for large thick works a little road-drift is introduced. All these substances get largely mixed with the sand of the floor, and lessen its binding quality, which is compensated for by occasional additions of new sand, and by using more moisture with the sand; as before extracting the patterns, the iron-founder wets the edges of the sand with a sponge, which has sometimes a nail tied to it to direct the water in a fine stream; for heavy works a watering pot is used.

The *green-sand moulds,* are made as in the brass-foundry, of the ordinary stock of old moist sand; these are often filled as soon as they have been made.

The *dry-sand moulds,* are made in the same manner, but with new sand containing its full proportion of loam; these moulds are thoroughly dried in a large oven or stove, and then blackwashed or painted, with thin clay water containing finely ground charcoal; this facing is also thoroughly dried before the moulds are poured.

The *loam moulds,* which are much used for iron castings and somewhat also for those of brass, are made of wet loam with a little sand, ground together in a mill to the consistence of mortar; the moulds are made partly after the manner of the bricklayer and plasterer, as will be explained: the loam moulds also are thoroughly dried, black-washed, and again dried, as from their greater compactness they allow less efficient escape for the vapour or air, and therefore they must be put into the condition not to generate much vapour when they are filled.

Iron moulds are also employed for a small proportional number of works which are then called *chilled castings;*

these were referred to at pages 258 and 259; and occasionally the methods of sand casting and chilling are combined, as in some axletreeboxes, which are moulded from wooden patterns in sand, and are cast upon an iron core. To form the annular recess for oil, a ring of sand, made in an appropriate core-box, is slipped upon the iron mandrel, and is left behind when the latter is driven out of the casting.

It would be a useless repetition to enter into the details of moulding ordinary iron-works; but from the horizontal position of the flasks it is necessary that the part of the work which is required to be the soundest, and most free from defects, should be placed downwards, as the metal is more condensed at the lower part, and free from the scoria or *sullage* which sometimes renders the upper surface very rough and full of minute holes. As the flasks almost always lie on the ground, it is also found the most convenient to retain them in contact by placing heavy 350 SIDE COKES USED BY THE IKON-FOUNDER.
weights upon them; the foundry should in consequence have an abundant supply of these.

The flasks require to be poured through a hole in the upper half, as seen at *r*, fig. 151, page 353, which hole is formed by placing a wooden *runner stick* in the top part A, whilst it is being rammed; and a small channel is afterwards cut sideways into the mould. Sometimes two, three, or even half-a-dozen or more runners are put to one single casting, either when it requires a great weight of metal, or when it is large but slight, as in trelliswork, in which case the metal might cool before filling the mould if only introduced at one single runner.

When the runners are required to be lofty, either to supply pressure to the metal, or as a reserve to fill up the space left by its contraction in cooling, iron rings of six or eight inches diameter are piled up to the required height, to support the tube of sand contained within them. Small objects that are poured from one hole, are frequently moulded with two runners, that the metal may *flow* through the mould, and that there may be a sufficient supply to meet the shrinkage, and also to supply head or pressure; another advantage also results, as it assists in carrying off the scoria or *sullage.*

The iron-founder employs all the methods of coring explained at pages 335 to 338, and also others of an entirely different kind but little required in brass-works; namely, for lateral holes in the parts of the castings buried beneath the general surface of the mould, and which are explained by the figs. 180 to 183. Thus ISO represents the finished casting, 181 the model of the same, 182 the appearance of the bottom flask or drag when the pattern is first removed, and 183 the flask and cores when closed ready for pouring; the moulds are inverted, and the same letters of reference refer to similar parts of all these figures.

The core print a, would deliver from the sand and leave the cavity at *a,* fig. 182, to be afterwards filled by the core shown black in fig. 183, the same as formerly explained at fig. 164, p. 335.

But the core print *b,* fig. 181, (which has reference to the blackstud *b,* fig. 183,) would tear away the sand above it in withdrawing the pattern; therefore the print *b,* should, like *d,* fig. 181, extend to the face of the pattern, or the parting line, represented by *e,* fig. 183. This being the case, the pattern would leave the space denoted at *d,* fig. 182; the core is put down sideways to the bottom of the reccss, and extends entirely across the same; the small open space above, is made good with the general surface, as shown by the shade lines in fig. 183, and this filling. in, at the same time fixes the core precisely where denoted by the print *d,* which latter has a mark to show to the moulder where the core is to end. The circular hole requires the core print shown at c, fig. 181; the cores themselves are made in the core-boxes, 165 and 106, before explained at page 336.

Fig. 185 represents the model and core-print, from which the finished casting shown at fig. 184 might be made from a solid pattern in a two-part flask; it would be inverted, and the parting would be made upon the line, *x.* The prints for the four holes *a a,* would be placed in the top flask, and those for the great apertures or panels *d,* would be made in a core-box of the express form, and as thick as the pattern and core-print measured together. The core would be deposited edgeways into the core-print, and the upper corners of the mould would be made good, as explained in fig. 183.

By the same method, a mortise wheel, or one with spaces around its edge, as at *m m,* fig. 186, to be filled with wooden cogs, might be made with a series of core-prints, as at *c,* brought up flush with the parting of the mould; if every print were filled with a core such as fig. 187, made in an appropriate core-box, the matter would be accomplished with great facility and truth.

The iron-founder makes frequent use of flasks which divide in three or four parts; this is done in many cases simply to increase the depth of the contained space; in which case when wooden flasks were employed, they admitted of being temporarily fixed together by dogs, or large iron staples, driven a little way into the neighbouring flasks, but the modern iron flasks are fixed by cotters. The following examples will show the nature of some other uses to which the flasks with several partings are applied. Figs. 188. 189.

A casting, such as fig. 1S8,".which represents the top of a sliding-rest for a lathe, might be moulded in a very deep tiropart flask, if the parting were made upon the dotted line *a a;* but there would be very great risk of tearing down the mould in drawing out the pattern, and from the depth, there would be scarcely a possibility of repairing it, and the metal would probably be strained. It would be also possible to mould it with the joining upon the line *b,* provided several cores were employed; but the following mode, which is generally adopted, is more convenient than either of these.

When the pattern is made in two parts, and the flask in three, as in fig. 189, A and B are first united and partly filled with sand, the pattern is knocked in as represented, and the whole well rammed, especially in the groove, the parting being made on the line, 1, 1, and dusted. C is now put on, filled, and struck off level, a board is put above it, and ABC are all turned over together, A becoming the top.

A is now removed, and the sand is cut away to make the second parting on the line 2, 2, after which, A is replaced, and the runner-stick is inserted to make the runner, *r.* On removing the pattern, the runner-stick *r* is first taken out, A, or the top part of the flask is lifted oft", and the white part of the pattern is drawn out; B, or the middle part, is then lifted, and the last or shaded piece of the pattern, is drawn out of the mould, which is now put together again, and poured through *r;* so that the top surface of the pattern, as seen in both views, becomes the *face,* from being cast downwards, or upon the lowest piece C, of the flask, called the drag.

The part *c,* fig. 188, might be cast with a chamfer in three different ways; although, in small castings, it is more usual to cast it square and plane it out of the solid. First, the pattern might be moulded square, and the top A, after removal, might be worked to the angle by aid of the trowel and a chamfered slip of wood, used as a gage; or secondly, by the employment of a core, the print of which is represented by the dotted lines terminating at the angle *d,* fig. 188; or thirdly, by having a loose slip on the pattern sliding on the line c, fig. 188, so as to be drawn off when the top A, had been lifted. The last method is analogous to that represented in fig. 190, also intended for a sliding rest; and which might be cast in a two-part flask, if the two chamfers, *c c,* were fitted loosely upon slides, as shown; but a three-part flask is more convenient, as explained by fig. 191, in which the pattern is inverted.

The lowest piece C, or the drag, is parted upon the line 1 1, but its sand extends upwards between the two sides of the pattern, as shown by the shade-lines. The middle piece 13, is parted through the line 2 2; and lastly A, the top is filled up level, the runner-stick at r being inserted at the time, A is first lifted, and all the pattern is then removed, excepting the chamfered bars and their slides, which are represented black; this vor. i. A A pattern delivers its own cores for the circular mortises *m m,* the sand forming them being a part of that in B, or the middle flask; lastly, B is lifted, and the chamfer-slips are picked off from C. This pattern may consequently be moulded without turning over the flask, and every part of the mould is quite accessible for repair.

The pedestal of the swage-block, fig. 128, page 231, is another good example of mouldiug in a three-part flask. The model is made with the upper fillet loose, also with the sides solid, or without the holes, and the object is moulded as it stands. The top part of the flask opens at the upper mouldiug, and which latter is then removed from the pattern; the middle flask divides at the plinth or flange, so that when this has been lifted, the pattern also may be withdrawn, leaving a square pedestal of sand, as large as the interior of the model, standing upon the bottom part or drag, as in 191. The panels are made by means of a core-box of the kind fig. 167, p. 336, the box is exactly as thick as the metal to be cast; and the circular cores are then fixed upon the pedestal of sand by means of a few wires or nails, after which the flask is put together ready for pouring.

If the fig. 128, here referred to, had four fluted columns at the four angles, either with a large cap to each, or with a square entablature connecting the whole of them, the object might be also cast in one piece, if moulded in a three-part flask. After removing the top flask, the entablature and capitals would be first withdrawn, the columns being divided through their *smallest* diameters; the mould would be then turned over, and upon lifting off the drag, or bottom-piece, the remainder of the pattern could be drawn, either in one single piece, or if the pillars were loose, the five parts could be more safely extracted; the threepart mould would be put together again, and reversed for pouring. In this general manner, by making either the mould, or the pattern, or both, in different pieces, and by the judicious employment of cores and drawbacks, objects apparently the most untractable are cast with very great perfection.

The iron-founders are likewise very dexterous in making castings in some respects different from the patterns from which they are moulded; thus if the pattern be too long, or that it be temporarily desired to obliterate some few parts, the mould is made of the full size and *slopped-off,* additional sand being worked into the mould by aid of the trowel and some temporary piece of wood to represent the imagined termination of the pattern. On the other hand, any simple enlargement or addition, is not always added to the pattern, but it is frequently *cut out* of the mould with the trowel, in a similar manner.

Many common works, such as plates, gratings, parts of ordinary stoves, and simple objects, are made to *written* measures, and without patterns, as a few parallel slips of wood to represent the margin of the casting, are arranged for the purpose upon a flat body of sand, which is modelled up almost entirely by hand; but for all accurate purposes and for machinery, good and wellmade patterns are indispensable, and to some particulars of which a little attention will be now devoted.

SECT. II. REMARKS ON PATTERNS FOR IRON CASTINGS.

Tiie construction of patterns for iron castings, requires not only the observance of all the particulars conveyed on pages 327 to 329, but in addition, the large size of the models, the peculiar methods employed in moulding them, and the nearly inflexible nature of the iron castings when produced, call for some other and important considerations; and which should not be entirely overlooked, even in works of comparatively small size, or it may lead to failure and disappointment.

Thus, it becomes necessary to make patterns in some degree larger than the intended iron castings, to allow for their contraction in cooling, which equals from about the ninety-fifth to the ninety-eighth part of their length, or nearly one per cent. This allowance is very easily and correctly managed by the employment of a *contraction rule,* which is made like a surveyor's rod, but one-eighth of an inch longer in every foot than ordinary standard measure. By the employment of such contraction rules, every measurement of the pattern is made proportionally larger without any trouble of calculation.

When a wood pattern is made, from which an iron pattern is to be cast, the latter being intended to serve as the permanent foundry pattern; as there aretwo shrinkages to allow for, a *double* contraction rule is employed, or one the length of which is one quarter of an inch in excess in every foot. These rules are particularly important in setting out alterations in, or additions to, existing machinery; the latter is measured with the common rule, and the new patterns are set out to the same nominal measures, with a single or double contraction rule as the case may be, the three being made in some respects dissimilar to avoid confusion in their use; the entire neglect of contraction rules, incurs additional trouble and uncertainty.

Patterns for iron castings are much more frequently divided into several parts than those for brass; for instance, the division into two equal parts after the manner of fig. 175, p. 339, (but without reference to the under-cutting,) is very common, as both the pattern and flask separate when the top part is lifted, and the halves of the pattern can then be drawn out from the halves of the flask with much less risk of tearing down the sand.

Referring to p. 319, fig. 141, if small, would be moulded as represented with false cores or drawbacks; but if it were a large fluted column, the iron founder would employ a solid two-part flask; the shaded parts would together represent the body of sand in the drag, and the pattern would be made in *three* parts something like a boot-tree. When the top flask had been lifted, the central slice of the pattern, extending from the two upper to the two lower angles would be withdrawn vertically, and the two outer pieces would be released sideways. The general rule is to divide the circumference of the pattern into six equal parts, and to let the central slice equal one of them in width.

The figures, 1S9 and 191, representing two parts of a slide-rest, and the pedestal, 128, are some amongst many of the common examples of the division of the patterns; and with which may be associated, the numerous subdivisions of the mould instead of the pattern by the employment of cores, many applications of which have been also explained. All these matters display much interesting and ingenious contrivance, resorted to either The contraction of brass is nearly three sixteenths of an inch in every foot, but from the small size of brass castings, the contraction rule is less required for them as the differences may be easily allowed for without it.

Iron castings weigh about fourteen times as much as the ordinary deal and fir patterns from which they are made, that being nearly the ratio of the specific gravities of those materials. All these matters are entered into in the author's pamphlet, "A New System of Scales of Equal Tarts," and his paper "On a scale of Geometrical Equivalents for Engineering and other Purposes," Lond. and Edin. Philos. Mag., July, 183S, wherein are described numerous applications of scales of equal parts to the purposes of drawing and calculation, and to the comparison and conversion of all kinds of measures, weights and other quantities.

to render possible the operation of moulding, or to facilitate its performance.

To lessen the distortion of castings from their unequal contraction in cooling, it is important that the models should be nearly symmetrical. For example, bars or rods of all the sections in fig. 140, p. 319, may be expected to remain straight; perhaps *g* is the most uncertain, but if the lower fins of *e* aud / were removed, their flat surfaces then exposed to the sand, would become rounding or convex in length, from the contraction of the upper rib being unopposed by that of a similar piece on the under side. Bars and beams, the sections of which resemble the letter I are of the most favourable kind for general permanence, and also for strength, and large panels may be cut out from their central plates to diminish their weight without materially reducing their stability. They are much used, not only in building, but also in the framing of machinery, which is in a great measure based upon the same general rules.

It is also of great importance, especially in castings of large size, that the *thickness* of the metal should be nearly alike throughout, so that it may cool at all parts in about the same time. Should it happen that one part is set or rigid whilst another is semi-fluid, or in the act of crystallising, there is great risk of the one part being altogether torn from the other and producing fracture. Or should the disturbing force be insufficient to break the casting, it may strain the metal nearly to its limit of tenacity or elasticity; so that a force far below that which the casting should properly bear, may break it in pieces.

An example of this is seen in wheels with very light arms and heavy rims or bosses. The arms sometimes cool so quickly as to tear themselves away from the still hot rim or nave; or when the arms are solidified without fracture, the contraction of the rim mayso compress the spokes endways, as to dish the wheel, (in the manner of an ordinary carriage wheel.) and thereby strain the casting nearly or quite to the point of fracture. The arms are sometimes curved like the letter S, instead of being straight and radial; the contraction then increases their curvature with less risk of accident than.to straight arms. A more elegant It appears to be often desirable to supersede the straight diagonal braces of iron castings, by curved lines, which are both more ornamental, and better disposed to yield to compression or extension by a slight alteration in their curvature.

358 INEQUALITY OF STRENGTH AND THICKNESS. way of avoiding the mischief was invented by Mr. Isaac Dodd, of the Horsley Iron Works, by placing the spokes as *tangents* to the central

boss, in which case the contraction of the rim makes a small angular change of position in the boss; for the rim in thrusting the spokes inwards, causes the boss to twist round a little way with far less risk of fracture.

The destructive irregularity of thick and thin works is partly averted by uncovering the thick parts of the casting, or even cooling them still more hastily, by throwing on water from watering-pots. In wheels this has been done by a hose, the axis of which is concentric with the wheel, the arms being all the time surrounded by the sand to retard-their cooling; but it is the most judicious in all patterns, to make the substance for the metal as nearly uniform throughout as circumstances will admit, so as not to require these modes of partial treatment, which often compromise the ultimate strength of the casting.

Another mode sometimes adopted for avoiding the fracture of wheels, from the great dissimilarity of their proportions, is by inserting *wrought iron arms* in the mould, but they do not always unite kindly with the iron of the rim and the nave. The same inconvenience occurs when iron pins are inserted in the ends of either of iron or brass castings, to serve for their attachment to their respective places: in iron castings it frequently produces the effect of chill casting, so as to render the works difficult to be turned or filed at the junction, and there is risk of the casting becoming blown or unsound in either case. When the pins are heated before being placed in the mould, they become nearly cold before the metal can be poured, and they also endanger the presence of a little steam or vapour, which is detrimental; therefore they are more generally put in cold, notwithstanding the sudden check they then give to the fluid metal.

The patterns for iron castings, of large size, are necessarily very expensive, especially those for hollow cylinders and pans, many of which are so large, that it would be impossible to find solid pieces of wood from which the patterns could be made; either with sufficient strength for present use, or with the Mr. Dodd had to contend with the shrinking of the *nave,* which was the last to cool; the accidents therefore occurred from the tension, instead of the compression of the spokes; this equally fatal effect was completely remedied by placing the arms as tangents.—Trans. Soc. of Arts, Vol. LI., p. 66.

necessary permanence of form-for a subsequent period, as they would be almost sure either to break, or to become distorted from the effects of unequal shrinking, as explained by the diagrams, 13, 14, 15, p. 49. Such patterns, therefore, require to be made of a great many thin layers or rings of wood, each consisting of 6, 8, or 12 pieces, like the felloes of wheels, so that in all parts the grain may be nearly in the direction of a tangent.

As they are glued up, every succeeding layer is connected with the former by glue and wooden pins or dowels, and the whole is afterwards turned to the tubular or hemispherical form, as the case may be. As the castings are generally required to be rather thin, such models are not only very expensive, but also very liable to accident; and besides it frequently occurs that only one or two castings of a kind may be required, which makes the proportional cost of the patterns excessive.

It fortunately happens, however, that this case, which is one of the most costly and uncertain by the employment of ordinary wood or metal patterns, becomes exceedingly manageable by a peculiar and simple application of the art of turning, (the one great center oT the constructive arts, to which these pages arc intended immediately and collaterally to apply:) and by which process, or one branch of *loam moulding,* to be explained in the following section, patterns are not generally required.

SECT. III. LOAM MOULDING.

Figs. 102, 193, and 194, are intended to illustrate this process as regards a steam cylinder. Fig. 192 is the entire section of the mould in its first stage; figs. 193 and 194 are the half sections of the second and third stages, preparatory to burying the mould'in the pit in which it is to be filled.

The inner part of the loam mould is called the *core* when small, but the *nowel* when large; the outer is called the *case* or the *cope.* Each part is built upon an iron loam-plate, or a ring cast rough on the face, and with four ears by which it may be lifted. The mould is occasionally erected upon four shallow pedestals of bricks for the convenience of making a fire beneath it to dry the loam; at other times it is made upon a low truck upon which it may be wheeled into the loam stove, which is heated to about the temperature of 300 to 400 degrees Fahrenheit.

A vertical axis, *a,* is mounted in any convenient manner, 360 MOULDS TURNED IN LOAM.

frequently in two holes in the truck itself, or as shown in the figure, in a pedestal or socket erected upon the truck; at other times the axis is mounted in a hole in the loam-plate, and in any bearing attached either to the building or its roof.

Figs. 192. 193. 194.

The first step is to fix upon the spindle the templet *b b,* at the distance of the radius of the cylinder, either by one or two *clutches,* with various binding screws. An inner cylinder of brickwork is then built 'up, plastered by the hands with soft loam, (which is represented black in all the figures,) and scraped into the cylindrical form by the radius board, which is moved round on its axis by a boy. When the surface is smooth and fair it is thoroughly dried, after which it is brushed over with blackwash, aud again dried. The charcoal dust in the blackwash serves as a parting, to prevent the succeeding portious of the loam mould from adhering to the first.

The templet *c c,* fig. 193, cut exactly to the external form of the cylinder, is now attached to the axis at the distance from the core /required for the thickness of the metal; some additional loam is thrown ou to form the *thickness,* which is smoothed in the saruc careful manner as the center, after which the templet and spindle arc dismounted, and the thickness, which is represented white in figs. 103 and 194, is also dried aud blackwashed.

The ring for the outer case or cope is uow laid down, aud its position is denoted either by fixed studs or by marks; and the outer case represented in fig. 194, is built up of bricks and loam, with an inner facing of loam worked very accurately to the turned thickuess. The new work, or the cope, is also thoroughly dried, and afterwards lifted off very carefully by means of the crane and a cross beam with four chains. This process likewise drags off the thickness, which usually breaks in the removal; its remains are carefully picked out of the cope; both parts of the mould are repaired, and again blackwashed and dried.

When the cylinder requires *ports* at the ends, or the short tubes with flanges for attaching the steam passages, models of the tubes are worked into the cope, and are afterwards withdrawn; the cores are made in core boxes, and are partly supported by the outer extremity, and partly upon *grains,* or two little plates of sheet iron connected by a central wire, the whole being equal to the thickness of the metal at the part. When steam passages are wanted, either along the side, or around the cylinder, they are worked up in clay upon the thickness, and duly covered in by the cope; their cores are supported, partly by their loose ends, and partly by grains, which become entirely surrounded by, and fixed in the metal, when it is poured.

The mould is now put together in a pit sunk in the floor of the foundry, and the two iron plates are screwed together; the surrounding space being rammed hard to prevent the mould from bursting open, but the inner part is left much more loose for the escape of the air. The top edges of the mould are covered over with a *loam-cake,* (which has been previously made and dried,) or a ring three or four inches thick, strengthened with iron bars amidst the clay, the joining being made air-tight by a little cow's hair, and by the pressure of a quantity of iron weights; the loam-cake is generally perforated with many holes as shown at *d,* for the entry of the metal and the escape of the air. But provision must always be made in casting thin cylinders, boxes, and such like forms, for the breaking up of the core as soon as the metal is set, to prevent the metal *scoring* or rending from its contraction upon a rigid unyielding center.f There U always some uncertainty of the sound union of the grains, or other pieces uf iron, with the cast-metal. Some cist them in iron and file them quito bright, others also tin them, apparently to preserve them from rust, as the tin must be instantly dissipated by the hot metal. Grains should always present clean metallic surfaces, and when used for very thin castings, to prevent them from dropping out, the wires are nicked with a file that they may be keyed in the metal. It is however better to avoid the use of grains, which may be generally done by giving the core *tand btarinjs,* and afterwards plugging up the holes in the casting.

t The largest cylinders, such as those of the Cornish pumping engines, of 80, 90, and 95 inches bore, and 12 or 14 feet in length, and the blowing cylinders of blast 302 LOAM MOTJIXS FOR PANS, PIPES, ETC.

Large pans, and various other circular works, are moulded precisely in the same way as cylinders; except that curved templets are used, and that towards the conclusion, the apertures through which the spindle passed are filled in and worked by hand to the general surface.

Water-pipes are made much in the same mode, but the cores for these are turned upon an iron tube pierced full of holes, which is laid horizontally across two iron trestles with notches, and is kept in rotation by a winch handle at the end; there is also a shaper-board or scraper fixed parallel with the axis; this primitive apparatus is called & *founder's lathe.*

The perforated tube, (serving as the mandrel,) is first wound round with haybands, then covered with loam, and the core is turned, dried and blackwashed; the thickness is now laid on and also blackwashed, after which the object is moulded in sand. The thickness is next removed from the core, which latter is inserted in the mould, and supported therein by the two prints at the extremities, and by grains with long wires, the positions of which may be seen by the little bosses on the pipe, the metal being there made purposely thicker to avoid any accidental leakage at those parts. When pipes are cast in large quantities, they are moulded from wooden patterns in halves, so that it only *becomes* necessary to turn the core, and this, when made in the above manner, is sufficiently porous for the escape of the air.

The moulds for crooked pipes and branches are frequently made in halves, upon a flat iron-plate. An iron bar or templet of the curve required is fixed ilown, and a semicircular piece of furnaces, sometimes 105 inches bore, are made without the employment of the thickness. The case or cope is built up in the pit, and turned *inside* with a radius bar, and the core is erected on a plate on the floor, and turned *outside* to a gage; when dried it is lowered into the other by the crane. The cylinders are cast one foot or upwards longer than required, to serve as a head of metal and make the top edge sound, and thus much is cut off before they are bored.

To enable the mould to resist the great pressure of the lofty column of fluid metal, (equal at the base to near 60 pounds on every square inch), the core U strengthened by diametrical iron bars entering slightly into the brickwork: the outer cylinder is surrounded at a small distance by iron rings piled on the other, the interval being rammed with sand; and stays are placed in all directions from the rings to the sides of the pit, which is either lined with brickwork, or when liable to be inundated with water, it is made of iron like a water-tight caisson.

Small cylinders are moulded in sand from wooden models, and only the cores are turned in loam; for cylinders of the smallest size the cores are made of sand in core boxes as already explained.

wood, called a *strickle,* is used for working and smoothing the half core; next a larger strickle is used for laying on the thickness, the two halves are then fixed together by wires, and moulded from in the sand flask; the thickness is now stripped off the core, which is fixed in the mould by its extremities, and if needful, is supported also upon grains. By the employment of these means, although the loam work requires time for the drying, yet with ordinary care an equality of thickness may be maintained, notwithstanding the complexity of the outline, and without the necessity for wooden patterns.

Very many of the large works in brass are also moulded in loam, the management being in most respects exactly the same as for iron, except that in some ornamental works wax is more or less employed, and is melted out of the moulds before the entry of the metal; a very slight view of the methods will serve as a sequel to the subject of brass founding.

Large bells are turned in almost the same manner as iron cylinders or pans, by means of wooden templets, edged with metal find shaped to the inner and outer contour of the core and thickness. The inscription and ornaments are either impressed within the cope, the clay of which is partially softened for the purpose, or the ornaments are moulded in wax, and fixed on the clay thickness before making the cope. Less generally the whole exterior face of the bell, or indeed its entire substance, is modelled in wax, and melted out before pouring. In any case, the concluding steps in filling up the apertures where the spindle passed, are to attach a dissected wooden pattern of the central stem and of the six *cannons* or ears by which the bell is slung, which parts are moulded in soft loam; and then, the parts having been dried and replaced, and the iron ring for the clapper inserted, the whole is ready for the pouring pit. The heaviest bells are moulded within the pit the same as huge cylinders.

Brass guns are also moulded in loam, and in a somewhat peculiar manner; a taper rod of wood much longer than the gun, is wound round with a peculiar kind of soft rope, upon which the loam is put for making the rough casting model of the gun, which is turned to a templet; the work is executed over a long fire to dry it as it proceeds, and the model is made about one-third longer than the gun itself. The model when dried and blackwashed all over, is covered with a *shell* of loam, not less than three inches thick, secured by iron bauds; the shell is also carefully dried; after this the taper.bar is cautiously driven out from its small end, the coil of rope is pulled out, and so likewise is every piece of the clay model of the gun.

The parts for the cascable and trunnions, which should have been worked separately upon appropriate wooden models, are then attached to the shell. Should the gun have dolphins, or any other ornamental figures, they are modelled in wax and fixed on the clay model before the shell is formed, and are then melted out to make the required space for the metal.

When all is ready and dried, six, eight, or more of these loam cases, or shells, are sunk perpendicularly in a pit at the mouth of the reverberatory furnace, and the earth is carefully rammed around them; at the same time a vertical runner is made to every mould, to enter either at the bottom or not higher than the trunnion: the upper ends of the runners terminate in the bottom of a long trough or gutter, at the far end of which is a square hole, to receive the excess of metal.

In casting brass guns, tapping the furnace is rather a ceremony, and certainly an imposing sight: the middle and the end of the trough are each stopped by a shovel or gate *held* across the same; and the runners are all stopped by long iroii rods, held by as many men. When all is pronounced to be *ready,* the stopper of the furnace is driven inwards with a long heavy bar swung horizontally by two or three men, and the metal quickly fills the trough; on the word of command, "*number one, draw,*" the metal flows into the first mould, and fills it quickly but quietly from the bottom; the mould being open at the top, no air can be accidentally enclosed. Numbers two, three, and four are successively ordered to draw. The first shovel is then removed from the great channel, and now the guns, five to eight or ten, as the case may be, are similarly poured and filled to the level of the trough; after which the last shovel is withdrawn, and the residue of the metal is allowed to run into the square bed or pit prepared for it. The flow of metal from the furnace is regulated by the tapping bar, the end of which is taper, and is thrust more or less into the mouth of the furnace as required; the trough and ruuners are thus kept exactly full, which is an important point in most cases of pouring, as it prevents a current of air being carried down along with the metal.

Large bells are poured much in the same manner, except that the runners are at the top, and the metal runs from the great channel, through smaller gutters to every sunk mould, the stoppers for which are successively drawn. For quantities of brass intermediate between the charge of an ordinary crucible, and such as require the reverberatory furnace, the large ladles or shanks of the iron founder are used; the contents of four or six crucibles being poured into the shank as quickly as possible, and thence in one stream into the mould.

SECT. IV.—MELTING AND POURING IRON".

Iron is usually melted in a blast furnace, or, as it is more commonly called, a cupola; although the cupola or dome leading The author of the article Founding, in the Encyclopedia Metropolitan, minutely describes three ways of casting large hollow statues, which are briefly as follows.

First; a rough model of the figure is made in clay, but somewhat smaller than its intended size, it is covered over with wax which is modelled to the required form, or the wax is worked up in separate pieces and afterwards attached: various rods or cylinders of wax to make the apertures for the runners and air holes are fixed about the figure and led upwards. The whole is now surrounded with a coating of loam and similar materials, the inner portion of which is ground very fine and laid on with a brush like paint; and the outer part is secured with iron bands. When all has been partially dried a fire is lighted beneath the grating on,which the figure is built, to cause the wax to run out through one or more aperttfres at the base, which are afterwards stopped, and all is thoroughly dried and secured in the pit, after which the charge of the furnace is let into the cavity left by the wax.

Secondly; the finished figure is modelled in clay, and stuck full of brass pins just flush with its surface, which surface is now scraped away as much as the thickness required in the metal; the reduced figure is now covered with wax mixed with pitch or rosin, which is worked to the original size with all the exactness possible. The other stages are the same as in the foregoing; the metal studs or pins prevent the mould and core from falling together, and they afterwards melt, becoming a part of the metal constituting the figure.

Thirdly; the finished figure is modelled in plaster, and a piece mould is made around it, the blocks of which consist internally of a layer of sand and loam 1J inch thick, and externally of plaster one foot thick. The mould when completed is taken to pieces, dried, and rebuilt in the casting pit; it is now poured full of a composition suitable for the core, the mould is again taken to pieces, the core is dried and scraped to leave room for the metal, and all is then put together for the last time, secured in the pit, and the statue is cast.

The first plan is the most wasteful of metal, the third the least so, although it is the most costly when the time occupied is also taken into account; but it has the advantage of saving the original work of the artist.
to the chimney, from which it would appear to have derived its name, is frequently omitted, the two or three furnaces being often built side by side in the open foundry.

At the basement there is a pedestal of brickwork about 20 to 30 inches high, upon which stands a cast iron cylinder from 30 to 40 inches diameter, and 5 to 8 feet high; this is lined with road-drift, which contracts its internal diameter to 18 or 24 inches. The furnace is open at the top for the escape of the flame and gases, and for the admission of the charge, consisting of pig-iron, waste or old metal, coke and lime, in due proportion. The lime acts as a flux, and much assists the fusion; chalk is considered to answer the best, but oystershells are very commonly used where they are abundant.

At the back of the furnace there are three or four holes one above the other for the blast, which is urged by bellows or by a revolving fan. No crucible is used, and as the fluid metal collects at the bottom of the furuace, the blast pipe is successively removed to a higher hole, and the lower blast-hole is Stopped with sand, which partly fuses and secures the blast-hole very effectually.

The front aperture of the furnace through which the metal is allowed to flow into the ladles or trough, is i sually made sufficiently large for the purpose of clearing or raking out rapidly the fuel and slag, as the process is most laborious owing to the excessive heat. This aperture is closed by a *guard-plate,* fixed on by staples attached to the iron-case of the furnace, in the center of which plate the tappiug hole is made: during the time the metal is fusing the tap hole is closed by sand well rammed in, and this, if well done, is never found to fail.

Many iron furnaces are made octangular, and in separate parts bound together by hoops, so that in the event of the charge becoming accidentally solidified in the cupola, the latter may be taken to pieces.for its removal, and thus avoid the necessity of destroying the furnace. There is frequently a light framing or grating above the furnace, upon which the small cores are placed that require to be dried.

In some foundries the cupolas are built just outside the moulding-shop, beneath one or more chimneys or shafts, which carry off the fumes; in such cases the fronts of the furnaces are accessible through an aperture in the foundry wall, with which FUUXACES, AND PROPORTIONINQ THE CHARGE.
307 they are nearly flush; when the furnaces are lofty there is a feeding stage at the back, from which the charge is thrown in.

For heavy iron castings, which sometimes amount to thirty tons and upwards in one piece, reverber'atory or air furnaces are also commonly used; the ordinary charge for these is four to six tons of iron, and five or six furnaces are commonly built close together, so that they may be simultaneously tapped in the production of such enormous works.

For melting iron in the small way, good air furnaces may be used, and also some of the black-lead furnaces, which are blown with bellows, but this is one of the processes that is not successful upon a limited scale.

Considerable judgment is required in proportioning the *charge* for the iron furnace, which always consists of at least two, and often of half-a-dozen kinds of new pig-iron mixed together, (as adverted to in the foot note, page 185,) and to which new iron, a small proportion of old cast-iron is usually added. The kinds and quantities used are greatly influenced by local and other circumstances, so that nothing can be said beyond a few general remarks.

When the principal object is to obtain sound castings with a very smooth face, as for ornamental works not afterwards wrought, the soft kinds of iron containing most carbon, which are most fusible and flow easily, are principally used. But such A few of the modern cupolas greatly exceed the air furnaces in effect, as they are calculated to contain upwards of twelve tons of melted iron. One of these, at the works of Messrs. Nasmyths' near Manchester, is six feet two inches diameter externally, and lined with Stourbridge bricks. It has three sheet iron tuyeres, nine inches diameter at the mouth, the blasts from which enter the furnace at three points of the circle, and they may be slid like telescope-tubes to either of the four series of holes, as the furnace becomes gradually filled.

There are three other furnaces progressively smaller, arranged beside the first; all of which may be used separately or in combination according to circumstances. The blast, which is under corresponding control, is obtained from two revolving fans, live feet diameter, making above 1000 revolutions per minute.

Messrs. Acramans, of Bristol, have likewise enormous cupola power; they have a seriei of four cupolas, in which collectively from forty to forty-five tons of iron may ba melted at one time.

In some cupolas the top is contracted by a cone made of iron plate; in Yates's patent, a brick trunk is built upon the cupola, with narrow arches crossing the trunk at right angles: this economises the heat by causing the flame and gaseous matters to be retarded, from pursuing a serpentine course in their escape.
metal would neither possess sufficient hardness, durability, nor strength, for many of the castings employed in the construction of edifices and machinery.

If the cupola contained a little hard pig-iron, but were in great measure filled with old cast-iron, which had been repeatedly melted, and had become successively harder from the loss of carbon at every fusion; such castings would be brittle, and sometimes so hard as scarcely to admit of being cut, these would be equally unfit for the generality of machinery from the opposite causes.

But the same mixture of iron will be found to differ very much according to the size of the objects in which it is cast; iron, which in a plate one-fourth of an inch thick may be quite brittle and hard, will mostly be of good, soft, and useful quality, in a stout bar or plate of t wo or three iuches thick. Thick castings are necessarily slow in cooling, and are seldom very hard, unless intentionally made so.

Between the extremes, (say three parts of pig-iron to one of old, or three parts of old iron to one of pig-iron,) various qualities may be selected; in casting for machinery the general aim is to obtain a strong, sound, and tough iron; mixtures of this nature which are used for iron ordnance, are called gun-metaJ amongst the gun-founders.

The fireman, or the individual having the management of the furnace, therefore always employs the scales in mingling the different kinds of iron, according to the magnitude and character of the works to be cast; and until the sorts in use arc familiarly known, it is partly a matter of trial, and requires the same attention as the making of alloys properly so considered.
When the management of cast-iron was less efficiently understood, it was occasionally alloyed with five or six per cent, of shreds of copper, thrown into the ladle full of iron to produce a close, sound, strong metal, suitable to three-throw cranks for pumping machinery, and other purposes. It is said that ten per cent, of copper renders cast-iron malleable, and that alloyed with copper or tin it is less disposed to *nust:* all these alloys may be now viewed as matters of experiment alone.
It is much to be regretted that no protection has yet been found to prevent the conversion of cast-iron into plumbago, or the carburet of iron, from long immersion in sea-water, or the water of copper mines, sewers, and other places. This, which is a most serious inconvenience in dock works, sea walls and mines, arises, says Dr. Faraday, from the circumstance that the protoxide of iron, formed beneath salt water, is soluble, and becomes washed away, thus robbiug the original mass of its iron; whereas the peroxide, or ordinary rust formed by exposure to

When enough iron is melted, (the common charge being twoand-a-half to four cwt., but sometimes above twelve tons,) the cupola is tapped in front, at a hole close to the bottom, which allows the whole contents to run out, either into ladles, or, in very large works, into channels leading directly to the moulds.

In pouring iron, the means of conveying the melted metal to the flasks, differ with the quantity. One man will carry from fifty to seventy pounds in a hand-ladle; three to five men will carry from two to four cwt. in a double hand-ladle, or a *shank:* larger quantities, amounting sometimes from three to six tons, are carried in the crane-ladle. These all possess one feature in common, namely their handles or pivots are placed but slightly above the center of gravity of the ladles; they may therefore be tilted very readily, as their fluid contents, in obeying the law of gravitation, are almost neutral in the operation of tilting, which they scarcely assist or retard, unless by mismanagement the ladle is over-filled, and thus rendered top-heavy.

All these ladles are coated with a thin layer of loam, and every time before use, they are brushed over with black-wash, and carefully dried. The hand-ladle has a handle three or four feet long, with a *crutch* or cross piece at the end, which is mostly held in the left hand; frequently the contents of half a dozen or more hand-ladles are poured simultaneously into the same flask. The shank has a single handle on the one side, and one made in two branches at the other, and together they measure six to eight feet in length; the tilting is completely under the command of the one or two men at the double handle.

The crane-ladle is carried from the furnace to the mould by the swinging and traversing motions of the crane, which is similar to those used at the iron forges, &c. (see p. 196,) and in very the air, is insoluble, and serves partly as a defence to the metal beneath. When first raised from the sea-water, the plumbago becomes exceedingly hot from the action of the atmosphere; it may be cut with a knife like an ordinary pencil.— Minutes of Conversation, Inst. Civ. Engineers, 8th Feb. 1842.
The furnace is not unfrequently tapped whilst the charge of metal is being melted, and in such cases when the required quantity has been removed into the ladles, the fireman re-stops the tap-hole, by a conical plug of clay on the end of a wooden bar; the process is called *hotting,* and requires a dexterous hand, or the whole contents of the furnace may escape. VOL. L n B large foundries, the plan of the building is divided into imaginary squares with a crane in the center of every square, so that the ladle is walked from one to the other, even to the far end of the shop, with great facility and expedition.
The *bail,* or handle of the crane-ladle, is fixed in its perpendicular position by

the *guard,* a simple bolt which prevents the ladle from being overset by accident until it has reached its destination. Two long handles, terminating in forked branches, are now fitted by their square sockets upon the swivels or pivots of the crane-ladle, and secured by transverse keys, after which the guard is withdrawn; and then two men at the ladle, two others at the crane, and one to skim the dross from the lip of the ladle, commonly suffice to manage two or three tons and upwards of fluid iron, with great ease and dexterity.

The observations offered on p. 344, respecting the temperature of the metal suitable to different brass works, might be here in a great measure repeated: namely that the smallest castings require very hot metal, and a gradually lower temperature is more suitable to works progressively heavier, to avoid their becoming sand-burned or rough on the face, from the partial destruction of the mould.

When cast-iron is very hot, the metal scintillates most beautifully, far more vividly than a mass of wrought-iron raised above the welding heat; as the metal cools, the sparks become intermittent, and at last the metal remains entirely quiet, excepting a multitude of lines vibrating in all directions, as if the surface were covered with thousands of wire-worms in great activity; this effect lessens until the metal solidifies. The softest iron shows most of this play of lines, or is said to *break* the best.

Iron castings are generally much heavier than those of brass, and the melting heat of the metal being considerably higher, the quantity of gas generated is very much greater; additional care is consequently required to provide for its escape, or the explosions are much more violent. The sand is punctured at many Mr. Nasmyth has added to the pivot of the large crane-ladle, a tangent-screw and worm-wheel, by which it may be gradually tilted by one man standing directly in front at any convenient distance; and another man skims the metal by a kind of throttle-valve coated with clay, which sweeps into the lip of the ladle, and keeps back the a ullage: the axis of the skimmer is continued as a long rod, at right angles to the first, and also terminating in a cross. By these arrangements any precise quantity of metal can be delivered, and the risk of accident scarcely exists.
places with a fine wire, before the removal of the patterns; sometimes also more coarsely as soon as the metal has become solidified. The gases issuing from the filled moulds are often lighted, either by the red-hot skimmer, or by a torch of straw with which the moulds are flogged: this lessens the accumulation of gas and the consequent risk of accident.

The pouring of very large objects in *open* moulds, such as plates, beams and girders, is a very beautiful and grand sight. The metal is led from the furnace, through a gutter lined with sand, into a large trough or *sow,* the end of which is closed with a *shuttle;* when the sow is full, the shuttle is raised; this allows the metal to flow very quickly into the mould, but enables it to be kept back should it be unnecessarily hot; the castings made in open moulds are generally covered up with sand as soon as the metal is set.

Thc above, and the casting of smaller objects, such as fiat plates in open moulds, may appear amongst the most certain modes of procuring sound castings; but unless the air be well *drawn* from the lower surfaces, they will become honey combed or full of air-bubbles. This defect is avoided by making the sand-bed sufficiently porous, and pricking it with many holes just below the surface, to serve as horizontal air-drains.
In casting lead, tin, &c. on a flat metallic plate, the formation of air-bubbles is lessened by placing a sheet of dry paper on the plate; it appears to keep down any little babbles of air or vapour, and to provide a thin channel for their escape.

But the most perfect example of a porous mould is that invented by Lord Oxmantown, to avoid the formation of air-bubbles in those speculums which are cast in open moulds. The plate, or bed of the mould, consists of a great number of slips of hoop iron placed edgeways and in contact; they are screwed tight within a frame, and are then turned in a lathe to the required curve: by this arrangement interstices exist at nearly every point, through which the air may escape downwards.

Speculum metal is perhaps the most untractable of any of the alloys, and it serves to illustrate in a most striking manner many of the effects that occur in castings generally. Small speculums are cast in sand; as soon as they are set, the sand core is pushed out of the aperture in such as are intended for Gregorian telescopes, to enable them to contract without fracture, and the red-hot disk is surrounded by ignited wood-ashes or any very bad conductor, to delay the cooling.

These precautions entirely fail with large speculums, as the margins solidify the first, and from the absence of ductility, the central parts tear away in the act of contraction, and the mass becomes rent or flawed. Lord Oxmautown considered this fracture would be avoided by cooling the speculums in *uniform layers* from below upwards, or as it were in infinitely thin lamina;, aud he therefore first employed iron moulds which were cooled by a stream of water projected against their under FILLING CLOSE MOULDS.

A far greater number of works are cast in *close* moulds, and in the horizontal position; the proportionate quantity of metal is carried to them in ladles, skimmers are held to the lips of the moulds at the time of pouring, to keep back all the sullage or dross. The number, position, and height of the runners are determined by circumstances; generally not less than two apertures are provided, the first for the entry of the metal, the second for the escape of the air, and to allow the metal to *flow through* the mould and carry off the sullage.

Sometimes in heavy castings, in addition to the runners one or more large heads or *feeds* are made at the upper part, to supply fluid iron as the metal shrinks in the act of solidifying; and in some such cases the feed is *pumped,* by moving an iron rod up and down iu the feed to keep the metal in motion, so that for a time the metal may freely enter and the air escape to increase the general soundness of the mass. The pumping should however be discontinued, the moment the metal begins to stiffen and clog the iron rod, or in other words to crystallise, otherwise mischief instead of benefit will accrue.

Works which are required to be particularly sound, as some cylinders, pipes, shafts and plungers, are cast vertically, the moulds are sunk in the earth, and well rammed to enable them to withstand the great pressure of the fluid column, without becoming strained or bursting open. Such objects are moulded and poured with a head, or an' additional portion about onethird the length of the finished casting, as mentioned in respect to brass guns.

In pouring cylinders of tolerably large size, the metal is surfaces; this partially answered with small speculums, but with those of 18 inches diameter it almost always failed, as the mould cracked before the metal was concealed. A general source of failure was the non-escape of air; this caused the lower surface to be full of air-bubbles, which it was tedious to grind out.

The plan ultimately adopted was the porous hoop-iron mould, with a marginal ring of sand; the mould was heated to about 212 F.J it was filled very quickly, and the moment the metal was solidified, it was drawn into an annealing oven previously heated to about the same temperature as the casting; so that for small reflectors of nine inches diameter, the cooliug might be extended over about three days, and for large ones of thirty-six inche diameter over about fourteen days; with these precautions the process was uniformly successful. *Sec* Trans. Royal Soc., 1840, pp. 510, 511. The whole of the paper "On the Reflecting Telescope,' pp. 503 to 527, is quite a fctudy for those interested in the construction of telescopes, and possesses nearly the same interest for the general mechanist.

The reader is referred to the Appendix of this volume, note F, page 402.

conducted from the sow through two sunk passages with side branches, entering the mould in the direction of tangents about one-third from the bottom, these keep the metal in circulation, and assist the rise of the sullage; cylinders are also poured through holes in the loam cake, other apertures being always provided in it for the escape of the air. Beneath the iron plate upon which the mould is built, is placed a central mass of haybands, in order that the air may have free passage to collect, and then to escape upwards to the surface of the earth, through one, two, three, or more internal or external tubes, as the case may be. The thick cylinders for hydrostatic presses are closed at one end, and those cast with the mouth downwards, require an air tube bent at each end, to lead from the core beneath the casting to the surface of the earth; the gas drives out in a stream, and is immediately ignited like a great torch: others prefer casting them with the mouth upwards, in order that less risk may exist of locking up air within the casting.

For the very heaviest works the three or four furnaces are usually tapped at the same moment, the stream from every one is conducted through a sand trough, and they all unite in one great trunk leading to the mould.

In pouring some of the largest cylinders, the trough is led entirely round the top of the loam mould, and from the circular channel, sometimes as many as thirty runners, every one of which is stopped by a shovel held by a man or a boy, descend to the mould, and as many air-holes are made between the in gates. When the foreman sees that all the furnaces are in full run, and that the channels are well supplied, he gives the word, "*up shovels,"* they rise at the instant, and allow the molten stream to deposit itself in its temporary resting-place.

At the time the cylinder is poured, all the precautions explained in the note, p. 362, are necessary to give the mould sufficient strength to resist the pressure of the fluid metal; but as soon as it becomes set, the conditions are altered, and this resistance must be removed from the inner surface, that the cylinder may shrink in cooling without restraint or fracture. Accordingly, after three or four hours' time, all the diametrical iron stays are knocked away by a vertical weight or monkey, and men descend by iron ladders into the cylinder, to break down the brick core. The heat is so terrific that they can only endure it for a minute or so at a time, but still the precaution is imperative: and even in comparatively small castings of hollow objects, such as cylinders, pans and boxes, it is desirable to break down the cores, to prevent the castings from *scoring* or breaking.

Although some iron castings employed for bridges, girders, and even for machinery, require the enormous quantities of iron referred to, on the other hand this useful metal is employed for exceedingly light and beautiful castings, abundant examples of which may be seen in the Berlin ornaments and chains. The links of most of the Berlin chains are connected with wroughtiron wire, but figs. 195 and 196 represent a chain made entirely by the process of casting.

Fig. 195. Fig. 196.

6 *b*

Its length is 4 feet 10 inches, it consists of about ISO links, and weighs If oz. avoirdupois. It was thus made, the larger links *a a,* were first cast separately, a solid model of the chain about 8 inches long, with core prints as in fig. 196, was then moulded; the links *a,* previously smoked to prevent the adhesion of the metal, were first laid in the mould, and afterwards the sand cores *b b,* and a separate runner was made to every one of the small links *c c,* so as to unite the whole when poured.f Through the kindness of John Taylor, Esq., F.R.S. , T.G.S., &c., I am possessed of some of these gems of art in the condition in which they left the moulder's hands, and also a portion of the sand employed: notwithstanding the minute size of the castings, some of them are quite hollow as if stamped out of thin steel metal.

Professor Ehrenberg says, that the iron employed for them is made from a bog iron ore, and that the sand is a kind of tripoli, also containing iron: both are entirely constituted of various kinds of animalcules, several of which are found, both in the fossil and recent states, in the neighbourhood of Berlin. Vide Paper read by Prof. Ehrenberg, at the Royal Academy of Sciences, Berlin, July 7, 1836; and also "ScientiBc Memoirs, Vol. I., part 3." (*Wilkinson's Engines of War,* p. 219.)

T Threo such chains were cast by a German workman in the employment of Messrs. Harvey and Co., of the Haylo Foundry, Cornwall, although in the same establishment they occasionally cast twenty tons in one piece. A large and a small link of the chain, weigh together *eight grains!*

See note AA, Appendix, vol. ii. page 974.

The concluding duty of the iron-founder, is to remove the castings from the mould and to break off the runners; after this all the loose sand, (which is reserved for making the partings of future moulds,) is scraped off with iron shovels and wire scratch brushes, and the seams are smoothed off with chisels and old files.

The skin or crust of a casting made in a sand mould, is in general harder than that of a loam casting; this appears to occur from the former being partially chilled by the moisture of the sand. In some cases, as in the teeth of wheels, it is desirable to retain this hard sand coat, on account of its greater durability; but when the crust is partially removed from thin or slight works it constantly happens that they spring or become distorted whilst under the treatment of tools from the general balance of strength being disturbed by the partial removal of the crust. This gives rise to continual interferences, which come however under the consideration of the mechanician rather than of the founder.

The crust of the casting which always retains some sand, is very destructive to the tools unless they can be sent in deep enough to penetrate to the clean metal beneath; when but little is to be removed from the casting, or that they are wrought with expensive tools and circular cutters, it is desirable to *pickle* the works, or to undermine the sand by dissolving a little of the metal with some acid.

Iron castings are pickled with sulphuric acid diluted with about twice as much water, the castings if small are immersed in a trough lined with lead; or else the acid is sprinkled over them; in two or three days a thin crust, like an efflorescence, may be washed off with the aid of water and slight friction.

Brass and gun-metal, when pickled, require nitric acid diluted with four to six times as much water, otherwise the rough coat should be removed with an old file, or a triangular scraper, but which is less effective than the dilute acid; this acid liquor should be also kept in leaden vessels, or in those of well-glazed earthenware or glass. The yellow brass is much improved by a good but *equal* condensation with the hammer, and in fact to whatever action the metals are subjected, whether natural in the mould, or artificial under the hammer and tools, it is of primary importance that all parts should be treated as nearly alike as possible.

CHAPTER XVIII.

WORKS IN SHEET METAL, MADE BY JOINING.

SECT. I. ON MALLEABILITY, ETC. J DIVISION OF THE SUBJECT.

The process of casting which has been recently considered under so great a variety of forms, is one of the most valuable courses of preparation to which the metallic materials are submitted. In the foundry, the metals are made to assume an infinitude of the most arbitrary shapes, but which are in general more or less *thick* or *massive.* It is now proposed to consider, a few of the methods and principles of another very extensive and serviceable employment of the malleable metals and alloys, which (excepting iron,) are cast into thick slabs or plates, and then laminated into *thin sheets* between cylindrical rollers.

Rollers have been used for a considerable period in the manufacture of sheets of malleable iron, steel and copper, when in the red-hot state, but most others of the metals and alloys are rolled whilst cold; and which economic application of power often nearly supersedes the use of the hammex, as it performs its functions in a more uniform and gradual manner; and at the same time increases to the utmost, the hardness, tenacity, elasticity, and ductility of such of the metals and alloys, as are submitted to this and similar courses of preparation for the arts. As stated at the beginning of the twelfth chapter, these processes of condensation cannot be carried to the extreme, without frequent recurrence at proper intervals to the process of annealing; and in rolling the thinnest sheets of metal, several are frequently sent through the rollers at the same time: but as in the instances of tin-foil, gold and silver leaf, and some others, the hammer is again resorted to after the metals have been rolled as thin as they will economically admit of, in this process of part-manufacture.

None of these preparations of the metals can go on without a material internal change of their substance, to which the celebrated Dr. Dalton thus refers: "Notwithstanding the hardness of solid bodies, or the difficulty of moving the particles one amongst the other, there are several that admit of such motion without fracture, by the application of proper force, especially if assisted by heat. The ductility and malleability of the metals, need only be mentioned. It should seem the particles glide along each other's surface, somewhat like a piece of polished iron at the end of a magnet, without being at all weakened in their cohesion."

This gliding amongst the particles of metals is exemplified by the action of thinning them by blows of the hammer: likewise by the actions of laminating rollers and the draw-bench, in which cases the external layers of the metals are retarded or kept back as it were in a wave, whilst the central stream or substance continues its course at a somewhat quicker rate. The necessity for annealing occurs, when the compression and sliding have arrived at the limit of cohesion; beyond this the parts would tear asunder, and produce such of the internal cracks and seams met with in sheet-metal and wire, as are not due to original flaws and air-bubbles, which have become proportionally elongated in the course of the manufacture of these materials.

A sliding or gliding of a very similar nature occurs also in every case in which the metals are bent; and this differs only in degree, whether we consider it in reference to a massive beam, a permanently flexible spring, a piece of thin sheet metal, or a film of gold leaf. For instance, the curvature of a cast-iron beam originally straight, is produced by the stretching or extension of the lower edge, and the shortening or compression of the upper edge, the central line or the *neutral axis,* remaining unaltered during the process. In like manner a spring derives its elasticity from the extension and compression of its opposite surfaces at every flexure; and the spring remains permanent, or endures its work without alteration of form, when the bending is not carried beyond its limit of elasticity: but when it is bent beyond a certain point, the spring either retains a permanent *set* or distortion, or it will break. In the same manner the beam when only bent to the limit of its elasticity, returns to its original form when the load is relieved, and the constant study of the engineer is so to proportion the beam, that it may Dr. Dalton's New System of Chemical Philosophy, 1808, p. 209.
never be required to exceed, nor even to arrive at the limit of its elastic force. For those parts of mechanism exposed to sudden shocks and strains, he will employ wrought-iron, the cohesive strength of which is considerably greater than that of cast-iron, although less than that of steel, which is the strongest and most permanently elastic of all metallic substances.

The thin metals also possess some elasticity, but this dies away before they reach the tenuity of leaf gold, in which however, the bending cannot be accomplished without a similar change in the arrangement of its opposite sides, although the difference is beyond the reach of our physical senses.

If we desire to wrap a piece of gold leaf around a cylinder of half an inch diameter, so small is the resistance that the least puff of breath suffices; a piece of thin tinfoil offers no more resistance than writing-paper; thin latten-brass, or China tea-lead, is bent more easily than a card; brass and iron the thirtieth or fortieth of an inch thick, could be readily bent with a wooden mallet; but metal of one-eighth of an inch thick would call for smart blows of a hammer, and in iron and steel the further assistance of heat would be likewise required, because in the last case a very considerable amount of the sliding motion of the metal would be called into play.

For example, the piece of metal of an inch thick, was originally flat and of the same size on its opposite surfaces; whereas now, neglecting any alteration of thickness, the inner part would equal the circumference of a circle J an inch diameter, and the outer that of a circle of inch diameter; or it would become 1 and 2i inches long respectively on its opposite surfaces. To produce this change of dimensions, would necessarily require far greater force than the bending of the gold leaf, the internal and external measures of which, viewed as a cylinder, could be ascertained alone by calculation, and not by ordinary means. On the other hand, the sliding of the thick sheet of metal would be illustrated most distinctly, if several pieces of writing-paper, equal to the original metal individually in surface and collectively iu thickness, were wrapped around the same cylinder. The inner paper would exactly meet, the outer would present an open seam inch wide. The metals possessed of the malleable property, undergo a nearly equal change in their arrangement; but the unmalleable or brittle metals break.

DIVISION INTO SINGLE AND DOUBLE CURVATURE. 379

Several of the processes of working the sheet metals are closely analogous to those employed in forging ordinary works in iron and steel; the differences being mainly such as arise from the thin and thick states of the respective materials, and their relative degrees of rigidity, or resistance; the illustrations will be selected indiscriminately from various trades in which the sheet metals are employed. It appears desirable however, to separate the subject into two principal parts: namely, the formation of objects *some* lines of which are straight; and the formation of objects *no* lines of which are straight.

The first division comprehends all objects with plane, cylindrical or conical surfaces, such as may be produced in pasteboard, by cutting out the respective sides, either separately or in clusters, and combining them in part by bending, and in part by cement. Similar works in metal are often produced by the precisely analogous means, of cutting, bending, and uniting, and which call for increase of strength in the methods, proportioned to the rigidity of the materials.

The second division comprehends all objects with surfaces of double curvature, including spherical, elliptical, parabolical, and arbitrary surfaces; as in reflectors, vases, and a thousand other things: none of which forms can be produced in stiff pasteboard, because this material is incapable of being extended or contracted in different parts, in the manner of sheet-metal; this is easily shown, by the following case amongst others.

Terrestrial globes are covered with *thin paper,* upon which the delineation of the surface of the earth has been printed; the paper may be cut into twelve gores, or fish-shaped pieces, all including thirty degrees from pole to pole. But the same gores cut out of *pasteboard* could not be applied to the surface of a globe, as pasteboard does not admit of that degree of gradual extension and contraction, required for the production of spherical and similar *raised* forms, from pieces originally flat, but will become abruptly bent and torn in the attempt.

On the contrary, a round disk of metal may be beaten into a hemisphere, or nearly into a sphere; but even thin paper is only A globe ia usually covered with 26 pieces of paper, namely 2 *pole papers,* or circles including 30" around each pole; and 24 *gores* meeting at the equator. Sometimes the gores extend from the pole to the equator; every gore has then a narrow curved central notch extending 30 from the equator.

possessed of this quality in a very limited degree, for the globe could not be smoothly covered with so few as two, three, or four pieces of the *thinnest* paper without its puckering up, showing that some parts of the material are in excess. The gliding property, or that of malleability and ductility, possessed by the metals, is indispensable to adapt the flat plate to the sphere, by stretching the central portion and gathering up the marginal part, an action, that admits of some comparison to the extension or compression of the slides of a telescope, except that the metal becomes *thicker* or *thinner* instead of being duplicated on itself.

SECT. II. WORKS IN SHEET METAL, MADE BY CUTTING, BENDING AND JOINING.

Every one in early life, has made the first step towards the acquirement of the various arts of working in sheet metal, in the simple process of making a box or tray of card; namely, by doubling up the four margins in succession to an equal width, then cutting out the small squares from the angles, and uniting the four sides of the box, either edge to edge, by paste, sealingwax, or thread, or in similar manners by lapped or folded joints. A different mode is to make the sides of the box as a long strip, folded at all the angles but one; or lastly, the bottom and sides may be cut out entirely detached, and united in various ways.

In the above, and also in the most complicated vessels and solids it is necessary to depict on the material the exact shape of *every* plane superficies of the work, as in the plans and elevations of the architect; and these may be arranged in any clusters which admit of being folded together, so as to constitute part of the joints by bending the material. Thus, a hexagonal box, fig. 197, can be made by drawing first the hexagon required for the bottom, as in fig. 198, and erecting upon every side of the same a parallelogram equal to one of the sides, which in this case are all exactly alike; otherwise the group of sides can be drawn in a line, as in fig. 199, and bent upon the joints to the required angle, or 120 degrees. Either mode would be less troublesome than cutting out seven detached pieces and uniting them; the addition of one more hexagon, dotted iu fig. 199, would serve to complete the top of the hexagonal prism, by adding a cover or top surface.

The same mode will apply to polygonal figures of all kinds, regular or irregular; thus fig. 200 would be produced when the group of sides in 201 were bent around the irregular octagonal base; or that the sides of 202 were separately turned up.

The cylinder, is sometimes compared with a prism of so many sides, that they melt into each other and become a continuous curve; and if the hexagon in fig. 199 were replaced by a circle, and the group of sides were cut out of equal length with the circumference of that circle, and in width equal to the height of the vessel, any required cylinder could be produced. And in like manner any vessels of elliptical or similar forms, or those with, parallel sides and curved ends, and all such combinations, could be made in the manner of fig. 201, (provided the sides were perpendicular,) by cutting out a band equal in length to the collective margin of the figure, as measured by passing a string around it; or the sides might be made of two, or several pieces, if more convenient, or if requisite from their magnitude.

All prismatic vessels require parallelograms to be erected on their respective bases; but pyramids require triangles, and frustrums of pyramids require trapezoids, as will be explained by figs. 204 and 205, which are the forms in which a single piece of metal must be cut, if required to produce fig. 203. Every one of the group of sides, must be individually equal to one of the sides of the pyramid, whether it be regular or irregular, and 203 being an erect and equilateral figure, all the sides in 201 and 205 are required to be alike, and would be drawn from one templet: an irregular pyramid would require all its superficies to be drawn to their absolute forms and sizes.

205.

The cone is sometimes compared with a pyramid with exceedingly numerous sides, (as the cylinder is compared with the prism,) and fig. 206, intended to make a funnel or the frustrum of a cone of the same proportions as 203, illustrates this case. The sides of the cone are extended until they meet in the center o, fig. 203, and then with the slant distances *o a,* and *o b,* the two arcs, *a a,* and *b b,* arc drawn with the compasses, from the center *o;* and so much of the arc *a a,* is required as equals the circumference *a,* of the cone: the margins *a b, a b,* are drawn as two radii. When the figure is curled up until the radial sides meet, it will exactly equal the cone, and the similitude between figs. 205 and 206, is most explanatory, as 206 is just equal to the collective group of the sides required to form the pyramid.

It will now be easily seen that mixed polygonal figures, such as figs. 207, 209, and 211, may be produced in a similar manner, provided their sides are radiated from the square, the hexagonal or other bases, in the manner of figs. 208, 210, 212, but the sides of the rays not being straight, it is no longer possible to group them by their edges, as in figs. 199, 201, and 205. The object with plane surfaces, fig. 207, is only the meeting of two pyramids, at the ends of a prism, and when unfolded, as in fig. 208, the center *a,* is equal to the base *a,* of the object; the sides *b,* radiate and expand from the hexagon at the angle of the faces of the inverted or lower pyramid *b,* and their vertical height in the sheet is equal to the slant height in the vessel; the superficies *c,* are those of a prism, therefore they continue parallel, and have the vertical height of the part *c,* of the figure; lastly, the sides (/, again contract as in the original, and at the same angle as the sides of the six upper faces; in a word, the faces *b, c, d,* are identical in the vase and in the radiated scheme.

Should the vessels, instead of planes, have surfaces of *single curvature,* as in figs. 209 and 211, the method is nearly as simple. The object is drawn on paper, and around its margin are marked several distances, either equal or unequal, and horizontal lines or ordinates are drawn from all to the central line. The radiating pieces for constructing the polygonal vases are represented in figs. 210 and 212, in which the dotted lines are parallel with the sides of the hexagons or the bases, and at distances equal to those of the steps, 1, 2, 3, to 8, around the curve of the intended vases; the lengths of these lines, or ordinates, 1 1, 2 2, 3 3, are in regular hexagonal vessels exactly the same in the radiated plans as in the respective elevations, because the side of the hexagon and the radius of its circumscribing circle are alike.

In all other regular polygonal vessels, the new ordinates will be reduced for figures of 8,10,12 sides, in the same proportions as the sides of these respective polygons bear to the radii of their circumscribing circles, and the ordinates for 3, 4, and 5 sided figures will be similarly increased.

It would have been easy to have extended these particulars to numerous other figures, such as the regular geometrical solids, oblique solids, and many others,f but enough has been advanced to explain the cases of ordinary occurrence, and in the delineations of which, the tinman, coppersmith, and others are very expert. Much of that which has been stated, as it will eventually appear, has been partly advanced in elucidation of the next chapter, on the less apparent methods practised in making similar All the above cases could be accurately provided for without any calculation, by the employment of a very simple scale represented in fig. 213, in which the angle 3 o/, shall contain 120 degrees, or the third of a circle; 4 *o f,* 90 degrees, or the fourth; 5 *o f,* 72, or the fifth; 6 *o f,* 60, or the sixth; and 8,10, and 12, respectively, the Sth, 10th, and 12th parts of a circle. The circular arcs are struck from the center o, and may be the 6th, 8th, 10th of an inch, or any small distance apart.

To learn the altered value of any ordinate, as for constructing a vase like the several figures 207, 209, 211, but with 10 sides; we will suppose the original ordinate to reach from o to x on the radius *o f,* the required measure would be the length of the arc x x, where intersected by the line 10, or that for a decagon; but it would be more convenient to make the angle half the sizeas then the new ordinate would be at once bisected, ready for being set off on each side the central line of the radiated plan. When one side had been carefully formed, a curved templet or gage would be made to the shape, by which all the other sides could be drawn.

For polygonal vessels with unequal sides, such as fig. 214, the curvatures of the edges of the rays will be identical, notwithstanding the differences of the sides. For example, the octagon drawn in the one corner shows that the figure resembles the regular octagon as far as the angles are considered: and that the regular octagon may be considered to be cut into four quarters and to be removed to the four corners, by the insertion of the two pairs of intermediate pieces *a a,* and 4 *h,* which latter would necessarily be parallel. In the like manner a pyramidal vessel built upon the same base, would require equal angles for all its sides.

t See "An Appendix to the Elements of Euclid, in Seven Books, containing Forty-two moveable Schemes for forming the various kinds of Solids and their Sections," *Slc.*, by John Lodge Cowley, F.R.S., 1759. The schemes include the five regular solids, and various irregular solids, prisms, pyramids, and frustrums thereof; all of which are cut out iu the plates, and may be folded up so as to become exact models of the solids. TOOLS FOR WORKING IN SHEET METALS. 385 forms out of flat plates, by the process called *raising;* this is done with the hammer alone, by stretching some parts of the metal and contracting others (the *drawing* and *upsetting* of the blacksmith), a process not required in any of the foregoing figures, the whole of which might be made in pasteboard, a material, that as before observed, does not admit of being raised or bulged into figures of double curvature. To conclude this chapter, it only remains to advert in the same general manner to the modes of bending and uniting the edges of the works represented.

The various works having been drawn upon the sheet of metal, the first process is to cut them out; this is almost always done with the shears; sometimes however for thick metal, the cold-chisel and hammer are used, the work being laid upon the bare anvil or upon a cutting-plate, as in forging: occasionally the metal is fixed in the jaws of the tail-vice, and cut off with the cold-chisel applied in contact with the vice; the edge of the chisel is placed nearly parallel with the jaws, which serve as a guide. In some cases very long vices with a screw at each end, are used in a similar manner, for the thick iron plates employed for boilers; f but the shears are the most generally convenient.

Although the tools used in working the sheet metals are extremely various as regards their sizes and specific forms, they may, with the exception of the shears and soldering-tools, be principally resolved into numerous varieties of hammers, anvils, swage-tools, and punches. Figs. 215 to 229 represent some few of the most common of these tools, which are used alike both in bent and raised works, and their close resemblance to those for ordinary forging in iron and steel will not escape observation. The most remarkable points of difference are in their greater height and length, which enable them to be applied to the interior of large objects, and also iu their square shanks, by which they are fixed in holes in the wooden blocks and benches.

The hammers are nearly alike at both ends; many of them have circular faces, either flat or convex; others resemble the straight or cross panes of ordinary hammers, and are also either flat or convex; and those used in finishing, are exceedingly See the Chapter on Shears, Vol. II., especially the figures on page 915. t See Plate 60. Buchanan, on Mill-Work. Edited by Rennie. 1841.

VOL. I. CO bright, in order that they may impart their own degree of polish to the work, which process is called *planishing.*

When thin metal is struck between tools both of which are of metal, it is invariably more or less thinned; and should the blows be given partially, such parts will become stretched or cockled, and will distort the general figure. It is therefore usual, whenever admissible, to employ wooden hammers of the forms described, and also wooden blocks or anvils when metal hammers are used; reserving the employment of tools *both* of metal, either for the concluding steps, or for those cases, where from the substance of the metal and the nature of the work, the wooden hammers would be ineffective, or a greater definition of form is required than wooden tools could give.

The anvil used by the coppersmith and similar workmen is usually square, say from six to eight inches on every side; and the smaller anvils which are called *stakes,* and also *teesls,* are of progressively smaller sizes, down to half an inch square, and even less. Some of them have one edge rounded like 21S; others have rounded faces as 219 and 220; a few assume the form of a rounded ridge, like fig. 222; and many have bulbs or buttons, as if turned in the lathe, as in fig. 223.

The *beak-irons* are also very unlike those used by the smith; they are seldom attached to the anvil, and are often exceedingly long, as in fig. 216; some few, for more accurate purposes, are turned in the lathe to the conical form, like 221, these are held in the vice, the jaws of which enter grooves in the shank; and mandrels four to six feet long, used for making long pipes, are attached to the bench by long rectangular shanks and staples.

Fig. 215, the *hatchet-stake,* is from two to ten inches wide; it is very much used for bending the thin metals, in the same manner as the rectangular edge of the anvil is used for those which are thicker; a cold-chisel fixed in the vice forms a small hatchet-stake; 224 is the *creasing-tool* for making small beads and tubes; 225 is the *seam-set* for closing the seams prepared on the hatchet-stake; 226 is a hollow and 227 a solid punch, the cutting edge of the former meets at about the angle of fifty degrees, the latter is solid at the end for small holes; both are struck upon a thin plate of lead.or solder laid upon the stake; 228 is a *riveting-set* or punch for the heads of rivets; and 229 is the *swage-tool,* a miniature of the tilt-hammer, to which a great variety of top and bottom tools, or *creases,* are added, which greatly economise the labour of making different mouldings and bosses; the stop is used to retain the parallelism of the mouldings with the edge of the metal, and a similar stop is also at times applied to the hatchet-stake, 215.

The sides of the vessels represented in figs. 197 to 212, if the metal were thin, would be bent to the required angles by laying the metal horizontally upon the hatchet-stake, with the lines exactly over the edge of the same, and blows would be given with the mallet (or with the hammer for more accurate angles), so as to indent the metal with the edge of the stake; it would be then bent down with the fingers, unless the edge were very narrow, as for a seam, when the mallet would be alone used. Thicker metal is more commonly bent over the square edge of the anvil, as in fig. 107, p. 216, a square set or hammer being held upon its upper surface; and sometimes the work is pinched fast in the vice, and it is bent over with the blows of a flat-ended punch or set, applied close in the angle, and then hammered down MODES OF BENDING.
square with the hammer; very strong metal is seldom bent in this manner, but the sides of objects are then made separately, and united in some of the ways which will be explained.

In bending thin metals either to circular or other curves, they are held on the one edge in the hand, and curled on the opposite edge over beak-irons or triblets with the mallet; when the metal is too stubborn or too narrow to be thus held in the hand (as the coppersmith scarcely ever uses tongs, except at the fire,) the metal is driven into a concave tool to curl up the edges. For instance, the crease, fig. 224, is frequently employed for making small tubes or edging; the strip of metal is laid over the appropriate groove, and an iron wire is driven down upon it with the mallet, this bends it like a waggon-tilt; the edges are then folded down upon the wire with the mallet, and it is finished by a top tool, or a punch, fig. 225, having a groove of similar concavity or radius to that in the crease.

For half-round strips, the crease together with the round wire suffice, or they would be more quickly made in the swage-tool, 229, and which might in this manner be made to produce any particular section or moulding, and that at any distance from the edge by means of the stop or gage. Large tubes are always finished upon beak irons, such as fig. 216, the round ends of which serve for curvilinear, and the square ends for rectilinear works.

Figs. 230. 231. 6 *b*

All the sheet metals up to the thickest boiler plate are treated much after the same general methods, large cast-iron moulds of various sweeps are employed, the stout iron being heated to redness, and set into them with set-hammers struck with the sledge. When a circular bend is wanted in the center of a long piece, it is conveniently and accurately done by bending it over a ridge, such as a parallel plate with a rounded edge, or a triblet, the ends of the work serving for a purchase, or as levers. Thus fig. 230 shows the common mode of bending thick plates to the form of the piece *d, b, 6,* for the internal flues of marine boilers; the plate is heated to redness in the middle, and pressed down until *a, c* assume the positions *d, 6*

In a similar manner, to bend long strips into easy curves, such as for cylindrical vessels, the tinmen use a *former,* fig. 231, a cylindrical piece of wood from two to four inches diameter, and two feet long, turned with a pivot at the one end; the pivot is laid upon the edge of the bench, and the man rests his chest against the other extremity of fig. 231, to support it in the horizontal position. The tin plate is first stretched in the hands by the two corners *a,* c, and rubbed over the *former* diagonally, to bend it at every part, this is repeated across the other diagonal to flatten the plate; it is afterwards folded round the stick, and rubbed forcibly down with the hand, as at *d,* to give it an easy bend approaching to the required curvature. Should the vessel require a bead at the upper edge, it is usually made by the swage tool, fig. 229, before the plate is curled up; the work is then much more rigid, and requires additional force to bend it.

Fig. 232 is intended to explain a very simple and useful machine, first employed by the tinmen for rolling up the cylinders for spring window-blinds, the sides of culinary vessels and similar works, and now also by the boiler-makers and others for the strongest plates. It has two cylindrical rollers *a b,* and *d,* which are connected by toothed wheels so as to travel in opposite directions, thus far exactly the same as a pair of laminating rollers for making the sheet metals; the third roller, *c,* is just opposite the two, and is free to move on its pivots, as it is unconnected with *a b* and *d;* and the third roller c, is capable of vertical adjustmeut.

390

B ENDING AND FLATTENING.

When therefore the metal is moved along by the carrying rollers *a b* and *d,* it strikes against the edge of the bending roller c, and is curled up to enable it to pass *over* the same; and as this bending occurs in an equal degree at *every* point of the sheet of metal, it assumes a circular sweep, the radius of which is dependent on the place of *c.* In the central position, the sheet would assume the circle e, /, *g;* and when *c,* is more raised as to the upper position, the metal would follow the dotted circle, the radius of which is much less; and when the bending roller *c* is placed out of level, the works are thrown into the conical form.

Fig. 233 shows Mr. Roberts' original application of the bending rollers to boiler plates; none of the rollers, o, *b, c,* touch each other, and *b,* is under adjustment for the different curvatures.

In the last four figures the same principle is employed, namely, the application of three forces, as in a lever of the first order, or as in bending or breaking a stick across the knee. The schoolboy's problem of " drawing a circle through three given points" is thoroughly exemplified in fig. 233; and in 232, the one force is the grip of the plate on the line of centers of *a, b, d;* the roller c curls the plate partly around the roller *a,* and the point at which the plate leaves *a b,* may be called the second force, or *b;* the third is the point of contact on *c.*

One of the most useful applications of the *bending* machines, is in *straightening* the metals, which may at first appear to be a misapplication of words, but in truth by the depression of *c,* to about the position c, it only bends the plate for the moment, just to the *limit* of its elasticity.. It results that when it has been passed through twice, or with each side alternately upwards, the elastic reaction just suffices to convert the figure temporarily given, or that of the arc of an *enormous circle,* into a plane or true surface; and as this is done without any blows, which produce partial condensation at such spots, the plate is less subject to after changes than if it had been hammered flat; as by the rollers *every part* of the plate is bent exactly to the limit of its permanent elasticity. In the tinmen's bending rollers, *d, c,* fig. 232, p. 389, are often turned with half-round grooves, to receive the thickened edge which contains the wire employed to stiffen the tops of the vessels; sometimes also the rollers are used for preparing the seam to contain the wire. Grooved rollers, (similar to those shown on pages 187 and 188) are very extensively employed likewise in other works in the arts besides the manufacture of iron, to which they are there more immediately referred.

The use of the plain cylindrical roller at *h,* page 187, is so simple as to be immediately apparent; rollers with curvilinear edges, such as at *i,* have been long employed for bending the steel and brass plates for fenders; similar rollers on a smaller scale and of numerous patterns, many of them chased and ornamented, are used in making jewellery, as for producing mouldings, headings, and matted, checkered, or other works.

SECT. III. ANGLE AND SURFACE JOINTS.

The next steps to be considered, appear to be the methods of uniting the edges of the vessels after they have been cut and bent to meet in angles, curves or plane surfaces; the principal modes of accomplishing this are represented in figs. 234 to 256, which are grouped together for the convenience of comparison.

Figs. 234 and 235 are for the thinnest metals, such as tin, which require a film of soft-solder on one or other side. Sheetlead is similarly joined, and both are usually soldered from within.

Figs. 236 and 237 are the *mitre* and *buit-jomts* used for thicker metals with hard-solders; sometimes 237 is dovetailed together, the edges being filed to correspond coarsely; they are also partly riveted before being soldered from within. These joints are very weak when united with soft-solder.

Figs. 234. 235. 236. 237. 233. 239. 240. 241. 242. 243. 241. 245.

Fig. 238 is the *lap-joint,* the metal is creased over the hatchet-stake. Tinplate requires an external layer of solder; spelter solder runs through the crevice and need not project.

Fig. 239 is folded by means of the hatchet-stake, the two are then hammered together, but require a film of solder to prevent them from sliding asunder.

Fig. 240 is the *folded angle-joint,* used for fire-proof deed boxes, and other strong works in which solder would be inadmissible, it is common in tin and copper works, but less so in iron and zinc, which do not bend so readily.

Fig. 241 is a *riveted* joint which is very commonly used in strong iron plate and copper works, as in boilers, &c.: generally a rivet is inserted at each end, then the other holes are punched through the two thicknesses with the punch 227, on a block of lead. The head of the rivet is put within, the metal is flattened around it, by placing the small hole of the riveting set 228 over the pin of the rivet, and giving a blow; the rivet is then clenched, and it is finished to a circular form by the concave hollow in another riveting set. When the works cannot be laid upon an anvil or stake, a heavy hammer is held against the head of the rivet to receive the blow; in larger works the holes are all punched before riveting, and the heads are left from the hammer.

Figs. 242 and 243; the plates *a a,* are punched with long mortises, then *b b,* are formed into tenons, which are inserted and riveted; but in 243 the tenous have transverse keys to enable the parts to be separated.

Fig. 244, the one plate makes a butt-joint with the other, and is fixed by _ formed rivets or screw-bolts , the short ends are generally riveted to the one plate, even when screwed nuts are used. This mode is very common for cast-iron plates, as in stove work.

Fig. 245 is the mode universally adopted for very strong vessels, as for steam boilers, in which the detached wrought-iron plates are connected by angle-iron, rolled expressly for the purpose (see /, fig. 76, p. 187). The rivet holes are punched in all the four edges, by powerful punching engines furnished with travelling stages and racks, which ensure the holes being in line, and equidistant, so that the several parts when brought together may exactly correspond. The rivet *r,* which may be compared to a short stout nail, is made red-hot, and handed by a boy to the man within the boiler, who drives it in the hole; he then holds a heavy hammer against its head, whilst two men quickly clench or burr it up from without: between the hammering, and the contracting of the metal in cooling, the edges are brought together into most intimate and powerful contact.

Bolts and nuts *b,* may be used to allow the removal of any part, as the man-hole of the boiler.

For the curved parts of the boilers, the angle-iron is bent into corresponding sweeps, and for the corners of square boilers, the angle-iron is welded together to form the three tails for the respective angles or edges which constitute the solid corner: this when well done is no mean specimen of welding.

It frequently happens that several plates are required to be joined together to extend their dimensions, or that the edges of one plate are united as in forming a tube; these joints are arranged in the figures 246 to 256, similarly to those for angles previously shown, from which they differ in several respects.

Fig. 246 is the *lap-joint,* employed with solder for tin plates, sheet lead, &c. , and for tubes bent up in these materials.

Fig. 247, the *butt-joint,* is used for plates and small tubes of the various metals; united with the hard solders they are moderately strong, but with tin solder the junctions are very weak from the limited measures of the surfaces.

Fig. 248 is the *cramp-joint;* the edges are thinned with the hammer, the one is left plain, the other is notched obliquely with shears, from one-eighth to three-eighths of an inch deep; each alternate cramp is bent up, the others down, for the insertion of the plain edge; they are next hammered together and brazed, after which they may be made nearly flat by the hammer, and quite so by the file. The crampjoint is used for thin works requiring *strength,* and amongst numerous others for the parts of musical instruments. Sometimes also 246 is feather-edged; this improves it, but it is still inferior to the cramp-joint in strength.

Fig. 249 is the lap-joint without solder, for tin, copper, iron, &c.; it is set down flat with a seam-set, fig. 225, and used for 247.
248. 249. 250. 251. 252. 253. 254. 255. 256. smoke pipes, and numerous works not required to be steam or water-tight.

Fig. 250 is used for zinc works and others; it saves the double bend of 249; it is sometimes called the *patent strip overlap.*

Fig. 251 is the *roll-joint* employed for lead roofs, the metal is folded over a wooden rib, and requires no solder: the water will not pass through this joint until it exceeds the elevation of the wood. The roll-joint is less bent when used for zinc, as that material is rather brittle; the laps merely extend up the straight sides of the wooden roll, and their edges are covered by a halfround strip of zinc nailed to the wood.

Fig. 252 is a hollow crease used for vessels and chambers for making sulphuric acid, the metal is scraped perfectly clean, filled with lead heated nearly to redness, and the whole are united by *burning,* with an iron heated also to redness. Solder which contains *tin* would be acted upon by the acid, whereas, until the acid is very concentrated, the lead is not injured: this method is however now superseded by the mode of autogenous soldering. The concentration of sulphuric acid and some other chemical preparations, is performed in vessels made of platinum.

Figs. 253 and 254 are very commonly employed either with rivets or screw-bolts; the latter joint is common in boilers, both of copper and iron, and also in tubes: copper works are frequently tinned all over the rivets and joints, to stop any minute fissures. Fig. 253 is the flange joint for pipes.

Fig. 255, with rivets, is the common mode of uniting the plates of marine boilers, and other works required to be flush externally.

Fig. 256 is a similar mode; used of late years for constructing the largest iron steam ships; the ribs of the vessels are made of T iron, varying from about four to eight inches wide, which is bent to the curve by the employment of very large surface-plates cast full of holes, upon which the wood model of the rib is laid down, and a chalk mark is made around its edge. Dogs or pins are wedged at short intervals in all those holes which intersect the curve; the rib, heated to redness in a reverberatory furnace, is wedged fast at the one end, and bent around the pins by sets and sledge-hammers, and as it grows or yields to the curve, every part is secured by wedges until the whole is completed.

CHAPTER XIX.

WORKS IN SHEET METAL, MADE BY RAISING.

SECT. I. CIRCULAR WORKS SPUN IN THE LATHE.

The former examples have only called into action so small an amount of the malleable or gliding property of the metals, that all the forms referred to could be produced in pasteboard, a material nearly incapable of extension or compression. The raised works now to be considered, call for much of this gliding or malleable action, which may be compared with the plastic nature of clay as an opposite extreme. Thus a lump of clay is thrown on the potter's horizontal lathe, a touch of the fingers shapes it into a solid round lump, the potter thrusts his clenched hand into the center, and it rises in form something like a bason; by applying the other hand outside to prevent the material from spreading, it will rise as an irregular hollow cylinder, and a gentle pressure from without, and a sustaining pressure from within, will gather up or contract the clay into the narrow mouth suited to a bottle, and which is made somewhat in this manner almost by the fingers alone.

A similar and parallel application, due to the *malleability* of the metals, and one which also requires the turning-lathe, is very extensively practised; namely, the art of *"spinning or burnishing to form"* thin circular works in several of the ductile metals and alloys, as for teapots, plated candlesticks, the covers of cups and vessels, the bell mouths of musical instruments, and numerous other objects required in great numbers, and of *thin* metals. Plated candlesticks are thus formed of several parts soldered together, or retained in position by the fittings of their edges, the whole being strengthened by a central wire, and by filling the entire cavity with a resinous cement. The figures 257 and 258, are intended to show the mode of spinning the body of a Britannia metal teapot from one unperforated disk of metal.

The wooden mould or chuck *a,* fig. 257, is turned to the form of the lower part of the teapot, and a disk of metal *b,* is pinched tight between the flat surfaces of *a* and *c,* by the fixed center screw *d* of the lathe, so that *a, b,* and c, revolve with the mandrel: and now by means of a burnisher *e,* which is rested against a pin in the lathe rest, as a fulcrum, and applied near the center of the metal; and a wooden stick /, held on the opposite side to support the edge, the metal is rapidly bent or swaged through the successive forms 1, 2, 3, to 4, so as to fit close against the curved face of the block and to extend up its cylindrical edge.

The mould *a,* is next replaced by *g,* fig. 258, a plain cylindrical block of the diameter of the intended aperture; one of various forms of burnishers *(h, i,* some bent, others f form, and so on, the surfaces of which are slightly greased) arc used together with the hooked stick or rubber,/, first to force the metal inwards as shown at 5, 6, 7, and also to curl up the hollow bead which stiffens the mouth of the finished vessel, 9. Sometimes the moulds are made of the entire form of the inside of the work, but of several pieces each smaller than the mouth; so that when the central block is first removed, the others may be successively taken out of the finished vessel, like the parts of a hat-block or of a boot tree.

It is of importance during the whole process, to keep the edge exactly concentric and free from the slightest notches, for which purpose it is occasionally touched with the turning tool during the process of spinning. The operation is very pretty and expeditious, and resembles the manipulation of the potter who forms a bottle or vase with a close mouth in a manner completely analogous, although the yielding nature of his material requires the fingers alone, and neither the mould, stick, nor burnisher.

The lenses of optical instruments are often fixed in their cells Fig. 259. by similar means; *a,* fig. 259, shows

"-- in excess the form of the metal when

Jj B turned, and *b,* the thin edge when TM curled over the glass by means of a burnisher applied whilst the ring revolves in the lathe.

Much of the cheap Birmingham jewellery, is also spun in the lathe, but in a different manner; for instance to make such an object as the ring represented black in fig. 260; a steel mandrel is turned upon a lathe to the same form as the ring, but less in diameter. The metal is prepared as a thin tube, it is soldered and cut into short pieces, each to serve for one ring, and these are spun into shape almost in an instant, between the arbor and the milling tool or roller, as seen in the front view fig. 261; it is clear that unless the arbor were smaller than the work, the latter from being undercut could not be released: sometimes only one broad milling tool is employed, at other times two or more narrow ones. This process is most distinctly a modification of two rollers, which travel by surface-contact instead of by toothed wheels, and differs but little from the embossing or matting rollers exployed by jewellers and others for long strips instead of rings; extending the same application to the millingtool upon a solid body such as milled nut, the interior metal supplies the resistance given by the arbor, in the last figure.

SECT. II.—WORKS RAISED BY THE HAMMER.

In raising the metals by the hammer, we have to produce similar effects to those in the spinning process, not however by the gradual and continued pressure of a burnisher, on one *circle* at a time, but by *circles of blows,* applied much in the same order, and as far as possible with the same regularity of effect.

The art consists, therefore, of two principal points: first, so to proportion the original size and thickness of the metal disk that it shall exactly suffice for the production of the required object; neither with excess of metal, which would have to be cut off with shears and thrown aside, wasting a part both of the metal and labour, nor with deficiency of metal, which would be nearly a total loss: secondly, that the work shall be produced with the *smallest possible number of blows,* which sometimes tend to thin, and at other times to thicken, the metal; whereas the finished works should present an uniform thickness throughout, and which is, in many cases, just that of the original metal when in the sheet.

For instance, a hollow ball six inches diameter is made of two circular pieces of copper, each seven and a half inches diameter: now calling the original circumference of the disk twenty-two and a half inches, this line eventually becomes contracted to eighteen inches, or the circumference of the ball; although at the same time the original diameter of the disk, namely a line of seven and a half inches, has become stretched to that of nine inches or the girth of the hemisphere.

This double change of dimensions, accomplished by the malleability or gliding of the metal, occurs in a still more striking manner in the illustration of spinning the tea-pot, in which the disk, originally about one foot diameter, becomes contracted to two or three inches only at the mouth. The precise nature of the change is seen on inspecting figs. 207 and 209, p. 383, in connection with the radiated pieces, 208 and 210, required for the formation of such polygonal vases, when bent up and soldered at their edges.

The same vases wrought to the circular figure from round plates, either by spinning or by the hammer, would not require disks of metal so large as the boundary circles in figs. 208 and 210; as the pieces between the rays would be entirely in excess, they would cause the vessels to rise beyond their intended sizes, and would require to be pared off. But the original disks for making the vases should be of about the diameters of the inner circles, as then the *pieces d,* beyond the inner circles, would be nearly equal to the *spaces e,* within these circles, which would leave the vessel of uniform thickness throughout, and without deficiency or excess of metal supposing the conversion to be performed with mathematical truth.

The first and most important notion to be conveyed in reference to raising works with the hammer, is the difference between those which may be called *opposed,* or *solid* blows, that have the effect of stretching or thinning the metal; and those which may be called *unopposed,* or *hollow* blows, that have less effect in thinning than in bending the metal; in fact, it often becomes thickened by hollow blows, as will be shown.

Figs. 262. 263. 264.

For example, the hammer in fig. 262 is directly opposed to the face of the anvil, or meets it face to face, and would be said to give a solid blow; one which would not jar the hand grasping the plate, were the latter ever so thick or rigid: and this blow would thin the metal by its sudden compression between two hard surfaces, the face of the hammer being represented at /.

The hammer in fig. 263 is not directly opposed to the anvil, or rather to that point of it which sustains the work, consequently this would be called a hollow blow, one which would jar the hand were the plate thick and rigid; and it would bend the plate partly to the form of the supporting edge, by a similar exhibition of the forces *a, b, c,* referred to in the diagrams, figs. 230 to 233, pages 388 and 389; not however by the quiet pressure therein 400 SOLID AND HOLLOW BLOWS.

employed, but by impact, or by driving blows. The band situated at *a,* fig. 263, would be insufficient to withstand the blows of the hammer at c, but for the great distance of *a b,* compared with *b c,* and the thin flexible nature of the material.

From these reasons the coppersmith and others never require tongs for holding the metal, the same as the blacksmith, except at the tire, as in annealing and soldering; in hammering thin works, a constant change of position is required, and which can be in no way so readily accomplished as by the exquisite mechanism given us by nature, the unassisted hand. When however the works are too rigid or too small to be thus held, the anvil is made to supply the two points *a, c,* as in fig. 264, and the blow of the hammer is directed between them

We will now trace the effects of *solid* and *hollow* blows given partially on a disk of metal *a a,* fig. 265, supposed to be twelve inches diameter; first within a centra circle *c c,* of three inches diameter; and then around the margin *a b,* to the width of three inches, leaving the other portions untouched in each case; the thickness of the metal is greatly exaggerated to facilitate the explanation.

The *solid* blows within the circle c c, would thin and stretch that part of the metal, and make it of greater superficial extent; but the broad band of metal *a c,* would prevent it from expanding beyond its original diameter, and therefore the blows would make a central concavity, as in a cymbal, or like fig. 266. And the more blows that were given, either inside the bulge upon a flat anvil, or outside the i i i j bulge upon an anvil or head of a globular form, the more would the metal be raised, from its being thinned aud extended: and thus it might be thrown into the shape of a lofty cone or sugar-loaf.

The *hollow* blows given within the same limited circle, would also stretch the metal and drive it into the hollow tools employed, such as fig. 264; thus producing the same effect as in 266, but by stretching the metal as we should the parchment 272.

SOLID AND HOLLOW BLOWS. 401 of a drum, by the pressure of the hand in the center, or by a blow of the drumstick.

The *solid* blows around the three-inch margin, would thin the metal and cause it to increase externally in diameter; but the plate would only continue flat, as in fig. 267, if every part of the ring were stretched proportionally to its increased distance from its first position. Were the inner edge towards *b,* thinned beyond its due amount, its expansion, if resisted by the strength of the outer ring a, would throw part of the work into a curve, and depress the metal, not as in the cymbal, but in the form of a gutter as in fig. 268; it would however more probably happen, that the inner edge alone of the marginal ring would be expanded, leaving the outer edge undisturbed, and producing the coned figure, 269.

The *hollow* blows given around the edge, as in fig. 263, would have the effect of curling up or raising the edge, first as a saucer 270, and then into a cylindrical form 271; provided that by the skilful management of the hammering, the metal could be made to slide upon itself without puckering, so as to contract the original boundary circle of the disk of twelve inches, into six inches, or the measure of the edge of the cylinder resulting from the drawing in of the three-inch margin.

In this process the metal would become proportionally thickened at the upper edge, because each little piece of the great circle, fig. 272, when compressed into a circle of half the diameter, would only occupy half its original length, as it could not be altogether lost; and the metal would therefore increase in thickness in a proportional degree. The remainder of the circle serves for the time as effectually to compress the metal in the direction of the tangent, as if the radii were the sides of an unyielding angular groove dotted in fig. 272: this contraction produces in fact the same effect, as the *jumping* or *upsetting* by endlong blows in smith's-work. Theoretically, the thickness of the upper edge of the cylinder would be doubled, and the lower edge would retain its original thickness, as in 271; whereas in extending the margin of the disk by *solid* blows as in fig. 267, the thinned edge would be found to taper away, also in a straight line, from the full thickness even to a feather edge if sufficiently continued, but neither of these cases would be admissible, as the general object is to retain a uniform substance.

VOL. I. D D

In equalising the thickness of the cylindrical tube, fig. 271, the solid blows would thin the metal, but at the same time throw it into a larger circle; it would then require to be again driven inwards, which would again slightly thicken it. So that in reducing the metal to uniformity, two distinct and opposite actions are going on; and upon the due alternation, combination, or proportioning of which, will entirely depend the ultimate form: that is, whether the metal be allowed to continue as a cylinder; to expand or to contract, either as a cone or as a simple curve; or to serpentine in any arbitrary manner, according as the one or other action is allowed to predominate with the gradual development. The treatment of such works with the hammer, is unlike spinning the teapot, at those parts of the work where the metal is folded down in close contact with the solid revolving mould therein employed; but in completing the upper part on the small block, fig. 258, the burnisher and rubber may be considered equivalent to the two antagonistforces, which lead the hammered vessel inwards or outwards, at the will of the operator.

This subject is too wide to enable anything more to be offered than a few general features, and I shall therefore proceed to trace briefly the practice in some examples.

Fig. 273 represents the first stage of making the half of a copper ball; the metal is first driven with a mallet iuto a concave bed, generally of wood, in which it is hastily gathered up to a sweep of about the third part of a sphere, as *a, a,* fig. 274; but this puckers up the edge like a piece of fluted silk, or the serpentine margin of many shells, in the manner represented at *fff* fig-275, which is of twice the size of 274.

The next step is to remove the flutes or puckers by means of blows of the raising hammer, applied externally as indicated by the black lines at *h,* fig. 275; and in fig. 276 are represented on a still more enlarged scale, the relative positions of the hammer, RAISING HEMISPHERES.

403 anvil, and work. Thus A represents the globular face of the anvil, B the rounded edge of the raising hammer, which like the pane of an ordinary hammer, stands at right angles to the handle, and *a* 1, shows the work, *a* being the edge, and 1 the point of the flute. The blows of the hammer are made to fall nearly on the center o, of the anvil, and at a small angle with the perpendicular, the hand being on the side *a.* A few blows are given as tangents, or directly across the point of the flute, and when it exceeds the width of the hammer, oblique blows are given to restore the pointed character, to be followed by other bl»ws parallel with the first, as shown at *h,* fig. 275. These hollow blows cause the sides of the flutes to slide into one another, almost as when two packs of cards, placed like the ridge of a house, penetrate into each other and sink down flat: in a manner somewhat resembling that by which the original and extreme margin in fig. 272, page 400, becomes, by the successive blows, contracted to the inner circle; but in the present case the plait slides down to the general curve of the spherical dish.

If however the puckers of a large globe were entirely removed by hollow blows, the central lines of the flutes would become thickened, and therefore solid blows are mingled with them, or rather the one blow partakes of the two natures. Thus from the curvature and oblique position of the hammer, fig. 276, its face is solid at , to that part immediately below it, but towards *h,* it rather bends than thins: the flatter the curves of the two surfaces, the greater the extent of the solid or thiuning blows. The plaits are not however entirely gathered up, as the dish *a, a,* fig. 274, always opens a little from the metal becoming stretched under the treatment for removing the flutes.

Throwing the work into flutes as described is not imperative, for the hemisphere might be entirely raised, as in the succeeding 404 RAISING HEMISPHERES. step, by blows on the outer surface upon a convex tool or head, but the flutes quicken the process, and speedily give a concavity which is convenient, as it makes the work hang better on the rounded face of the anvil.

The outer curve *a a,* fig. 274, p. 402, which represents the copper dish when the puckers have been removed, will not be sent into the hemispherical form, or the inner line *d d,* at one process, hut will progressively assume the curvatures *b b, c c,* and sometimes many others: neither will the work be changed from the curve *a* a, to that of *b b,* at one sweep, nor as with the burnisher in spinning, even by one consecutive ring or wave. The hammer must necessarily operate by successive blows arranged in circles, the proximity of which circles, will at length include within their range the entire sweep *a a,* or *b b,* each of which is called a *course:* and before proceeding from one course or sweep to the next, the metal requires to be annealed.

Figs. 276 and 277 explain the transition or conversion from the first sweep *a,* to the second sweep *b;* the black lines represent the metal after a few circles of blows have been given. Fig. 277 shows the narrow edge of the raising-hammer, in the act of descending upon the center of the head or stake, and as a tangent to the circle; it first throws in a little rim

at 1, which connects the new and old sweeps by a curve or ogee; then another little circle 2, will be similarly gathered in, then 3,4, 5, and so on, up to the edge. Now the artifice consists in making the intervals both of the great sweeps, *a, b, c,* fig. 274, and of the little waves 1, 2, *3,* of fig. 277, as large as practicable, RAISING HEMISPHERES.
405 provided they do not cause the exterior metal to pucker or become in plaits, as this would endanger its ultimately cracking at those places, where the metal might have become plaited.

In thus *raising-in* the metal, it necessarily becomes thickened from its contraction in diameter, but as in fig. 276 the hammer at *h,* gives a hollow blow and bends, whilst the part *s,* gives a solid blow and thins, the two effects are thus combined; and when they are duly proportioned, by a hammer more or less round, and blows more or less oblique, the true thickness as well as the desired change of figure are both obtained.

It is easier to get the hemisphere by a little excess of thinning, or by a superfluity of blows: so that the less skilful workman will use a piece of copper of seven inches diameter, with additional blows, for a six-inch hemisphere: but the more skilful will take a piece of seven and a half inches diameter, and obtain the work with less labour. Occasionally, when the work is common and thin, from three to six hemispheres or other pieces are hollowed together, the outer piece is cut as a hexagon or octagon, and its angles are bent over to embrace the inner pieces, before the process of hollowing is begun, and which scarcely consumes more time than for one only. This is a general practice in hollowing tin-works, such as the covers of saucepans, as the number of thicknesses divide the strength of the blows; the several pieces are then twisted round at intervals, so as to arrange them in a different order, which mixes the little imperfections, and tends to their mutual correction: the raising process represented in fig. 277 is also performed upon two or three pieces at a time, when they are sufficiently thin to permit it.

One of the most conspicuous and remarkable examples of raised works, is the ball and cross of St. Paul's Cathedral, London. The old ball consisted of sixteen pieces riveted together; the present, also 6 feet diameter and £ inch thick, was raised in *ttco* pieces only, and may therefore be considered to mark the improvement in the coppersmith's art in making large works, such as sugar-pans, stills, &c.
The metal was first thinned and partly formed under the tilt-hammer at the copper-mills, or sunk in a concave bed; the raising was effected precisely as explained in fig. 277. and with hammers but little larger than usual; the two parts were riveted together in their place, and the joint is concealed by the ornamental band.

All the work is modern, and is mostly hammered up, except the cast gun-metal consoles beneath the ball, which formed part of the original metallic edifice; a name to which it is justly entitled, the height being 29 feet, and the weight of copper 3i tons. The new ball and cross were erected in 1821, by Messrs. Kepp, of

Having conveyed the full particulars for raising a hemispherical shape, the modifications of treatment required for various other forms will be sufficiently apparent. Thus, below the dotted lines *a d,* in fig. 278, the sweeps are exactly the same as in fig. 274, but the metal rises higher from having been originally larger: in the courses *g h,* it is first kept rather thicker on the edge, and towards the conclusion, it is thinned on the edge to the common substance, and curled over by hollow blows from within, although the whole figure might be produced by external blows, but which would be a more tedious method.

On the other hand, by the continuance of the *raising-in,* explained by diagram 277, the metal would be gathered into a smaller diameter through the steps *ij k I,* in the latter of which the metal would become thickened, unless the solid or thinning blows were allowed to predominate. If enough metal had been given in the first instance, when the mouth had been contracted as to the form of a teapot, it might be extended upwards as a cylindrical neck, in the manner explained in fig. 271, p. 400, and curled over at the top, as on the opposite side of fig. 278, at *h.*

To lessen the labour of raising works from a single flat plate, soldering is sometimes resorted to; thus the teapots, figs. 207 and 209, p. 383, might be made in two dished pieces, and soldered at the largest diameter; the lofty vase or coffeepot, fig. 211, could be made from a cylinder of midway diameter soldered up the side, the bulge being set-out by thinning the metal, and the contraction above being drawn-in by hollow blows.

Vases in the shape of an earthen oil jar, or of the line *I dn,* fig. 278, could be made from a cone such *as op,* with a bottom soldered in: these preparations would save the work of the hammer, although such forms and others far more difficult could be raised entirely by the hammer from a flat piece of metal.

Should any of the above vessels require a solid thickened edge or lip, beyond that which would result from the drawing-in of the metal, it would be necessary to select a piece of metal of smaller diameter but thicker, and to retain the margin of the full thickness by directing all the blows within the same; someLondon; they are strengthened by a most judicious inner framing of copper and wroughtiron bars, stays, bolts, and nuts, extending through the arms and downwards into the building; thus adding about 2 tons of iron to the load of copper, and to the 38 ounces of gold used in its decoration.
HAlSlNG COMPLEX WORKS. 407 times on the other hand, works require to be thinned on the edge, these are then cut out proportionally smaller than their intended sizes, as illustrated by the following example, which is considered the most difficult of its kind.
The bell of a French horn, together with the first coil of the tube, are made of a flat strip of metal about 4 feet long and 2 inches wide; for making'the bell of the instrument, there is an enlargement at one end of the strip, in the form of the funnel piece 206, page 382, and of the width of 16 to 22 inches, the small-

er piece being adopted when the bell is required to be very thin. The narrower piece of metal, when first bent up, much resembles the butt end of a musket, terminating in a small tube; the metal is united and soldered down the edge with a cramp or dovetail joint, fig. 218, page 393; it is next thrown into a conical form of about five inches diameter, and expanded from within, first with blows of a wooden mallet upon a wooden block, and then with those of a hammer on iron stakes.

When nearly finished, or about one foot diameter, it is hammered very accurately upon a cast-iron mould turned exactly to the form of the bell, which is thus rendered much thinner than the general substance, and remarkably exact; the band containing the wire for stiffening the edge of the bell, is attached by dexterous hammering and without solder. To bend the tube to the curve without disturbing its circular section, it is filled with a cement, principally pitch, which allows the tube to be bent to the scroll of the instrument, without suffering the metal to be puckered or disturbed from its true circular section; and in bending similar tubes to smaller curves, they are filled with lead. These materials serve as flexible and fusible supports, which are easily removed when no longer required.

Should any of the raised works have ornamental details, such as concave or convex flutes, or other mouldings, they would be mostly overlooked until the general forms had been given; and then every little part would be proceeded with upon the same principles of solid and hollow blows. Each of the series of flutes would be first slightly developed all around the object, then more fully, and so on until the completion: when, however, the details are so large as to form what may be considered integral partsj it is necessary to prepare for them at an earlier stage.

Thus, to take an excellent, familiar, and *agreeable* example, let 408 RAISING COMPLEX WORKS.

fig. 279 represent plans, 280 sections, and 281 elevations of jelly moulds, many of which require the greatest skill of the coppersmith. The general outline is that of a cylinder *abed,* upon a larger cylinder *efg h,* as a hase. The twelve large and deeply indented flutes or finials, rise perpendicularly to a great height from the plane surfaces *a c* and *e b,* and yet the whole is hammered out of one flat plate.

The first step is to raise the summits of the flutes *i* or *k,* preparatory to the general formation of the upper cylinder *abed,* and then the two are worked up together, leaving for a time the expanded base *efgh,* but ultimately the whole receive a general attention in common. If the flutes were polygonal, and terminated in ornaments like spires or finials, as at *k,* they would be first treated as if for the more simple or generic form *i,* and the details would be subsequently produced.

I have before me a mould which consists of two series of polygonal flutes, such as *k,* rising one above the other, and ending in pyramids which altogether present the appearance of a beautiful and symmetrical group of crystals; this was produced in the same general manner, by external blows, and almost without the employment of compasses or measuring instruments.

Those moulds which require an inner tube, (the Turk's cap of the confectioner,) although for economy usually made in two parts, may be made in one. In this case the tube would be first raised in the ceuter of a flat plate, beginning from a very small hole, the edge of which would be first curled up perpendicularly, and then the flat disk would be driven downwards by blows within the angle, throwing part of its central substance into the body of the tube; after the completion of which the surrounding parts would be raised as before described, but the COMPARISON BETWEEN RAISING AND STAMPING. 409 process would be difficult, from the narrow space for the stakes or inner tools. The skill called for in such works is greatly enhanced by the attention which is required to preserve a nearly uniform thickness in the metal, notwithstanding the apparent torture to which it is submitted; and this is only endured in consequence of a frequent recurrence to the process of annealing, which reinstates the malleable property.

In cases of extensive repetition, or where large numbers of any specific shape are required, expensive dies of the exact forms are employed; but these are only applicable to objects in small relief, and to those in which the parts are not quite perpendicular. Dies would be entirely inapplicable to objects such as the jelly moulds, fig. 280, although a common notion exists that they are rapidly made by that method, but which is in general utterly impossible when such objects are made in one piece. In all such cases, the metal has to undergo the same bendings and stretchings between the dies as if worked by the hammer, and which unless gradually brought about, are sure to cut and rend the metal; the production of many such forms with dies is therefore altogether impracticable.

For example, the patera or moulding, *z,* fig. 282, is only in small relief, and yet the flat piece of metal *a,* would be cut in two or more parts if suddenly compressed between the dies A, B; as the edges *i,j,* would first abruptly bend and then cut the metal, without giving it the requisite time to draw in, or to ply itself gradually to the die, beginning at the center as in the process of hammering.

In fig. 283, the successive thicknesses obliterate the effect of the acute edges of the bottom die B; the face and back of every thickness differ, as although parallel they are not alike, but they become gradually less defined, so that in fig. 283, the top die A, requires nothing more than a flowing line with slight undulations. no STAMPING. PECULIARITIES OF THE TOOLS

Therefore, when two or three dozen plates are inserted between the dies A, B, the transition from *a* to *z* is so gradual, that the metal can safely proceed from *a* to *b,* from *b* to *c,* and so on, and it will be progressively drawn in and raised without injury. When one or two pieces alone are required, they are *blocked-down* to fit the mould, by laying above them a thick piece of lead, which latter is struck with the mallet or hammer; by the yielding resistance the lead opposes, the thin metal is drawn into the die with much less risk of accident, than if it were subjected to the blows without the intervention of the lead.

In producing many pieces, however, one piece *a,* is added at the top, between every blow, and one piece *z,* is also removed from the bottom; occasionally two, three or more are thus added aud removed at one time, and generally as the concluding step, every piece is struck singly between dies, such as fig. 282, which exactly correspond. In general the process of annealing must be also resorted to once or more frequently during the transition from *a* to . For the best works the bottom die is mostly of hardened steel, sometimes of cast iron, or hard brass; the top die is also of hardened steel in the best works, but in very numerous cases lead is used, from the readiness with which it adapts itself to the shape required.

Stamping is very common for many works in brass, but which would be inapplicable if the pieces had perpendicular and lofty sides, as in fig. 280, page 408: such lines, although rounded by the successive thicknesses of metal, would still present perpendicular sides, and therefore render this mode of treatment with dies impracticable, without reference to cost. Thimbles are raised at five or six blows, between as many pairs of *conical* dies successively higher, but the metal requires to be annealed every time. See note A C, page 974, Appendix, Vol. II.

SECT. III. PECULIARITIES IN THE TOOLS AND METHODS.

Before concluding the remarks on raised works, it may be desirable to revert to some of the principal and distinguishing features of the tools employed in these arts. As a general rule, it will be observed that all these manifold shapes are the more quickly obtained, the more nearly the various tools assimilate to the works to be wrought. For instance, the several dies and wage tools quickly and accurately produce mouldings, of the USED FOR THIN METAL WORKS. 411 specific forms of the several pairs of dies; but it is utterly impossible to extend this method to all cases, and the *progressive* changes required, from the flat'disk, the cylinder or cone, as the case may be, to the finished object; and therefore certain ordinary forms of tools can alone be employed, and they are continually changed as the work proceeds.

For hollow works with contracted mouths, the inner tools are required gradually to decrease in bulk and to increase in length, in order to enter the cavities; but they can be rarely the exact counterparts of the transient forms of the works, nor is it always desirable they should be so. The tools are often required to be bent at the end, to extend within a shoulder or gorge; the small stake in the tool, fig. 220, p. 386, is an example of this; the dotted line represents the work, such as the perforated cover of a cylinder, or the top of a teakettle: the strong wrought-iron arm or *horse,* fig. 220, carries the small steel tools, and which latter, may be also fixed by their shanks either in the bench or vice, according to circumstances.

There are many curious circumstances respecting the modification of the *materials* for, as well as thc forms of, the hammers and anvils, if the use of these terms may be extended to the various contrivances, by the action and re-action of which thin metal works are produced; and the concluding examples are advanced to bring some of these peculiarities of method into notice.

The plated metals have so thin a coating of silver, that they require more expert hammering than similar works in solid silver, otherwise the removal of the bruises left by the hammer, by scraping and polishing, might wear through the silver and show the copper beneath. The bruises are therefore driven to the copper side, by hammering upon the silver or the face, with a very smooth planishing hammer, and covering the anvil or bottom tool with *cloth.* On account of the elasticity thus given, the blows become so far hollow that all the little bruises descend to the copper side, or that which is exposed to the cloth, and the face becomes perfectly smooth.

When the inside of a vessel is required to be smooth, it is the hammer that is covered with cloth, stretched over it by an iron ring, and the polished stake or head within the vessel is left uncovered; and in those cases in which the work is required to be good on both sides, the faces both of the hammer and anvil are each muffled; this gives them some of the elasticity of wooden tools, but with superior definition of figure.

Plated works are generally furnished with an additional thickness of silver at the part to be engraved with a crest or cypher, in order that the lines may not penetrate to the copper, (see page 282;) should it, however, be requisite to remove the engraved lines for the substitution of others, the following mode is resorted to.

The object is laid upon the anvil over a piece of sheet lead and it is struck with a bare hammer *upon the engraved lines,* these latter are therefore *hollow* as regards the face of the hammer; in consequence of which, the reaction of the lead causes it to rise in ridges corresponding with the engraved lines, and to drive the thin plated metal before it. The device is thus in great measure obliterated from the silver face and thrown to the copper side, so as to leave much less to be polished out; this ingenious method is appropriately called *reversing.*

In making vases, such as figs. 209 and 211, page 383, the metal is first driven into concave *wooden* blocks with a wooden mallet, as in fig. 273, page 402, in order to gather up the metal into the fluted concave 275, page 403, but without making any sensible alteration in its thickness. In the next stage of the work, *metal* tools are alone employed, whether the object be made by raising-in with hollow blows, or by setting-out with solid blows as adverted to; and the sizes and curvatures of the tools require to be accommodated to the changes of the work.

Supposing the vases to have either concave or convex flutes, escutcheons, or similar ornamental details, they are now sketched with the compasses upon the plain surface of the vase; and if from the shape of the works swage tools similar to fig. 229, p. 386, cannot be employed for raising the projecting parts, they are *snarled-up,* by the method represented overleaf in fig. 284.

Thus at *v,* are the jaws of the tail vice, in which the *snarling-iron s,* is securely fixed; the extremity *b,* which is turned up must be sufficiently long to reach any part of the interior of the vessel, but yet small enough to enter its mouth. The work is held firmly in the two hands, with the part to be raised or set-out exactly over the end *b;* and when the snarling-iron is struck with a hammer at /(, the reaction gives a blow within the vessel, which throws the metal out in the form of the end of the tool, whether angular, cylindrical, or globular: except in small works, two individuals are required, one to hold and the other to strike.

k *rr*--0 'P285' 288'

Figure 285 shows the last stage of the work prior to polishing; thus in finishing the flutes and other ornaments after they are snarled-up, the object is filled with a melted composition of pitch and brick-dust, sometimes the pitch is used alone, or common resin is added; the ornaments are now corrected with punches or chasing tools of the counterpart forms of the several parts; some portions of the metal are thus driven inwards, whilst those around rise up from the displacement and re-action of the pitch. To avoid injuring the lower surface of the work it is supported upon a sand-bag *b,* like those used by engravers, and the perpendicular lines *p,* denote the usual position of the chasing tool.

Works in copper and brass are sometimes filled with lead at the time of their being chased, but the silversmiths and goldsmiths are studious to avoid the use of this metal, as if it gets into the fire along with their works, it is very destructive to them.

Pitch and mixtures of similar kind, are constantly used in the art of chasing in its mote common acceptation; from its adhesive and yielding nature it is a most appropriate support, as it leaves both hands at liberty, the left to hold the punch, the right for the small hammer used in striking it.

The pitch-block, fig. 286, is employed to afford the utmost choice of position, for works..from the smallest size. to those of six or eight inches long. The lower part is exactly hemispherical, and it is placed upon a stout metal ring or collar of corresponding shape, covered with leather. The mass of metal makes a firm solid bed to sustain the blows, and the ball and socket contact, allows the work to assume every obliquity, and to be twisted round to place any part towards the artist.

Large flat works in high relief are frequently sketched out and commenced from the reverse face, the prominent parts of the subject being sunk into tbe pitch, which after a short time must be melted away to allow the metal to be annealed; and this is frequently required when the works are much raised. In the concluding steps the artist works from the face side.

Many of the chased works are cast in sand moulds from metal models, which have been previously chased nearly to the required forms; the castings are first pickled to remove the sand coat, and in such cases, chisel and gravers are somewhat used in removing the useless and undercut parts.

The art of chasing may be considered as the sequel to that of forging (that is, setting aside the employment of the red-heat), but the various hammers and swage tools now dwindle into the most diminutive sizes, and are required of as many shapes as may nearly correspond with every minute detail of the most complex works. Some of them are grooved and checkered at the ends, and others are polished as carefully as the planishing hammers, that they may impart their own degree of perfection and finish to the works; in a similar manner that the polish and excellence of coins and medals are entirely due to that of the dies from which they are struck, the chasing process being, as it were, a minute subdivision of the action of the die itself.

SECT. IV.—THE PKINCIPLES AND PRACTICE OF FLATTENING THIN PLATES OF METAL WITH THE HAMMER.

I Have purposely reserved this subject for a distinct section, on account of its great general importance in the arts, and have placed it last, in order that the various applications of the hammer might have been rendered comparatively familiar; for Many ancient specimens of armour, gold and silver plate, vases and ornaments, are excellent examples of raised, chased, inlaid and engraved works, both as regards design and execution. In our own times, the Hungarian silversmith, Szentepeteri, has produced a very remarkable alto-relievo in copper, taken from Le Brun's picture of the battle of Arbela, in which some of the legs of the horses stand out and are entirely in relief from the background.

It would appear from a prior attempt, also exhibited, in which the artist had failed, as if the metal were cut through around the legs, and that the edges of the hole were drawn together to complete the background, whilst the edges of the removed piece were also stretched and curled backwards so as to unite at the hinder part of the limb. This singular chasing was exhibited in London in 1S38 and 1851; and the author of " Hungary and Transylvania," (1839,) who visited the artist, during the progress of the work, speaks warmly of his unpretending skill.

although the plane surface, may appear to be of more easy attainment than many of the complex forms which have been adverted to, such is by no means the case.

The methods employed are entirely different from that explained at page 247, in reference to flattening *thick* rigid plates, which are corrected by *enlarging the concave side,* with blows of the sharp rectangular edge of the hack hammer, applied *within* the concavity. A method which bears some analogy to that employed by the joiner in straightening a board which is curved in its width, namely, the *contraction of its convex side* by exposure to heat, as adverted to at page 51. In thin metal plates neither of these modes is available, as the near proximity of the two sides causes both to be influenced in an almost equal degree by any mode of treatment.

Thin plates are flattened by means of solid and hollow blows, which have been recently explained, but they require to be given with considerable judgment; and a successful result is only to be obtained by a nice discrimination and considerable practice. All therefore that can be here attempted is an examination of the principles concerned, and of the general practice pursued; as the process being confessedly one of a most difficult nature, success is only to be expected or attained by a strict and persevering regard to principle.

As respects thin works no figure is so easily distorted as the true plane, and this arises from the very minute difference which exists between the span or chord of a very flat arch, and its length measured around the curve. For example, imagining the span of an arch to be one inch, and the height of the same to be one-twentieth of an inch (a monstrous error as regards a flat plate), the curve would be only about one-200th of an inch longer than the span: and therefore, if any spot of one inch diameter, were stretched until, if unrestrained, it would become one inch and one-200th, in diameter, such spot would rise up as a bulge one-twentieth of an inch high. This trivial change of magnitude would be accomplished with very few blows of the hammer, and much less than this would probably distort the whole plate.

In general however, there would be not one error only, but several, the relationship of which would be more or less altered with nearly every blow of the hammer; thence arises the difficulty, as the plane surface cannot exist so long as any part of 416 FLATTENING THIN PLiTEB the plate is extended beyond its just and proportional size, and which it is a very critical point to arrive at.

There is another test of the unequal condition of flat works besides that of form, namely their equal or unequal states of elasticity, and which is an important point of observation to the workman. For instance, if we suppose a plate of metal to be exactly uniform in its condition, it will bend with equal facility at *every* point, so that bending a long spring or saw, will cause it to assume a true and easy curve; but supposing one part to be weaker than the remainder, the saw will bend more at the weak part, and the blade will become as it were two curves moving on a hinge. When such objects are held by the one extremity and vibrated, the perfect, will feel as a uniformly elastic cane; the imperfect, as a cane having a slight flaw, which renders it weak at one spot; and in this manner we partly judge of the truth of a hand-saw, as in shaking it violently by the handle, it will, if irregularly elastic, lean towards the character of the injured cane, a distinction easily appreciated.

A thin *plate* of metal can only be perfectly elastic, when it is either a true plane or a true curve, so that every point is under the same circumstances as to strength. Thus a hemisphere, as at *a,* fig. 287, possesses very great strength and rigidity owing to its convexity, but as the figure becomes less convex it decreases gradually in strength, and when it slides down to the plane surface, as at /, the metal assumes its weakest form.

A nearly plane surface will necessarily consist of a multitude of convexities or bulges varying in size and strength, connected by intermediate portions, which may be supposed to be plane surfaces; the whole may be considered as greatly exaggerated in the figure. The bulged parts are stronger than the plain flat parts, it follows that the bending will occur in preference at the plane or weak parts of the plate, precisely as in the injured cane.

When the bulges are large but shallow, they flap from side to side with a noise at every bending, as their very existence shows that they cannot rest upon the neutral or straight line; such parts are said to be buckled, their ready change of posi tion renders them flaccid and yielding under the pressure of the fingers, and they are therefore called *loose* parts, but at the same time it is certain that they are too large.

On the contrary, those parts which are intermediate between the bulges, feel *tight* and tense under the fingers, because they are stretched in their positions and rendered comparatively straight, by the strong edges of the bulged or convex parts: the flat portions are the hinges upon which the bulged parts move, and such flat parts are sensibly too small for their respective localities, the others being too large.

Now, therefore, in prescribing the rule for the *avoidance* of these errors, it is simply *to treat every part alike,* so that none may be stretched beyond its proper size so as to become bulged, and thereby to distort the whole plate. "When the mischief has occurred, the *remedy* is to *extend all the too-small parts,* or the hinges of the bulges to their true size, so as to put every part of the plate into equal tension, by allowing the bulged or too-large parts room to expand. Uniform blows should be therefore directed upon all the straight or too-small parts of the plate, the force and number of the blows being determined by the respective magnitudes of the errors, and the rigidity of the plate.

In flattening plates, the greater part of the work is done with solid blows upon a true and nearly flat anvil; the face of the hammer is slightly round, and its weight and the force of the blows are determined by the strength of the plate, the slighter plate requiring more delicate blows, and being more difficult to manage. In the commencement, the rectangular plate is hammered all over with great regularity in parallel lines beginning from one edge; it is generally turned over and similarly treated on the other side. Circular plates are hammered in circular lines beginning from the center, that is supposing the plates of metal to be soft, and in about the ordinary condition in which they are left by the laminating rollers; as the equable hammering gives a general rigidity, which serves as a foundation for the correctional treatment finally pursued. With a steel plate hardened in the fire, and which is already far more rigid than the soft plate, it is necessary to begin at once upon the reduction of the errors and distortions, which usually occur in the hardening and tempering.

The hammer should be made to fall

on one spot with the uni VOL. I. E E formity of a tilt-hammer, the work being moved about beneath it. As, however, the regularity of a machine is not to be expected from the hand, it is scarcely to be looked for that the work shall be at once flat. Whilst the errors are tolerably conspicuous or considerable, the man accustomed to the work will still keep the hammer in constant motion, and will so shift the work, as to bring *the tight parts alone* beneath its blows, hammering with little apparent concern just around the margins of the loose parts, or at the foot of every rise. As the plate becomes more nearly flat, it is necessary to proceed more cautiously, and to hold the plate occasionally between the eye and the light to learn the exact parts to be enlarged, the straight-edge is also then resorted to.

In many works, especially in saws which require very great truth, the elasticity is also examined; this is frequently done by holding the opposite edges of the plate between the fingers and thumbs, and bending them at various parts. As previously explained, all the portions which are technically called tight, or those lines upon which the loose enlarged parts appear to move as on hinges, are strictly the parts to be extended by gentle hammering. For instance, supposing that in the plate, fig. 288,

Figs. 288. 289.

there were only one central buckle *a,* the whole exterior portion would require to be stretched, beginning from the base of the bulge; but it must be remembered the extreme edges of the plate will yield with greater facility than the more central parts, and therefore require somewhat fewer blows, as the blows are all given as nearly as possible of the same intensity, and the number of them is the source of variation.

If, as it is more to be expected, there are two or more loose parts, such as *a,* and *b c,* the more quiescent part between them must be first hammered, as working upou any loose or bulged part only magnifies the evil. Where the intermediate space is narrow as at *d,* less blows will be needed, and such tight parts will soon, and sometimes very suddenly, become loose from the two bulges melting into one. It should be rather the general aim to throw the several small errors into a large oue, by getting the.plate into one regular sweep; dealing the blows principally between the dotted lines, not carelessly so as to increase the general departure from the plane surface, but with an acute discrimination to lead all the defects in the *same direction,* by making the plate as it were a part of a very great cylinder, as at *e* or /, fig. 289, but with as little curvature as possible.

When this is accomplished, and that the work is free from loose parts, it is hammered on the *rounding* side, in lines parallel with the axis of the imaginary cylinder; so that in *e,* the lines would be parallel with the edge from which the rise commences, and in /, or the plate which is bent diagonally, the lines of blows would be necessarily oblique, although as regards the curvature, the same as in *e.* The reason why any reduction of curvature should at all result from this treatment (action and re action being alike), is due to the greater roundness of the hammer than the anvil; the rounder hammer effects the change more rapidly, but also the more indents the work. *a b ii* Fig. 290. ft *a*

In a circular saw, the general aim is first to throw the minor errors into one regular concavity, which may be supposed to extend to *b, b,* in the imaginary section, fig. 290, and then the margin *a, b,* would be hammered in a proportional degree, to enlarge it until it just allowed the interior sufficient room to expand to the plane surface.

It may happen'in the course of the hammering that from *b* to *b,* becomes loose, whilst the extreme edge *a a,* is also loose, and that the intermediate part towards *b,* requires to be stretched. These minor differences cannot be told alone by bending the plate with the fingers, as errors frequently exist which are too minute to yield to their pressure, and then, the eye and straightedge are conjointly employed in the examination.

In a saw, the general aim is to leave the edge rather tight or 420 FLATTENING THIN PLATES. small, as then, the small amount of expansion it acquires when at work, from heat and friction, will enlarge the edge just sufficiently to bring the saw into a state of uniform tension. Otherwise, if before the saw is set to work the edge is fully large enough, when expanded by the heat it is almost sure to become loose on the edge, and to vibrate from side to side, without proper stability; so as to produce a wide irregular cut, and make a flanking whip-like noise, arising from the violent vibration of the buckled parts of the plate, in passing through the saw kerf; the sides of the wood will then exhibit ridges like the ripple marks on the sandy shore.

In hammering all plates, preference should in the like manner be given to keeping the edge rather small or stiff", to serve as a margin or frame to the more loose parts within; it gives a degree of stability somewhat as if the object had a thickened rim, and when a rim really exists, the process of flattening is comparatively easy.

If by undue stretching, the edge is made too loose, the whole piece becomes flaccid and very mobile, and we seem to lose the governing power, or those retaining points by which the changes of the plate are both influenced and rendered apparent; the edge should be therefore *always* kept somewhat tight, from being proportionally less hammered, especially as the edge more easily admits of expansion than the inner part.

As a general rule, it may be said that every part of the plate which is straight and tense, whilst others are curved and flaccid, denotes that every straight part is under restraint; and that its straightness is due to its being, as it were, stretched either lengthways or around its edges, by the other parts which are too loose, and therefore arched, and also strong. In such cases, the straight lines require to be extended in length, to allow sufficient room for the curves to expand to their proportional sizes. This refers not only to small local efrors towards the inner part of the plate, as explained by diagram, fig. 288, p. 41S; but should the one edge of a plate be tolerably straight, whilst the opposite is loose and flaccid, the rule also applies with equal truth, and the straighter side must be hammered; in this case the curved side is as it were a great bulge cut in two parts.

Should a circular saw have a sudden dent, such as at *g*, fig. 290 on the last page, standing the reverse way, and which may result from its having rested upon a small lump of coke whilst in the fire, the first hlows will be given on the *hollow* side, between the lines *i i*, to lessen the abruptness of the margin by stretching it to the dotted curve, and then it will be driven downwards by violent blows, to form a part of the general sweep or concavity; a little time is gained by these driving blows, over the mode of stretching by the hammer.

The foregoing descriptions have all referred to *solid* blows, upon the face of the hard anvil, but to expedite the process, recurrence is often had to *blocking*, which is only one application amongst many others of a wooden anvil or block with a narrow flat-faced hammer, such as fig. 263, page 899. In this case the blows arc to a certain extent *hollow*, as the wood immediately beneath the hammer-face yields to the blow, whereas the margin around the same does not. Such blows are therefore unquestionably hollow, and *bend* with very little stretching.

The blocking is considerably employed in saw-making, *after* the loose parts have been entirely removed, as the hollow blows correct anyslight errors of figure, by bending alone,and with little risk of *stretching* the plates, if the work be delicately performed.

Towards the conclusion, however, all the different modes of work are required to be used in combination, as the true condition of the plate is only the exact balancing of all the forces, or of the tension of the several parts; and it constantly happens that attention to one error causes a partial change and fluctuation throughout the whole. It therefore requires great tact to know when to leave the anvil for the block, and when to return to the anvil, and so on alternately; and also which side of the plate should be upwards for the time, which particular points should be struck, and the required force of the blows.

Of course, within certain limits, a thick plate is easier to hammer than a thin one, as the latter is difficult from its excessive mobility; also a soft plate of iron is more difficult than a hard plate of steel, although the latter requires more blows to produce the same effect; but when the works are very thick they become laborious, and the difficulty always increases rapidly with the size of the plate.

Those who may desire to practise this art should therefore commence with a plate some 4, 6, or 8 inches square, and moderately stout, and subsequently proceed to pieces larger and thinner. They will also find some advantage in raising the anvil to within about a foot of the eye, as the alterations can be then more easily seen whilst the work lies on the anvil, and the effect of any predetermined blows can be the better watched. One other observance is essential, namely, *patience;* as although the process is thoroughly reducible to system, and no blow should be struck in vain, the beginner will frequently find it necessary to pause, examine, and consider, especially as the errors decrease; whereas the accustomed eye will follow the fluctuations of the plate almost without intermission of the blows, and will also accomplish the task with the fewest possible number of blows, which is the great desideratum.

Indeed it may happen from hammering some parts of a plate excessively and improperly, that it is rendered so hard and rigid, as to make its correction very tedious, or indeed nearly impossible without previous annealing, as the plate might *burst* or crack from the extension being carried beyond the safe limit of malleability. As in raised works, the annealing is mostly done by a gentle red heat; but in hardened steel plates, a slight increase of temperature barely sufficient to discolour the plate, will make a perceptible difference; and this latter process is always the last step in making a saw, in order to restore, by a gentle heat, the proper elasticity which has been mysteriously lost in the grinding, polishing, and hammering required in its manufacture.

CHAPTER XX. PROCESSES DEPENDENT ON DUCTILITY.

SECT. I. DRAWING WIRES, ETC.

The ductility of many of the metals and alloys, or the quality which allows them to be drawn into wire, is applied to a variety of curious uses in the manufacturing arts, and the process may be viewed as the sequel to the use of grooved and figured rollers; but the ductile metals submit to this process with various degrees of perfection.

In drawing wire, the metal is first prepared to the cylindrical form, either directly by casting, or between rollers with semicircular grooves; and the process is completed by pulling the metal through a series of holes gradually less and less, made in a metallic plate, by which the wire becomes gradually reduced in size, and elongated; but as in rolling, the process of annealing must be resorted to at proper intervals.

In general,the draw-plates are made of hardened steel, and they are formed upon the same principle, whether for round, square, or complex sections, either solid as wires, or hollow as tubes; the substance of the metal is partly kept back, as in a wave, by a narrow ridge within the draw-plate, acting as a burnisher.

The section of the holes is explained by fig. 60, p. 173, which represents one of the jewel draw-plates patented by Mr. Brockedon, for gold, silver, and other fine wires; but the plates are generally made of hardened steel, or else of alloys of partly similar nature, which allow the holes to be contracted and repaired, by closing them with blows of a pointed hammer or punch around the hole.

The holes for round wires are sometimes ground out from both sides upon the same brass cone or grinder, the sides of which vary in obliquity from 10 to 30 degrees, according to the metal to be drawn; for the sake of strength the ridge is mostly nearer to the side on which the metal enters, and the sharp edge WIRE

DRAWING.

is also removed, either by wriggling the plate upon the grinder in order to round the inside, or in any other manner.

The end of the wire is pointed to enable it to be passed through the hole, and it is then caught by a pair of nippers, themselves at the extremity either of a chain, rope, toothed rack, or screw, by which the wire is drawn through by rectilinear motion. The nippers or *dogs* resemble very strong carpenters' pincers or pliers, the handles of which diverge at an angle; they are sometimes closed by a sliding ring at the end of the strap or chain, which slides down the handles of the nippers; there are some other modifications, all acting upon the same principle, of compressing the nippers the more forcibly upon the wire the greater the draught. It requires a proportionally strong support to resist the strain; and to avoid the fracture of the hardened steel drawplate, it is usually placed against a strong perforated plate of wrought-iron. In manufactories where large quantities of wire are made, the wire is more usually attached to the circumference of a reel, which is made to revolve by steam or other power.

It is necessary often to anneal the wire, but no general rule can be stated in respect to its recurrence; and before resuming the drawing process, the wire is invariably immersed in some acid liquor or pickle, to remove the slight coating of oxide, which would otherwise rapidly destroy the plates, (as many of these metallic oxides are used in polishing), in general some lubricating matter is applied to reduce the friction, as beer-grounds, starchwater or oil; and for gold and silver, wax is generally used. (See note A B, Appendix, Vol. II. page 974.)

Most of the wire is drawn upon reels, and is therefore met with in circular coils, and it is necessary, in almost every case, to straighten it before use. The soft or annealed wires, such as the copper wire used for bell-hanging, the soft iron *binding-wire* used in soldering, and others, are stretched and straightened by fixing the one end, and pulling the other with a pair of pliers; or short pieces of soft wire may be straightened by rolling them between two flat boards.f Sometimes the plate is made with three cones instead of two; the third cone is exaggerated in fig. 302, p. 429, which represents the arrangement for drawing tubes, the central cone being just equal to the wave, or the quantity the metal is reduced. Under any circumstances all the keen edges are removed, as they would tear instead of compress tho material.

t Soft steel wire for making needles is straightened by rolling or *rubbing:* it ia The hard-drawn and unannealed wires, used for making pins, bird-cages, blinds, and numerous other wire-works, are too elastic to yield to the above methods, and fig. 291 represents the mode employed *to take the spring out of them,* or in other words, to straighten these hard wires. The coil of wire on the reel /, which revolves on a pin, is drawn through the *riddle g,* by the pliers. The riddle is a piece of wood or metal with sloping pins, which lean alternately opposite ways, so as to keep the wire close down on the board, and yet to compel it to pursue a slightly zigzag, or rather serpentine course, which is considerably magnified in the figure.

The pins are equivalent to the three forces *a, b, c,* of the bending machine, page 389, several times referred to. Were the three first pins *critically placed,* they would suffice to bend the wire to the limit of its permanently elastic force, and would leave it perfectly straight; commonly however, five pins are used, and sometimes seven or nine. The same riddle will not serve for wires differing in diameter; and were this simple tool more expensive so as to render it desirable, a universal riddle out up in lengths of 4 or 5 inches, and arranged in cylindrical bundles, within iron hoops of 4 inches diameter; the rubber is a bar of cast-iron about two feet loug, narrow enough to lie between the rings. *See* Lardner's Cyclopedia, VoL 2, Manufacturers in Metal, chapters XIV. and XV. of which contain much information on wire-drawing and wire-working.

In actual practice, the riddle is made wider than represented, so as to contain about half a dozen rows of pins, suitable to as many sizes of wire; between every set of pins, and fixed close down to the board, is a straight wire about three time the diameter of the one to be straightened: very great importance is attached to this latter or central wire, being itself straight, it serves as a metallic bed for the small wire to run upon, and it thereby gets worn into furrows crossing it obliquely from pin to pin. The board is retained by two staples at the far end, which fit loosely on two studs or nails driven into the work-bench. might be made by placing the pins *b* and *d,* under a simple screw-adjustment; but in actual practice, a tap of the hammer is found sufficient to correct their positions.

It is necessary to be very particular in pulling the wire through, not to allow it to lean sensibly against either of the last two pins, or it will assume a curve; and in this manner, by drawing the wire designedly at different angles, it may be thrown into any required circular arc, instead of the right line.

The great bulk of wire is cynndrical, but draw-plates are also made of various other forms, as oval, half round, square, and triangular, for the wires figs. 293; and also of more complex forms, as for the production of steel of the sections of figs. 294, known as pinion wire, the whole of the illustrations being printed from the wires themselves. The largest of 294, serves for the pinions of clocks, and the smallest for those of watches; in these cases the entire arbor, (which carries one of the toothed wheels,) is made of the pinion wire, but the teeth are removed from every part, excepting that which works into the adjoining wheel of the train. The plates for pinion wire are exactly the same as the others in principle, and exhibit a remarkable degree of perfection Cylindrical shafts may be viewed as large wires, and when they are turned with ordinary care, in a slide-lathe with a back-stay, it becomes pretty certain that the shafts are circular, and of true diameters; but they are frequently more or lea crooked or bent when they leave the lathe.

In straightening the axes or shafts intended for the *Calculating Machine,* which were of steel, about 6 to 10 feet long, and $ to 1 inch diameter, Mr. Clements employed three half-round dies, fig. 292, *a, c,* fixed to the bed of a fly-press, and 6 to the screw of the same, which was so adjusted that 6, could only bend the portion of the shaft between *a, c,* to the limit of its elasticity; and therefore, by keeping the press constantly at work, drawing the rod through, and twisting it round so as to bend it at *every point* of its length, every shaft was made perfectly straight.

The straightening of black wrought iron shafts *p revion t* to turning, is now accomplished by three equidistant rollers, say a foot diameter and twelve feet long, similar to fig. 233, p. 889. The shaft is heated to redness, and tho center roller is raised at the moment of its introduction, and then a few turns are given to the whole; this straightens the shaft, and retains it so until partially cooled; the other end of the shaft, should it exceed the length of the rollers, is then heated, and treated in the same manner.

All these modes are highly useful, as they operate upon the materials without partially condescending any point, from which unequal treatment a loss of figure would be almost certain to occur, when any such condensed point is partially removed by the turning-tool or otherwise; as it appears to be quite impossible to prevent all sorts of perplexition, when, by any mode of operation, the one point of a material receives a different treatment from the remainder.
in their construction, as for every size there must be a series of many holes gradually assuming the form of a circular foliated Gothic window, with six, seven, eight, or more foils.

Some of the printed calicoes and muslins are also curious examples of the wire-drawing process; the pattern fig. 295, consists of no less than 205 different pieces of copper wire of various forms, fixed into a wood block; the surfaces of the wires when filed smooth, are printed from after the manner of printers' types; the few detached pieces, fig. 296, show some of the sections of such wires, and which may be combined in endless variety. In the same manner, the specimen of music fig. 297, is printed from the surfaces of detached wires and slips of copper fixed in a wooden block; this is only one amongst the many ingenious processes for printing music by letter-press or surface printing.

Fig. 298 represents the *double-plates* or *swage-bits* used for some of the pieces in figs. 295 and 297; the dies are fitted into n small frame or *cramp* with a side screw, (much the same as dies for cutting screws,) so that the metal may be gradually reduced by one pair of swage-bits. This method is very much employed by the silversmith and goldsmith for mouldings, the tools being much cheaper than rollers: the piece 299 was thus prepared for the edging of silver and gold boxes; it is bent round to the form of the box or cover, whether square, circular or oval, and the rebate on the straight side of the band serves for receiving the flat plate to constitute the cover.

The window lead, shown in section in fig. 300, does not admit of being drawn in tbe ordinary manner, from the softness of the material, nor of being rolled, because of its undercut section; the two principles are, therefore, curiously combined in the *glazier's vice.* It maybe conceived that the shade lines of 301, represent parts of two narrow rollers with roughened edges, (equal in thickness to the glass,) which indent the bot tom of the groove, and thereby carry the lead between the figured sidepieces *8 8,* one only shown. In some cases cutting is combined with drawing, cutters are then fixed to the draw-plate, this method has been adapted to making rules and similar rods; and in perfecting the flattened wire for the *reeds* used in looms, the edges are rounded by reeling the wire beneath a forked cutter, a process intermediate between turning and planing.

The process of wire-drawing is seldom practised by the general mechanician, and still less by the amateur; but when it is necessary to produce a wire either of some unusual section not prepared by the manufacturer, or that the equality of size requires more than usual exactness, the process may be accomplished in the small way, by fixing the draw-plate in the tailvice and drawing the wire through with the pliers or a handvice, or by a reel moved by a winch handle.

The most perfect example of this application of the drawing process is in the British Mint: two fixed rollers are employed after the manner of a draw-plate, and the long strip of gold or silver, when rolled very nearly to the thickness, is drawn through the *ttationary* rollers, by dogs attached to one of the links of an endless chain which is in continual motion from the steam-engine. It was found barely possible to make the surfaces of *revolving* rollers so truly concentric that the equality of thickness in the metal could be obtained with the rigorous exactness required, so as to dispense entirely with the necessity of scraping every piece individually, a mode still practised in some of the Continental mints.

The metal, when drawn, is tested by punching out one blank at each end; these are carefully weighed, and if found correct, the whole strip is punched into blanks; and such is the accuracy of the drawing and punching processes, that without the smallest after-adjustment, any fifty or one-hundred blanks weigh alike to the fraction of a grain. See further, page 938, vol. ii. This beautiful arrangement was invented by the late Sir John Barton, then Comptroller of H. M's. Mint.

SECT. II. DRAWING METAL TUBES.

The perfection of tubes is mainly dependent on the drawing process, conducted in a manner similar to that employed for drawing wire. Many of the brass tubes for common purposes, when they have been bent up and soldered edge to edge, as in fig. 247, page 393, are only drawn through a hole which makes them tolerably round and smooth externally, but leaves the interior of the tubes in the condition in which they left the fire after they were soldered, and nearly as soft as at first.

The sliding tubes for telescopes, and many similar works, arc "*drawn inside and out,"* and rendered very hard and elastic, by the method represented in fig. 302; the form of the plate *b,* being exaggerated to explain the shape. For example, the tube when soldered is forced upon an accurate steel cylinder or triblet, in doing which it is rounded tolerably to the form with a wooden mallet, so as to touch the mandrel in places; the end is set down with the hammer around the shoulder or reduction of the triblet, and on drawing the tube and triblet, by means of the loose key or transverse piece *a,* through the draw-plate *b,* the tube becomes elongated, and contracted close upon the triblet at every part, as the metal is squeezed between the mandrel and plate. The fluted tubes for pencil-cases, such as c, are drawn in this manner through ornamental plates, the triblets being in general cylindrical. Some of the drawn tube, called joint-wire, is much smaller than *d,* and is used by silversmiths, for hinges and joints. It is drawn upon a piece of steel wire, which being too small· to admit the shoulder for holding on the tube, the latter is tapered off with a file, and the tube and wire arc grasped together within the dogs, and drawn like a piece of solid wire. A semicircular channel is filed half way in both the 430

TUBES DRAWN INSIDE AND OUTSIDE.

parts to be hinged, and short pieces of the joint-wire are soldered in each alternately.

Triangular, square, and rectangular brass tubes, are in common use in Trance for sliding rules and measures; these are made in draw-plates with moveable dies, fig. 303, which admit of adjustment for size; the dies are rounded on their inner edges, and are contained in a square frame with adjusting screws, and the whole lies against a solid perforated plate.

Iu the general way, tubes of small diameters are completed at two draughts, sometimes three are used, and by this time the tube has received its maximum amount of hardness; therefore the first thickness of the metal and the diameter of the plates require a nice adjustment. The tube, when finished, is drawn oft' the triblet by putting the key through the opposite extremity of the same, and drawing the triblet through a brass collar, which exactly fits it; this thrusts off the tube, which will in general be almost perfectly cylindrical and straight, except a trifling waste at each end.

It requires a very considerable assortment of truly cylindrical triblets to suit all works: and when the tubes are used in pairs, or to slide within one another as in telescopes, it calls for a nice correspondence or strict equality of size, between the aperture of the last draw-plate, and the diameter of the triblet for the size next larger; and as these holes are continually wearing, it requires good management to keep the succession in due order, by making new plates for the last draught and adapting the old ones to the prior stages. Sometimes, for an occasional purpose, the triblet is enlarged by leaving a tube upon it and drawing the work thereupon; but this is not so well as the turned and ground surface of the steel triblet.

Tubes from TV inch internal diameter, and 8 or 10 inches long, up to those of 2 or 3 inches diameter, aud 4 or 5 feet long, are drawn vertically by means of a strong chain wound on a barrel by wheels and pinions as in a crane. In Messrs. Donkin's enormous tube-drawing machine, which is applicable to making tubes, or rather cylinders, for paper-making aud other machinery, as large as 20 inches diameter and 6J feet long, a vertical screw is used, the nut of which is turned round by toothed wheels driven by six men at a windlass.

All the tubes previously referred to are made of-sheet-metals turned up and soldered edge to edge, but lead and tin pipes for water, and other fluids, have for a long period been cast as thick tubes, some 20 to 30 inches long, and extended to the length of 10, 12 or 15 feet on triblets, which require to be very exactly cylindrical or they cannot be withdrawn from the pipes.

The brass tubes for the boilers of locomotive engines are now similarly made by casting and drawing without being soldered, and some of these are drawn taper in their thickness as described in Note A D, page 976 of the Appendix to Vol. II.

The ductility of tin is very great; it was from the ordinary tin tube of commerce, (which is cast about 2 feet long, £ inch thick, and drawn out to about 10 feet,) that Mr. Rand prepared his patent collapsible vessels for artists' oils and colours. Pieces 3 inches long were extended to 36 inches by drawing them through ten draw-plates, which are sometimes placed in immediate succession, the one to commence just as the other had finished. The tube seemed to grow under the operation, and it was thus reduced without annealing, from half an inch thick as cast, to the 170th of an inch thick, audit was stretched fully sixty times in length. This mode of making the tubes of patent collapsible vessels has been superseded by another, presenting far greater ingenuity and described in the Note A E, page 977 of the Appendix to Vol. II.

Some of the smallest tin tube of commerce, when removed from the ten-foot triblets, is drawn through smaller plates without any triblet being used; this reduces the diameter with little change of thickness, so that the half-inch tube becomes a nearly solid wire measuring about inch diameter externally, which is known as beading, and used to form the raised ledges around tables and counters covered with pewter.f Various patents have been taken out by Burr, Hague, Gethen, Hanson, and others, for making lead pipes without the necessity of drawing.—*See* London Journal of Arts and Sciences. In Hanson's patent (1837, Vol. XVI., p. 344) the melted lead is poured into a vertical cylinder, "and as soon as this metal is *set,* or becomes hard or solid, and before it becomes cold," the lead is forced downwards by hydrostatic pressure, through a short tube or mould, with a central triblet kept in position by four arms or thin fins, and the metal becomes a continuous pipe, as the four sections are effectually united by the great heat and the pressure employed.

\+ Before quitting the subject of tubes and wires, I will refer to an ingenious instrument connected with them; a pair of proportional compasses, tho arms of which are as 1 to?, so that the diameter of the tube being measured by tho shorter arms, tho longer denote the width of the strip of metal required to produce *iU*

CHAPTER XXI.

SOLDEEINO.

SECT I. GENERAL REMARKS, AND TABULAR VIEW.

Soldering is the process of uniting the edges or surfaces of similar or dissimilar metals and alloys by partial fusion. In general, alloys or solders of various and greater degrees of fusibility than the metals to be joined, are placed between them, and the solder when fused unites the three parts into a solid mass; less frequently the surfaces or edges are simply melted together with an additional portion of the same metal.

The chemical circumstances to be considered in respect to soldering, are for the most part set forth in the section on the fusibility of alloys, page 300 to 304, to which the reader is referred. It is there explained, that the solders must be necessarily somewhat more fusible than the metals to be united; and that it is of primary importance, that the metallic oxides and any foreign matters be carefully removed, for which purpose the edges of the metals are made chemically clean, or quite bright, before the application of the solders and heat; and as during this period their affinity for oxygen is violent, they are covered with some flux which defends them from the air, as with a varnish, and tends to reduce any portion of oxide accidentally existing.

The solders are broadly distinguished as *hard-solders,* and *softsolders;* the former only fuse at the red heat, and are consequently suitable alone to metals and alloys which will endure that temperature; the soft-solders melt at very low degrees of heat, and may be used for nearly all the metals.

The attachment is in every case the stronger, the more nearly the metals and solders respectively agree in hardness and malleability. Thus if two pieces of brass or copper, or one of each, are brazed together, or united with spelter-solder, an alloy nearly as tough as the brass, the work may be hammered, bent and rolled, almost as freely as the same metals when not soldered, beeause of the nearly equal cohesive strength of the three parts.

Lead, tin, or pewter, united with soft solder, are also malleable, from the near agreement of these substances, whereas when copper, brass and iron are soft-soldered, a blow of the hammer or any accidental violence, is almost certain to break the joint asunder, so long as the joiut is weaker than the metal generally; and therefore the joint is only safe when the surrounding metal from its *thinness* is no stronger than the solder, so that the two may yield in common to any disturbing cause.

The forms of soldered joints in the thin metals have been figured and explained in pages 391 to 394; and soldered joints in thicker works resemble the several attachments employed in construction generally. When the spaces between the works to be joined are wide and coarse, the fluid solder will probably fall out, simply from the effect of gravity; but when the crevices are fine and close, the solder will be as it were sucked up by capillary attraction. All soldered works should be kept under motionless restraint for a period, as any movement of the parts during the transition of the solder from the fluid to the solid state, disturbs its crystallization and the strict unity of the several parts.

In hard-soldering, it is frequently necessary to bind the works together in their respective positions; this is done with soft iron *binding-wire,* which for delicate jewelry work is exceedingly fine, and for stronger works is the twentieth or thirtieth of an inch in diameter; it is passed around the work in loops, the ends of which are twisted together with the pliers. The Asiatics seldom use binding-wire, see note AF, Appendix. Vol. II., page 977.

In soft-soldering, the binding-wire is scarcely ever used, as from the moderate and local application of the heat, the hands may in general be freely used in retaining most thin works in position during the process. Thick works are handled with pliers or tongs whilst being soft-soldered, and they are often treated much like glue joints, if we conceive the wood to be replaced by metal, and the glue by solder, (see page 57,) as the two surfaces are frequently coated or tinned whilst separated, and then rubbed together to distribute and exclude the greater part of the solder.

The succeeding "Tabular View of the Processes of Soldering" may be considered as the index to the entire chapter; which refers to the ordinary methods of soldering most metals. The chapter is arranged under three divisions, illustrated in distinct sections, preceded by one section on the modes of applying heat.

VOL L TABULAR VIEW OP THE PROCESSES OF SOLDERING. *Nott.*—To avoid continual repot itlon, references are mode to the pages of this volume which illustrate the resp.'ctive subjects, and also to the lists on the opposite page, in which some of the solders, fluxes, and modes of applying heat are enumerated. t.

HARD SOLDERING. 441. *Applicable to nearly all metalt less fusible than the solders; tJic modes 0/ treatment nearly similar throughout. The hard solders most commonly used are the spelter solders, and silver solders. The general Jlux is borax marked* A, *on next page; and the modes of heating are the naked fa-e, the furnace or muffle, and the blowpipe, marked* a, b, g. *2ote.*—The examples commence with the solders, (the least fusible first,) followed by the metals for which they are commonly employod.

Fine Gold, laminated and cut into shreds, is used as the solder for joining chemical vessels made of platinum.

Silver is by many considered as much the best solder for German silver.

Copper in shreds, is sometimes similarly used for iron.

Gold solders laminated, are used for gold alloys. See 275-6 and 444.

Spelter solders granulated whilst hot, are used for iron, copper, brass, gunmetal, German silver, &c., 263, 441-3.

Silver solders laminated, are employed for all silver works and for common gold work, also for German silver, gilding metal, iron, steel, brass, gunmetal, &c., when greater neatness is required than is obtained with spelter solder. 283 and 443.

White or button solders granulated, are employed for the white alloys called button metals; they were introduced as cheap substitutes for silver solder. 272-3.

SOFT-SOLDERING. 444. *Applicable to nearly all the metals; the modes of treatment very different. The soft-solder mostly used, is 2 parts tin and 1 part lead; sometimes from motive of economy much more lead is employed, and* li *tin to* 1 *lead is the most fusible of the group unless bismuth is used. The Jinxes* B *to* G, *and the modes of heating* a *to* i, *arc all used with the soft solders.*

Note.—The examples commenco with the metals to be soldered. Thus in the list Zinc, S, C,/, implies, that zino is soldered with No. 8 alloy, by the aid of the muriate or chloride of zinc, and the copper bit. Lead, 4 to S, F, *d, e,* implies that lead is soldered with alloys varying from No. 4 to S, and that it is fluxed with tallow, the heat being applied by pouring on molted solder, and the subsequent uso of the heated iron not tinned; but In general one only of the modes of heating is selected, according to circumstance.

Iron, cast-iron, and steel, 8, B, D, if thick heated by *a, b,* or *c,* and also by *g.* 447 and 448. Tinned iron, 8, C, D, /. 446.

Silver and Gold are soldered with puro tin or olso with 8, E, *a, g,* or *h.*

SOFT-SOLDERING—*Contin ued.*

Copper and many of its alloys, namely, brass, gilding metal, gun-metal, &c. 8, B, C, D; when thick heated by *a, b, c, e,* or *g,* and when thin by /, or *g.* 447-9.

Speculum metal, 8 B, C, D, the heat should be most cautiously applied, the sand-bath is perhaps the best mode.

Zinc, 8, *C,f.* 447.

Lead and lead-pipes, or ordinary plumbers' work, 4 to 8, F, *d,* or *e.* 445.

Lead and tin pipes, 8 D & G mixed, *g,* and also/. 449.

Britannia metal, 8, C, D, *g.*

Pewters, the solders njust vary in fusibility according to the fusibility of the metal, generally G, and *i,* are used, sometimes also G and *g,* or/. 449-50.

Tinning he metals, and washing them with lead, zinc, &c. 450-1.

SOLDERING PER SE, OR BURNING TOGETHER. 452. *Applicable to tome few of the metali only, and which in general require no /lux.*

Iron and brass, &c., are sometimes burned, or united by partial fusion, by pouring very hot metal over or around them, *d.* 452-4.

Lead is united without solder, by pouring on red-hot lead, and employing a redhot iron, *d, e,* 452, and also by the autogenous process, pages 454-6.

The table by H Gaulthiek Dk Claubrt, from which the present extract is derived, enumerate! 102 different alloys intended to be used, for the safety-plugs of steam boilers. In order that the fusion of the plug, and the consequent escape of the wo,(er, may occur when the steam exceeds any predetermined pressure, dependent on thermometric temperature. Bee *le Dictitmnaire de Vlndnstrie Manufaeturitre, Commercialt el Agricole; par A. Baudrimont, Blanquiahu tt autres.* Paris, 1833. Vol. I. p. 326.

The 13 proportions here given include the extremes of the original, and being arranged so that the tin commences at the minimum and ends at the maximum, the temperatures form two reversed series, No. 7 being the lowest unless bismuth is present. No 5 is the *'Plumbera" Sealed Bolder?* which is assayed after the manner of pewter, (nee p. 284,) and is then stamped by au officer of the Plumbers' Company. The table appears to be particularly useful as regards the pewter alloys and their appropriate solders; but it sheuld be observed that the temperatures, where comparable, are a few degrees lower than tbose given by English autherities.

SECT. II.—THE MODES OF APPLYING HEAT IN SOLDERING.

The modes of heating works for soldering are extremely varied, and depend jointly upon the magnitude of the objects, the general or local manner in which they are to be soldered, and the fusibility of the solders. It appears to be now desirable to advert to such of the modes of applying heat enumerated in the tabular view, as are of more general application, leaving the modes specifically employed in heating works in their respective sections.

In hard-soldered works, the fires bear a general resemblance to those employed in forging iron and steel, and already described; in fact, the blacksmith's forge is frequently used for brazing, although the process is injurious to the fuel as regards its ordinary use. Coppersmiths, silversmiths, and others, use a similar hearth, but which stands further away from the upright wall, so as to allow of the central parts of large objects being soldered; the bellows are always worked by the foot, either by a treadle, as in fig. 90, p. 203, or more commonly by a chain from the rocking-staff terminating in a stirrup.,

The brazier's hearth for large and long works, is a flat plate of iron, about four feet by three, which stands in the middle of the shop upon four legs: the surface of the,plate, serves for the support of long tubes and works over the central aperture in the plate which contains the fuel, and measures about two feet by one, and five or six inches deep. The revolving fan is commonly used for the blast, and the tuyere irons, which have larger apertures than usual, are fitted loosely into grooves at the ends, to admit of easy renewal, as they are destroyed rather quickly. The fire is sometimes used of the full length of the hearth, but is more generally contracted by a loose iron plate; occasionally two separate fires are made, or the two blast pipes are used upon one. The hood is suspended from the ceiling, with counterpoise weights, so as to be raised or depressed according to the magnitude of the works: and it has large sliding tubes for conducting the smoke to the chimney.

Furnaces are occasionally used in soldering, or the common Some parts of the remarks on forging iron and steel, p. 195 to 205, and also of those on hardening and tempering steel, 210 and 246, refer to similar applications of heat to those required in soldering.

fire is temporarily converted into the condition of a furnace from being built hollow, or by the insertion of iron tubes or muffles, amidst the ignited fuel, as already explained in reference to forging and hardening. For want of any of these means, the amateur may use the ordinary grate, or it is better to employ a hrazier or chafing-dish containing charcoal, and urged with hand-bellows blown by an assistant, as then both hands are at liberty to manage the work and fuel.

Fresh coals are highly improper for soldering, on account of the sulphur they always contain; the best fuel is charcoal, but in general coke or cinders are used. Lead is equally as prejudicial to the fire in soldering, as it is in welding iron and steel, or in forging gold, silver, or copper: as the lead readily oxidizes and attaches itself to the metals that are being soldered or weldedj preventing the union of the parts, and in almost all cases rendering the metals brittle and unserviceable.

There are many purposes in the arts which require the application of heat having the intensity of the forge fire or of the furnace, but with the power of observation, guidance, and definition of the artist's pencil. These conditions are most efficiently obtained by the blowpipe, an instrument by which a stream of air is driven forcibly through a flame, so as to direct it either as a well-defined cone, or as a broad jet of flame, against the object to be heated, which is in many cases supported upon charcoal, by way of concentrating the heat.

The blowpipe is largely used—namely, in soldering, in hardening and tempering small tools, in glass blowing for philosophical instruments and toys, in glass pinching with metal moulds made like pliers, in enamelling, and by the chemist and mineralogist, as an important means of analysis: the instrument has consequently received very great attention both from artizans and distinguished philosophers.

Most of the blowpipes are supplied with common air, and generally by the respiratory organs of the operator; sometimes by bellows moved with the foot, by vessels in which the air is condensed by a syringe, or by pneumatic apparatus with water pressure. In some few cases oxygen or hydrogen, or the same gases when mixed, are employed; they are little used in the arts.

TUley's Blowpipe. Trans. Soc. of Arte, VoL 31, p. 106; and Toft's Clow, pipe. See also Griffin on the Blowpipe.

The ordinary blowpipe is a light conical brass tube, about 10 or 12 inches long, from one-half to one-fourth of an inch diameter at the end for the mouth, and from one-sixteenth to one-fiftieth at the aperture or jet; the end is bent as a quadrant, that the flame may be immediately under observation.

Fig 304, represents the same instrument when fitted with a ball for collecting the condensed vapour from the lungs; it is seen by the enlarged section, fig. 305, that the tube is discontinuous, and any moisture within it, proceeding in the direction of the arrow, is arrested in the ball. There are several other blowpipes for the mouth, with various contrivances, such as series of apertures of different diameters, joints for portability! and for placing the jet at different angles, and projecting parts to support the instrument upon the table, but none of these are in common use.

The lungs may be used for the blowpipe with much more effect than might be expected, and with a little practice a constant stream may be maintained for many minutes, if the cheeks arc kept fully distended with wind, so that their elasticity alone shall serve to impel a part of the air, whilst the ordinary breathing is carried on through the nostrils for a fresh supply.

The most intense beat of the common blowpipe is that of the pointed flame; with a thick wax candle, and a blowpipe with a small aperture placed slightly within the flame, the mineralogist succeeds in melting small fragments of all the metals, when they are supported upon charcoal, and exposed to the extreme point of the inner or blue cone, which is the hottest part of the flame; that is, fragments of all metals which do not require the oxyhydrogen blowpipe.

Larger particles, requiring less heat, are brought somewhat nearer to the candle, so as to receive a greater portion of the flame; and when a very mild degree of heat is needed, the object is removed further away, sometimes as in melting the fluxes preparatory to soldering, even to the stream of hot air beyond the poiut of the external yellowish flame.

The first or the silent pointed flame, is used by the chemist and mineralogist for reducing the metallic oxides to the metallic state, and is called the *deoxidizing* flame; the second, or the The inventions of Bergman, Black, Gahn, Macgellan, Pepys, Tennant, Wollaston, &c.

noisy brush-like flame, is less intense, and is called the *oxidizing* flame.

The artizan employs in soldering a much larger flame than the chemist, namely that of a lamp the wick of which is from a quarter to one inch diameter, this must be plentifully supplied with oil; the blowpipe in such cases is selected with a larger aperture; it is blown vigorously, and held a little distant from the flame, so as to spread it in a broad stream of light, extending over a large surface of the work, which is in most cases supported upon charcoal. When any minute portion alone is to be heated, the pointed flame is used with a milder blast of air and a decreased distance.

Fig. 306 is an arrangement, the use of which is attended with no fatigue to the operator; it is much employed by the cheap jewellery manufacturers at Birmingham. A stream of air from a pair of bellows directs a gas flame through a trough or shoot, the third of a cylindrical tube placed at a small angle below the flame. Instead of a charcoal support, they employ a wooden handle, upon which is fixed a flat disk of sheet-iron, about three or four inches diameter, covered with a matting of waste fragments of binding wire, entangled together and beaten into a sheet, about three-eighths or half an inch thick; some few of the larger pieces of wire, extend round the edge of the disk to attach the remainder. The work to be soldered is placed upon the wire, which becomes partially red-hot from the flame, and retains the heat somewhat as the charcoal, but without the inconvenience of burning away, so that the broad level surface is always maintained. Small cinders are frequently placed upon the tool, either instead of, or upon the wire.

Sometimes, as in fig. 307, the gas pipe is surmounted by a square hood, open at both ends, and two blast pipes are directed through it; the latter arrangement is used by the makers of glass toys and seals; these are pinched in moulds something like bullet-moulds; the devices on the seals are produced by inserting in the moulds dried casts, made in plaster of Paris.

Makers of thermometers and other philosophical instruments, generally use a table blowpipe, with a shallow oval or rather a kidney-shaped lamp, fig. 308, with a loop placed lengthways upon the short diameter for holding the cotton, which is sometimes an inch long and half an inch wide. The wick is plentifully supplied with tallow or hog's lard, and a furrow is made through it with a wire to afford a free passage for the blast from the fixed nozzle, by the size of which, and its distance from the flame, the bitter is made to assume the pointed or brush-like character. This lamp is more cleanly and emits less smell than those supplied with oil; any overflow of the tallow is caught in the outer vessel or tray, and when cold, the fat solidifies. The forge, fig. 90, page 203, has also a blowpipe and lamp to enable it to be apphed to the arts in a similar manner, and a very cheap table blowpipe is described by Dr. Faraday, in his "Chemical Manipulation," page 120-169.. See also Appendix, note G, p. 462.

Many blowpipes have been invented for the employment of oxygen and hydrogen; the mixed gases were first used by Dr. Hare of Philadelphia, who has been followed in various ways by Clark, Gurney, Cumming, Hemming, Marcet, Leeson, and many others. Two subsequent modifications of gas blowpipes which have been invented for the workshop, will alone be here described, namely, Sir John Robinson's Workshop Blowpipe, intended for soldering, hardening, and other purposes;f and the Count de Richemont's Airo-hydrogen Blowpipe.J

The general form of the " workshop blowpipe" is that of a tube opeu at the one end, and supported on trunnions in a wooden pedestal, so that it may be pointed vertically, horizontally, or at any angle as desired. Common street gas is supplied through The construction ami management of nearly all the blow-pipes are described in Dr. Faraday's "Chemical Manipulation," 1830, pages 107 to 123. Also in "A Practical Treatise on the Use of the Blowpipe," by John Griffin. Glasgow, 1S27.

f ''The Workshop Blowpipe," which resembles a howitzer, is engraved and minutely described in the number of the Mechanic's Magazine for 2nd April, 1842.

£ See the figure and description, Sect. v., page 454 8.

the one hollow trunnion, and it escapes through an *annular* opening; whilst oxygen gas, or more usually common air, is admitted through the other trunnion which is also hollow, and is discharged in the center of the hydrogen through a central conical tube; the magnitude and intensity of the flame being determined by the relative quantities of gas and air, and by the greater or less protrusion of the inner cone, by which the annular space for the hydrogen is contracted in any required degree.

From amongst numerous other small applications of heat, Mr. Gill's portable blowpipe furnace may be noticed; it consists of a lump of pumice-stone three or four inches diameter, scooped out like a pan or crucible, and filled with small fragments of charcoal; sometimes a conical perforated cover is added; the inside may be intensely ignited, whilst the slow conducting power of the pumice-stone guards the hand from inconvenient heat.

SECT. III.—EXAMPLES OF HARD-SOLDERING.

It was mentioned in the tabular view, that the several works united with hard solders, receive nearly the same treatment; a few examples will therefore serve to convey a general idea of hard-soldering; a process commonly attended with some risk of partially melting the works, because the fusing points of the metals and their respective solders often approach very nearly together.

Several of the hard-solders contain zinc, which appears to be useful in different ways: first it increases their fusibility; in cases where the solder cannot be seen it serves as an index to denote the completion of the process, for when the solder is melted the zinc volatilizes, and burns with the well-known blue flame; and as at this moment some of the zinc is consumed, the alloy left behind becomes tougher, and more nearly approaches to the condition of the metal which it is desired to unite. The zinc may be therefore considered to act as a flux, and so likewise does the arsenic occasionally introduced into the gold and silver solders, as the arsenic is for the most part lost, between the processes of making and using the solders; but this metal being of a noxious quality, it is but little resorted to, and besides, it renders the other metals very brittle.

In every case of soldering, a general regard to cleanliness in the manipulation is important, and for the most part the edges of the metals are filed or scraped prior to their being soldered, as before observed; in those cases in which the red-heat is employed, filing or scraping are less imperative, as any greasy or combustible matters are burned away, and the borax has the property of combining with nearly all the metallic oxides and earthy bases, thereby cleansing the edges of the metals should that proceeding have been previously omitted.

The works in copper, iron, brass, &c. , having been prepared for *brazing,* (or soldering with a fusible brass,) and the joints secured in position by binding wire where needful, the granulated spelter and pounded borax are mixed in a cup with a very little water, and spread along the joint by a slip of sheet metal or a small spoon.

The work, if sufiiciently large, is now placed above the clear fire, first at a small distance so as gradually to evaporate the moisture, and likewise to drive off the water of crystallization of the borax; during this process the latter boils up with the appearance of froth or snow, and if hastily heated it sometimes displaces the solder. The heat is now increased, and when the metal becomes faintly red, the borax fuses quietly like glass; shortly after, that is at a bright red, the solder also fuses, the indication of which is a small blue flame from the ignition of the zinc. Just at this time some works are tapped slightly with the poker to put the whole in vibratiou, and cause the solder to run through the joint to the lower surface, but generally the solder *flashes,* or is absorbed in the joint, and nearly disappears without the necessity for tapping the work.

It is of course necessary to apply the heat as uniformly as possible, by moving the work about so as to avoid melting the object as well as the solder; the work is withdrawn from the fire as soon as the solder has flushed, and when the latter is set, the work may be cooled in water without mischief.

Tubes are generally secured by loops of binding-wire twisted together with the pliers; and those soldered upon the open fire are almost always soldered from within, as otherwise the heat would have to be transmitted across the tube with greater risk of melting the work, air being a bad conductor of heat; it is necessary to look *through* the tube to watch for the melting of the solder. Long tubes are rested upon the flat plate of the brazier's hearth, and portions equal to the extent of the fire are soldered in succession. The common Birmingham tubes for gasworks, bedsteads, and numerous other purposes, are soldered from the outside; but this is done in short furnaces open at both ends and level with the floor, by which the heat is applied more uniformly around the tubes.

Works in iron require much less precaution in point of the heat, as there is little or no risk of fusion; thus in soldering the spiral wires to form the internal screw within the boxes of ordinary tail vices, the work is coated with loam, and strips of sheet brass are used as solder, the fire is urged until the blue flame appears at the end of the tube, when the fusion is complete; the work is withdrawn from the fire and rolled backwards and forwards on the ground to distribute the solder equally at every part. Other common works in iron, such as locks, are in like manner covered with loam to prevent the iron from scaling off.

The finer works in iron and steel, those in the light-coloured metals generally, and also the works in brass which are required to be very neatly done, are soldered with silver-solder. From the superior fusibility of silver-solder, and from its combining so well with the different metals without *"gnawing them,* or *eating them away"* or wasting part of the edges of the joints, silver-solder is very desirable for a great many cases; and from the more careful and sparing manner in which it is used, many objects require but little or no finishing subsequently to the soldering, so that the more expensive solder is not only better, but likewise in reality more economical.

The practice of silver-soldering is essentially the same as brazing. The joint is first moistened with borax and water; the solder, (which is generally laminated and cut into little squares with the shears,) is then placed on the joint with forceps. In heating the work additional care is given not to displace the solder; and for which reason some persons *boil* the borax, or drive off its water of crystallization at the red heat, then pulverize it, and apply it in the dry state along with the solder; others fuse the borax upon the joint before putting on the solder.

Numerous small works united with the hard-solders, such as mathematical and drawing instruments, buttons, and jewellery, "Sheet iron may be soldered by filings of soft cast iron, applied in the usual way of soldering with borax, which has been gradually dried in a crucible and powdered, and a solution of sal-ammoniac." London Journal of Arts and Sciences, vol. v. 1823, p. 274.

444 SOLDERING GOLD AND SILVEK. 8OFT-SOLDEUING. are soldered with the blowpipe; in almost all cases the work is supported upon charcoal, and sometimes for the greater concentration of the heat it is also covered with charcoal. The management of the blowpipe having been explained, it is only necessary to add that the magnitude and shape of the flame are proportioned to those of the works.

In soldering gold and silver, the borax is rubbed with water upon a slate to the consistence of cream, and is laid upon the work with a camel's-hair pencil, aud the solders although generally laminated are also drawn into wire, or filed into dust; but it will be remembered, the more minute the particles of the granulated metals, the greater is the degree of heat required in fusing them.

In many of the jewellery works the solder is so delicately applied, that it is not necessary to file or scrape off any portion, none being in excess, and the borax is removed by immersing the works in the various pickling and colouring preparations to be adverted to. See Note A G, Appendix, Vol. II., page 978.

SECT. IV.—EXAMPLES OF SOFT-SOLDERING.

In this section the employment of the less fusible of the softsolders will be first noticed; the plumbers' sealed solder, 2 parts lead and 1 of tin, melts at about 440 F.; the usual or fine tinsolder, 2 parts tin and 1 of lead, melts at 340; and the bismuthsolders at from 250 to 270: the modes of applying the heat consequently differ very much, as will be shown.

The soft-solders are prepared in different forms suited to the nature of the various works. No. 5, p. 435, the plumbers'solder, is cast in iron moulds into triangular ingots measuring from 1 to 6 superficial inches in the section. No. 8, the fine tinsolder, is cast in cakes about 4 by 6 inches, and J to *h* inch thick; and this and the move fusible kinds, are traled from the ladle upon an iron plate or flat stone, to make slight bars, ribbons, and even threads, that the magnitude of the solder may be always proportioned to the magnitude and circumstances of the work.

It is very essential that all soft-soldered joints should be particularly clean and tree from metidlic oxides; and except where oil is exclusively used as the flux, greasy matters should be avoided, as they prevent the ready attachment of the aqueous fluxes. It is therefore usual with all the metals, except clean tinned plate, and clean tin alloys, to scrape the edges immediately before the process, so far as the solder is desired to adhere.

Lead works are first smeared or soiled around the intended joints, with a mixture of size and lamp-black, called *soil,* to prevent the adhesion of the melted solder; next the parts intended to receive the solder are scraped quite clean with the *shave-hook,* (a triangular disk of steel riveted on a wire stem,) and the clean metal is then rubbed over with tallow. Some joints are *wiped,* without the employment of the soldering iron; that is, the solder is heated rather beyond its melting point, and poured somewhat plentifully upon the joint to heat it; the solder is then smoothed with the cloth, or several folds of thick hed-tick well greased, with which the superfluous solder is finally removed.

Other lead joints are *striped,* or left in ridges, from the bulbous end of the plumber's crooked soldering-iron, which is heated nearly to redness, and not tinned; the iron and cloth are jointly used at the commencement, for moulding the solder and heating the joint. In this case less solder is poured on, and a smaller quantity remains upon the work; and although the stripedjoints are less neat in appearance, they are by many considered sounder from the solder having been left undisturbed in the act of cooling. The vertical joints, and those for pipes, whether finished with the cloth or iron, require the cloth to support the fluid solder when it is poured on the lead.

Slight works in lead, such as lattices, requiring more neatness than ordinary plumbing, are soldered with the *copper-bit* or *copper-bolt* represented in figs. 311 and 312; they are pieces of copper weighing from three or four ounces to as many pounds, riveted into iron-shanks, and fitted with wooden handles. All the works in tinned iron, sheet zinc, and many of those in copper and other thin metals, are soldered with this tool, frequently misnamed a soldering-*iron,* which in general suffices to convey all the heat required to melt the more fusible solders now employed.

If the copper-bit have not been previously tinned, it is heated in a small charcoal stove or otherwise to a dull red, and hastily filed to a clean metallic surface; it is then rubbed immediately, first upon a lump of sal-ammoniac, and next upon a copper or tin plate, upon which a few drops of solder have been placed; this will completely coat the tool; it is then wiped clean with a piece of tow and is ready for use.

In soldering coarse works, when their edges are brought together, they are slightly strewed with powdered resin contained in the box fig. 310, or it is spread on the work with a small spoon; the copper-bit is held in the right hand, the cake of solder in the left, and a few drops of the latter are melted along the joint at short intervals. The iron is then used to heat the edges of the metal, both to fuse and to distribute the solder along the joint, so as entirely to fill up the interval between the two parts; only a short portion of the joint, rarely exceeding six or eight inches, is done at once. Sometimes the parts are held in contact with a broad chisel-formed tool, or a hatchetstake, whilst the solder is melted and cooled, or a few distant parts are first *tacked* together or united by a drop of solder, but mostly the hands alone suffice, without the tacking.

Two soldering tools are generally used, so that whilst the one is in the hand, the other may be reheating in the stove; the temperature of the bit is very important; if it be not hot enough to raise the edges of the metal to the melting heat of the solder, it must be returned to the fire; but unless by mismanagement it is made too hot and the coating is burned off, the process of tinning the bit need not be repeated, it is simply wiped on tow, on removal from the fire. If the tool be overheated, it will make the solder unnecessarily fluid, and entirely prevent the main purpose of the *copper-hit,* which is intended to act both as a heating tool, and as a *brush,* first to pick up a small quantity or drop from the cake of solder which is fixed upright in the tray, fig. 309, and then to distribute it along the edge of the joint.

The tool is sometimes passed only *once* slowly along the work, being guided in contact with the fold or edge of the metal. This supposes the operator to possess that dexterity of hand, which is abundantly exhibited in many of the best tin wares; in these the line of solder is very fine and regular. The soldering-tool is then thin and keen on the edge, and the flux instead of being resin is mostly the muriate of zinc, with which the joint is moistened by means of a small wire or a stick prior to the application of the heated tool; sometimes the workman cools the part just finished, by blowing upon it as the bit proceeds in its course; and the iron, if overheated is cooled upon a moistened rag placed in the empty space of the tray containing the solder.

Copper works are more commonly fluxed with powdered salammoniac, and so is likewise sheet-iron, although some mix powdered resin and sal-ammoniac; others moisten the edges of the work with a saturated solution of sal-ammoniac, using a piece of cane; the end of which is split into filaments to make a stubby brush, and they subsequently apply resin; each method has its advocates, but so long as the metals are well defended from oxidation any mode will suffice, and iu general management the processes are the same.

Zinc is more difficult to solder than the other metals, and the joints are not generally so neatly executed; the zinc seems to remove the coating of tin from the copper soldering-tool; this probably arises from the superior affinity of copper for zinc than for tin. The flux sometimes used for zinc is sal-ammoniac, but the muriate of zinc, made by dissolving fragments of zinc in muriatic acid diluted with about an equal quantity of water, is much superior; and the muriate of zinc serves admirably likewise for all the other metals, without such strict necessity for clean surfaces as when the other fluxes are used.

The copper tool is only applicable to *thin* metals, because it requires such a degree of heat, as will allow it to raise the temperature of the work to be joined, to the melting point of the solder; and the excess of heat thus required for *stout* metals, is apt, either to burn off the coating of solder, or to cause it to be absorbed as a process of superficial alloying. It requires some tact to keep the heat of the tool within proper limits by means of the charcoal or cinder fire, but with the airo-hydrogen blowpipe, explained at page 454-6, it is easy to maintain any required temperature for an indefinite period.

Thicker pieces of metal, such as the parts of philosophical apparatus, gas-fittings, and others which cannot be conveniently managed with the copper-bit, are first prepared by filing or turning, and each piece is then separately tinned in one of the following ways. Small pieces, immediately after being cleaned with the file or other tool, and without being touched with the fingers, are dipped into a ladle containing melted solder, which is covered with a little powdered sal-ammoniac. The flux meets the work before it is subjected to the heat, and the tinning is then readily done; sometimes the work is in the first instance sprinkled with resin, or rubbed over with sal-ammoniac water; the latter is rather a dangerous practice, as the moisture is apt to drive the melted metal in the face of the operator.

Thin pieces of brass or of copper alloys, if submitted to this method, must be quickly dipped, or there is risk of their being attacked and partly dissolved by the solder. There is some little uncertainty as to iron, and especially as to steel, being well coated by dipping; sometimes a forcible jar or a hard rub will remove most of the tin, and it is therefore safer to rub these works with a piece of heated copper shaped like a file, imrae. diately on their removal from the melted solder, which makes the adhesion more certain.

Larger pieces of metal, or those it is inconvenient to dip into the ladle, are first moistened with sal-ammoniac water, or dusted with the dry powder or resin, and heated on a clear fire either of charcoal, coke, or cinders, until the strip of solder held against them is melted and adheres; as the lowest heat should be always used. Another cleanly way of applying the heat, and which is also employed in tempering tools, varnishing, and cementing, is to make red-hot a few inches of the end of a flat iron bar about two feet long, to pinch it in the vice by the cold part, and to lay the work upon that spot which is at a suitable temperature; the work can be thus very conveniently managed, especially as it may be likewise placed in a good light.

Until the two parts of the work are thoroughly tinned, they must be well defended from the air by the flux to prevent oxidation; they are next made a trifle hotter than is required for tinning, and placed in contact whilst the solder is quite fluid, and a little additional solder is also used; when practicable the two surfaces are rubbed together to perfect the tinning and spread the alloy evenly through the joint; the work is then allowed to cool under pressure applied by the hammer handle, the blunt end of a tool, the tail-vice, or in any convenient manner. The stages of this practice, are similar to those of the carpenter, who having brushed the glue over the two pieces of wood, rubs them together and fixes them with the hand-screws until cold, as before adverted to.

Small works are sometimes united by cleaning the respective surfaces, moistening them with sal-ammoniac water, or applying the dry powder or resin, then placing between the pieces a slip of tin foil, previously cleaned with emery paper, and pinching the whole between a pair of heated tongs to melt the foil; or other similar modifications combining heat and pressure are used.

Many workmen who are accustomed to the blowpipe, as jewellers, mathematical instrument makers, and others, apply the blowpipe with great success in soft-soldering; but as the methods are in other respects similar to those given, they do not require particular notice, except that in some cases there is no choice but to tie the works together with binding-wire as in hard-soldering; but the preference is always given to detached tinning and rubbing together.

The modern gas-fitters are remarkably expert in joining tin and lead pipes with the blowpipe; they do not employ the method of the plumbers and pewterers, or the *spigot* and *faucet* joint surrounded by a bulb of solder, but they cut off the ends of the pipes with a saw, and file the surfaces to meet in butt joints, in mitres, or in T form joints as required. In confined situations they apply the heat from one side only with the blowpipe and rushes; they employ a rich tin solder, with oil and resin mixed in equal parts as the flux; the work looks like carpentry rather than soldering. See also Appendix, Note G, page 462.

The pewterers employ a very peculiar modification of the blowpipe, which may be called the *hot-air blast,* and the names Fig. 313. for which apparatus are no less An ingenious workman assured me that he had employed this mode, for lead pipes measuring externally one inch and a half diameter and situated in angles, by placing pieces of slate against the floor and the perpendicular partition to defend them from the flame, the action of which was assisted by two pieces of charcoal inserted in the corners. And also that as a trial of skill, be had made fifteen joints in three quarter inch tin pipe, five of each kind, namely, plain, mitre, and T form, including the preparations, in the exceedingly short period of twenty-five minutes. 450 SOLDERING PEWTER WORKS, TINNING, ETC. of air, through the pipe *c,* from bellows worked by the foot. The pewter wares, many of which are circular, are placed on the *gentleman,* or a revolving pedestal which may be adjusted by the side screw to any height: the workmen dip the strip of solder in a little pot of oil, and apply it to the joint with the right hand, whilst they slowly revolve the work with the left. This, which is a very controllable application of heat, includes in its range a moderately large extent of the pewterer's work, and answers the purpose extremely well; by some, the rushes and mouth blowpipe are used for circular as well as for other articles in pewter.

The pewters bear nearly the proportion of the alloys Nos. 8 to 12, page 435: for the less fusible containing most tin, the solder No. 8, or 2 tin 1 lead, is used; for the more fusible containing most lead, the bismuth solders, 2 tiu 1 lead 1 bismuth, and others of similar low degrees of fusibility, are employed. The first solder is called by the pewterers, *hard-pale,* the last *softpale;* and to suitthe pewterersof intermediatedegreesof-fusibility, the two are mixed in variable proportions and called *middlingpale;* but the table on page 435, and especially the original from which the 18 terms there given are extracted, would enable the solders to be definitely proportioned, to their respective metals.

The flux always used by the pewterers, is Gallipoli oil; it is a second rate olive oil, of peculiar quality, rather thick, green, and unfit for the table, but its selection requires judgment.

Iron, copper, and alloys of the latter metal are frequently coated with tin, and occasionally with lead and zinc, to present surfaces less subject to oxidation; gilding and silvering are partly adopted from similar motives. As regards iron, the method of making the tin plate is strictly a manufacturing process which has been slightly noticed at page 284, and that of covering iron with zinc by Mallett's patent process at page 301, so that it principally remains to describe the ordinary method of tinning vessels and other objects of copper, brass, and iron, after they have been manufactured, and which is in general thus performed.

Copper and brass vessels are first pickled with sulphuric acid, mostly diluted with about three times its bulk of water; they are then scrubbed with sand and water, washed clean and dried; they are next sprinkled with dry sal-ammoniac in powder, and heated slightly over the fire; then a small quantity of melted block-tin is thrown in, the vessel is swung and twisted about to apply the tin on all sides, and when it has well adhered the portion in excess is returned to the ladle, and the object is cooled in water. When cleverly performed very little tin is taken up, and the surface looks almost as bright as silver; some objects require to be dipped into a ladle full of tin.

Iron presents rather more difficulty, the affinity of the tin being less strong for iron than for copper; but the treatment is in general nearly the same. Old works require that the grease should be removed with concentrated muriatic acid, before the other processes are commenced; and in cast-iron vessels the grease often penetrates so deeply, owing to the porous nature of the metal, that the re-tinning is sometimes scarcely possible, and it is often more economical to obtain a new vessel.

An alloy of nickel, iron, and tin, has been introduced as an improvement in tinning the metals. Mr. G. M. Braithwaite, one of the patentees, informs the author that, "the nickel and tin compound is harder than tin, and endures a much longer time; it is less fusible, and will not run or melt at a heat that would cause the ordinary tinning of pans to forsake the sides and lie in a mass at the bottom. Also, that as an experiment to show the tenacity of the nickel, a piece of cast-iron tinned with the compound, had been subjected by him for a few minutes to the white heat under a blast, and although the tin was consumed, the nickel remained as a permanent coating upon the iron."

There is also another method, that of *cold-tinning,* by aid of the amalgam of mercury, described at page 301; but this process when applied to utensils employed for preparing or receiving food, appears questionable both as regards effectiveness and "The proportions of nickel and iron mixed with the tin in order to produce the best tinning, are ten ounces of the best uickel, and seven ounces of sheet-iron, to ten pounds of tin. These metals are mixed in a crucible, and to prevent the oxidation of the tin by the high temperature necessary for the fusion of the nickel, the metals are covered with one ounce of borax and three ounces of pounded glass. The fusion is completed in about half an hour, when the composition is run off through a hole made in the flux. In tinning metals with this composition the workman proceeds in the ordinary manner." See specification of Messrs. Richardson and Braithwaite's Patent, 1840, in *The Inventorl't Advocate,* 1841,p. 197. The process was discovered by M. Budie, of the firm of Blaise & Co., Paris.

wholesomeness, and the activity of the muriatic acid must not be forgotten; it should be therefore washed carefully off with water. The tin adheres however, sufficiently well to allow other pieces of metal to be afterwards attached by the ordinary copper soldering-bit. SECT. V.—SOLDERING PER SE, OR BURNING TOGETHER.

This principally differs from ordinary soldering, in the circumstance that the uniting or intermediate metal is the same as those to be joined, and that in general no fluxes are employed.

The method of burning together, although it only admits of limited application, is in many cases of great importance, as when successfully performed the works assume the condition of greater strength, from all parts being alike. There is no dissimilarity between the several parts as when ordinary solders are used, which are open to an objection that the solders expand and contract by heat either more or less than the metals, to which they are attached. There is another objection of far greater moment: the solders *oxidize* either more or less freely than the metals, and upon which circumstances hinge some galvanic or electrical phenomena; and thence the soldered joints constitute galvanic circuits, which in some cases, cause the more oxidizable of the two metals to waste with the greater rapidity, especially when heat, moisture, or acids are present.

In chemical works this is a most serious inconvenience, and therefore leaden vessels and chambers for sulphuric acid must not be soldered with tin solder, the tin being so much more freely dissolved than the lead. Such works were formerly burned together by pouring red-hot lead on the joint, and fusing the parts into one mass, by means of a red-hot soldering-iron, as noticed at page 394; this is troublesome and tedious, and it is now replaced by the autogenous soldering, to be explained.

Pewter is sometimes burned together at the external angles of works, simply that no difference of colour may exist; the one edge is allowed to stand a little above the other, as in fig. 234, page 391, a strip of the same pewter is laid-in the angle, and the whole are melted together, with a large copper-bit, fig. 311, page 445, heated almost to redness; the superfluous metal is then filed off, leaving a well-defined angle without any visible joint.

Brass is likewise burned together; for instance the rims of large mural circles for observatories, that are five, six, or seven feet diameter, are sometimes cast in six or more segments, and attached by burning. The ends of the segments are filed clean, two pieces are fixed vertically in a sand mould in their relative positions, a shallow space is left around the joint, and the entire charge of a crucible, say thirty or forty pounds of the melted brass a little hotter than usual, is then poured on the joint to heat it to the melting point. The metal overflows the shallow chamber or hole, and runs into a pit prepared for it in the sand; but the last quantity of metal that remains solidifies with the ends of the segments, and forms a joint almost or quite as perfect as the general substance of the metal; the process is repeated for every joint of the circle.

Cast-iron is likewise united by burning, as will be explained by the following example: to add a flange to an iron pipe, a sand mould is made from a wood model of the required pipe, but the gusset or chamfered band between the flange and tube, is made rather fuller than usual, to afford a little extra base for the flange. The mould is furnished with an ingate, entering exactly on the horizontal parting of the mould, at the edge of the flange, and with a waste head or runner proceeding upwards from the top of the flange, and leading over the edge of the flask to a hollow or pit sunk in the sand of the floor.

The compensation balance of the chronometer and superior watches, is an interesting example of natural soldering. The balance is a small fly-wheel made of one piece of steel, covered with a hoop of brass; the rim consisting of the two metals, is divided at the two extremities of the one diametrical arm of the balance, so that the increase of temperature which weakens the balance-spring, contracts in a proportionate degree the diameter of the balance, leaving the spring less resistance to overcome. This occurs from the brass expanding much more by heat than steel, and it therefore curls the semicircular arcs inwards, an action that will be immediately understood if we conceive the compound bar of brass and steel to be straight, as the heat would render the brass side longer and convex, and in the balance it renders it more curved.

In the compensation balance, the two metals are thus united; the disk of steel when turned and pierced with a central hole, is fixed by a little screw-bolt and nut at the bottom of a small crucible with a central elevation, smaller than the disk; the brass is now melted and the whole allowed to cool The crucible is broken, the excess of brass is turned off in the lathe, the arms are made with the file as usual, the rim is tapped to receive the compensation screws or weights, and lastly the hoop is divided in two places, at opposite ends of its diametrical arm.

A little black lead is generally introduced between the steel and the crucible; and other but less exact modes of combining the metals are also employed.

The end of the pipe is filed quite clean at the place of junction, and a shallow nick is filed at the inner edge to assist in keying on the flange; lastly the pipe is plugged with sand and laid in the mould. After the mould is closed, about six or eight times as much hot metal as the flange requires, is poured through the mould; this heats the pipe to the temperature of the fluid iron, so that on cooling, the flange is attached sufficiently firm to bear the ordinary pressure of screw-bolts, steam, &c.

The method of burning is occasionally employed in most of the metals and alloys, in making small additions to old castings, and also iu repairing trifling holes and defects in new ones; it is only successful, however, when the pieces are filed quite clean, and abundance of fluid metal is employed, in order to impart sufficient heat to make a natural soldering. A process which is also, although differently accomplished, in plating copper with silver, (page 282,) as the two metals are raised to a heat just short of the melting point of the silver, and the metals then unite without solder by partial alloying.

To conclude the description of soldering processes, we have to refer to figure 314, which represents the *airo-hydrogen* blowpipe invented in France by the Count de Riehemont;t it is in a great measure converting the oxy-hydrogen blowpipe invented by Dr. Hare, to the service of the workshop, and it is done with great simplicity and safety. The elastic tube *h,* supplies hydrogen from the generator, and the pipe *a* supplies atmospheric air from a small pair of double bellows *b,* worked by the foot of the operator, and compressed by a constant weight *w;* the two pipes meet at the arch, and proceed through the third pipe *e,* to the small jet/, from whence proceeds the flame. All the connexions are by elastic tubes, which allow perfect freedom of motion, so that the portable blowpipe is carried to the work.

Steam and water-tight joints, in cast-iron works not requiring the power of after-separation, are often made by means of iron cement in the following proportions: 112 lb. of cast-iron filings or borings, 1 lb. of sal ammoniac, 1 lb. of sulphur, and 4 lb. of whitening. Small quantities of the materials are mixed together with a little water shortly before use.

For minute cracks the cement is laid on externally as a thin seam, or for larger spaces it is driven iu with caulking-irons. The edges of the metal and the cement shortly commence one common process of rusting, and at the end of a week or ten days, the joints will be found hard, dry, and permanent.

t The process is patented in this country by Mr. Delbruck, and is practised under his licence by his agents, Messrs. A. Clark & Co., of Southwark, and various others. THE AIRO-HTDROGEN BLOWPIPE. 455

In soldering by the autogenous process, the works are first prepared and scraped clean as usual, the hydrogen is ignited, and the size of the flame is proportioned by the stop-cock *h;* the air is then admitted through *a,* until the flame assumes a fine pointed character, with which the work is united after the

Fig. 814.

general method of blowpipe soldering, except that a strip of lead is used instead of solder, and generally without any flux.

This mode is described as being suitable to most of the metals, but its best application appears to be to plumber's work, and it has been adopted for such in our government dock-yards. The weight of lead consumed in making the joints, is a mere fraction of the weight of ordinary solder, which is both more expensive and more oxidizable, from the tin it contains. The gas soldering, as it is called, removes likewise the risk of accidents from the plumbers' fires, as the gas generator, which is in itself harmless, may be allowed to remain on the ground whilst the workman ascends to the roof, or elsewhere, with the pipe.

Lead is interposed as solder in uniting zinc to zinc, and it is also used in soldering the brass nozzles and cocks to the vessels of lead, and those of copper coated with lead, used as generators.

Another very practical application of the gas flame, is for keeping the copper soldering tool, fig. 315, at one temperature, which is done by leading the mixed gases through a tube in the handle, so that the flame plays on the back of the copper bit. This mode seems to be very well adapted to tin-plate and zinc works, especially as the common street gas may be used, thereby dispensing with the necessity for the gas generator, the construction and management of which alone remain to be explained.

The gas generator, fig. 314, bears some resemblance to Pepys gasometer. When it is first charged, the stopper 1, is unscrewed, and the lower chamber is nearly filled with curly shreds of sheet zinc, and the stopper is replaced. The cover is now removed, and a plug with a loDg wire is inserted from the top into the hole near 3; the upper chamber is next filled with dilute sulphuric acid,(1 acid and 6 water,) until it is just seen through the central hole to rise above the plate immediately beneath it. This measures the quantity of liquid required to charge the vessel without the risk of overflow. The plug is now withdrawn from 3, and the cocks 4, and *h,* being opened, the air escapes from the lower vessel by the pressure of the column of water which enters beneath the perforated bottom 5, upon which the zinc rests. The cocks 4 and *h* are now closed, and by the decomposition of the water, hydrogen is generated, which occupies the upper part of the lower chamber, and drives the dilute acid upwards, through the aperture 3, so as to place matters in the position of the engraving, which represents the generator about two-thirds filled with gas.

The gas issues through the pipe *h,* when both cocks are opened, but it has to proceed through a safety box 6, in which the syphon tube, dips two or three inches into a little plain water introduced at the lateral aperture 7; by this precaution the contents of the gasometer cannot be ignited, as should the flame return through the pipe *h,* it would be intercepted by the water in the safety box. After three or four days' constant work the liquid becomes converted into the sulphate of zinc, and is withdrawn through the plug 8; the vessel is then refilled with fresh dilute acid as already explained, but the zinc lasts a considerable time.

The generators are made of lead, or where portability and lightness are required, of copper washed with lead, and all the exposed parts of the brass work are washed and united with lead to defend them from the acid. Occasionally the air is likewise supplied by *aerometers,* or vessels somewhat resembling the gas generator, but which are only filled with common air, and therefore do not require the zinc or acid.

The great and unintended length to which the pages of the present volume have extended, is to be solely attributed to a constant desire to set forth in sufficient detail, the general principles and features of the numerous subjects which have been considered. In the fulfilment of this task, I have been induced very greatly to enlarge upon the original manuscript during its passage through the press, by the notice and explanation of additional illustrations, many of which have been indeed acquired during that period.

Yet notwithstanding this great extension, it may be satisfactorily added, that this has in no respect led to any departure from the arrangement first proposed in pages 10 to 12, of the introductory chapter; as all the processes forming the subject matter of the foregoing sheets, are accomplished nearly without the use of *cutting tools,* and are such as could not, consistently with the plan of this work, be so well placed elsewhere. The following is the broad difference between the airo-hydrogen and the oxy-hydrogen blowpipes. In the oxy-hydrogen blowpipe, the pure gases are mixed in the exact proportions of two volumes of hydrogen to one of oxygen, which quantities when combined constitute water, and in this particular case there is the greatest condensation of volume, and the greatest evolution of latent as well as of sensible heat.

The airo-hydrogen blowpipe, is supplied with common air and with pure hydrogen; this instrument is also the most effective when the oxygen and hydrogen are mixed in the proportions of 1 to 2; but the nitrogen, which constitutes four-fifths of our atmosphere, is now in the way and detracts from the intensity of the effect.

END OF THE FIRST VOLUME. APPENDIX.

During the period in which thit Volume hat been passing through the press, some new matters having relation to its pages have been puUislied; a few oj these are here noticed, and by aid of the references in Vie body of the work, these Notes will come under observation in their appropriate places.

Note A.—To follow the end of page 46.

The Patent Wood Carving. This is not accomplished in the usual manner, by cutting away the wood with chisels, but it is burned away, or rather converted into charcoal. The oak, mahogany, rosewood, horse-chesnut, or other wood, is steeped in water for about two hours; and the cast-iron dye or mould containing the device, is heated to redness or sometimes to a white heat, and applied against' he wood; either by a handle as a branding-iron, by a lever-press, or by a screwpress, according to circumstances; the moulds are made by the iron-founder from plaster casts of the original models or carvings.

Had not the wood been saturated with water it would be ignited, but until the moisture is evaporated it is only charred; it gives off volumes of smoke, but no flame. After a short time the iron is returned to the furnace to be reheated, the blackened wood is well rubbed with a hard brush to remove the charcoal powder, which, being a bad conductor of heat, saves the wood from material discolomtion; and before the re-application of the heated iron the wood is again soaked in water, but for a shorter time, as it now absorbs moisture with more facility.

The rotation of burning, brushing, and wetting is repeated ten or twenty times or upwards, until in fact the wood fills every cavity in the mould, the process being materially influenced by the character and condition of the wood itself, and the degrees in which the heat and moisture are applied. The water so far checks the destruction of the wood, or even its change of any kind, that the burned surface simply cleaned by brushing, is often employed, as it may be left either of a very pale or deep brown, according to the tone of colour required, so as to match old carvings of any age; or a very little scraping removes the discoloured surface. Perforated carvings are burned upon thick blocks of wood, and cut off with the circular saw.

The patent mode is considerably cheaper than ordinary carving, and the more so the greater the complexity and delicacy of the design. The date of the Patent granted to Messrs. A. S. Braithwaite and Co. for this novel process is Nov. 1840.

Note B.—To follow the Foot Note on page 116.

Subsequently to the extract from "*Br. Boucherie's Memoir on t)te Preservation of Woods*" having been printed upon pages 113—116 of this work, the subject came under the notice of the Institution of Civil Engineers; and in justice to the prior claim of Mr. Bethell, I have quoted the following paragraphs from the Minutes of Proceedings of that Institution for 1842, pages 88—9.

"Mr. Bethell remarked that the process described in Dr. Boucherie's pamphlet was identical with that patented by him July 11th, 1838, two years before Dr. Boucherie's was mentioned in Paris, which was in June 1840. The specification filed by Mr. Bethell stated 'that trees just cut down may be rapidly impregnated 460 APPENDIX.—NOTES C AND D.
with the solution of the first class, hereafter mentioned, (among which is included the pyrolignite of iron,) by merely placing the butt ends in tanks containing the.,solution, which will circulate with the sap throughout the whole tree; or it may be done by means of bags made of water-proof cloth affixed to the butt ends of the trees and then filled with the liquid.'—*See* Specification in Repertory of Patents, March, 1842.
"Mr. Bethell found that some solutions were taken up more rapidly by the sap and circulated with it more freely than others, and the pyrolignite of iron seemed to answer best; he had not hitherto introduced the process in England because it was much more expensive than the oil of tar, the pyrolignite costing from *dd.* to *Od.* per gallon, and the oil being delivered at 3d. per gallon."

"In answer to a question from Mr. Pellatt, Mr. Bethell stated that his experiments on the use of silicate of potash or soluble glass for rendering wood uninflammable were not yet concluded: he had proved its efficacy in this poiut—that as soon as the prepared timber was heated, the glass melted and formed a filmy covering over the surface, which protected it from the oxygen of the air and prevented its catching fire. The silicate also hardened the wood and rendered it more durable. This process was included in his patent of July Uth, 1838. "

Note C, (and also D, E,) to follow the Foot Note, page 234.

"On some peculiar Changes in the Internal Structure of Iron, independent of, and subsequent to, the several processes of manufacture." By Mr. Charles Hood, F.R.A.S., &c. This paper was read before the Inst. Civ. Eng., 21 June, 1842.

"The singular and important changes in the structure of iron, which it is the object of this Paper to explain, are those which arise in the conversion of the quality of iron, known by the name of 'red short iron,' which is tough and fibrous, into the brittle and highly-crystallised quality known by the name of ' cold short iron.' "—" The principle causes which produces this change are percussion, heat, and magnetism, and the author traces through a great number of practical cases of ordinary occurrence, the joint as well as the separate effects of these three causes; showing that the rapidity of the change is proportional to the combined action of these several causes, and that in some cases where all three causes are in operation at the same time, the change of structure is almost instantaneous; while in other cases, where this united operation does not occur, the change is extremely slow, extending over several years before it becomes sensible."

Various fractured specimens were shown to the Meeting, by which the coarse crystalline structure was principally ascribed to the cold hammering and planishing. See Minutes of Proceedings Inst. Civ. Eng. Pages 180—184; transcribed in Civil Eng. and Arch. Journal, 418.

Note D, page 234.

"Experiments on the Tenacity of Wrought Iron," by Mr. James Nasmyth, C.E.— Communicated to the Meeting of the British Association, Manchester, 1842.

Mr. Nasmyth commences by entirely objecting to a part of the opinion recently advanced by the French Commission, appointed to inquire into the circumstances of the recent lamentable accident on the Versailles railway; namely, that the axles of railway carriages, although they may be originally good, become gradually deteriorated from electrical or magnetical causes, owing to their revolving in contact with the rails. Mr. Nasmyth, on the contrary, ascribes the mischief principally to the workshop, and considers that the very best iron may be greatly deteriorated by the continuance of the hammering or swaging until the iron is cold, which, although it adds to the finish and smoothness of the forgings, at the same time detracts, and sometimes enormously so, from their "shock-resisting quality."

Mr. Nasmyth's experiments were made on 'parts of the same bar of the very best iron, 1J inch square, laid obliquely across the anvil, with the end overhanging the same 2 to 3 inches; the blows of a heavy sledge hammer were directed upon the bar, and immediately over the edge of the anvil.

Exp. 1. The bar, in the state in which-it left the manufacturer's hands, and at the temperature of 60, broke with nine heavy blows, "the fracture exhibiting that clear crystalline texture due to a good quality of iron *at tliat temperature.* "

Exp. 2. A piece of the same bar, heated red hot and hammered until cold, broke at 60, with one slight blow, "the fracture exhibiting a most beautiful *clou* crystalline grain, more like the fracture of steel than iron."

Exp. 3. The latter bar (exp. 2) heated to a dull red, and allowed to cool at its leisure, refused to break, it doubled upon itself, the outer part of the bend became three-fourths of an inch thinner from the extent to which it had been stretched, the inner part became thickened in the same degree from the compression; and after receiving 105 heavy blows, and being doubled or folded down flat, no evidence of fracture was visible.

Exp. 4. A portion of the original bar (1) warmed to 100, after receiving about fifty blows over the edge of tie anvil, curled into the form of a crook, and split open on the outer edge, the fracture being entirely *fibrous* like wood, of a fine lead gray colour, and totally free from the appearance of any sparkling crystals;" showing remarkable changes from experiment 1, simply by the addition of 40 degrees of temperature.

Mr. Jfasmyth thence argues, that temperature greatly influences both the tenacity and the appearance of the fracture in iron; and that the best practice undoubtedly is, to hammer the work until cold, and then to anneal it; "the curative process," being most simple and of insignificant cost. See Civil Eng. and Arch. Journal, page 285.

Note E, page 234.

A Machine for Forging Iron and Steel, has been lately invented by a gentleman at Bolton, named Rider. This machine which was exhibited in action at the Meeting of the British Association at Manchester, is quite portable, occupying a space of 3 feet by 4 feet. It may be worked by steam or water power, so as to make about 650 blows per minute.

It contains five or six sets of anvils and swages. The anvils are arranged in a row in the frame at the usual height from the ground, and every swage is fixed to the lower end of a vertical bar, moving between proper guides, so as to be capable of rising and falling through a small space above its anvil. A horizontal axis passes across the upper ends of these swage-bars, and has an eccentric for every swage, so that the uniform rotation of this axis causes every one of the swages to rise and fall periodically in order. The workman has merely to heat the bar in the fire, and hold it under the vibrating swage, turning it or otherwise changing its position according to the form he wishes to produce.

It is stated that the machine will perform the labour of three men and their assistants or strikers, and will complete its work in a very superior manner and with great rapidity. Thus a piece of round iron 1 inch in diameter, was reduced to a square of 5 inch, 2 feet 5 inches long at one heat.

This swage machine promises to be of particular use in forging such works as cotton spindles and others required in great numbers, as the diameters and lengths of their respective parts, can be determined with the same accuracy aa that of objects moulded in sand and cast in the ordinary way; and the forgings require but little adjustment beyond centering, to adapt them to the turning lathe.

Note F.—To follow the Foot Note, page 372.

The Six-foot Speculum. In casting this remarkable work, the Earl of Rosse, (formerly Lord Oxmantown,) used a porous metallic mould with a ring of sand, as before described. The mould was composed of slips of hoop-iron, four inches wide, one-eighth of an inch thick, wedged up tight in a frame, and turned coarsely to the curve on the bed of the grinding-machine, with a rude curvilinear slide-rest.

The diameter of the Speculum was six feet and a fraction, thickness in the center five inches, at the edge five and a half inches. Three tons weight of metal, which had been previously run into ingots, was poured from three cast-iron pots into the mould in four seconds of time; the mould was shortly after dragged into the heated annealing oven, and every aperture having been stopped off most carefully the annealing was extended over above two months, and was quite successful.

The Telescope will be erected in a lofty stone tower, with a stage at the top; the speculum will be supported in its tube upon a system of levers, to resolve its bearing to three points only, (as described in the Trans. Royal Soc. 1840, p. 524,) and at the distance of 54 feet, or its focal length, the plane mirror at 45, and the lateral eye-piece, will be attached to the tube, as in Newton's arrangement. The instrument will be situated in the plane of the meridian with an equatorial movement by clockwork for half an hour on either side, to extend observations for one hour.

The construction of the Reflecting Telescope has been long a favourite pursuit with amateurs, but this instrument is an enormous extension upon any hitherto attempted. The Karl of Rosse does not consider that he has arrived at the mechanical limit of size, but that no further optical advantage is to be anticipated from a larger instrument, owing to atmospherical interferences. This arduous undertaking is expected to be completed in the course of 1843.

Since the above lines were written, the Earl of Rosse has completed his gigantic undertaking in a most satisfactory manner, no alterations from the original intentions having appeared necessary. The observations made with the new telescope, have been found to realise every expectation that had been formed of its powerful and surprising definition.

Note G.—To follow the second paragraph, page 440. A very convenient *portable torch,* which is much used by the gas-fitters and pewterers, consists of three or four dozen rushes such as are used for rush-lights; they are expressly prepared by the tallow-chandlers with only a slight coating of tallow, and are retained in a paper sheath. This torch *may be carried to the vvrk,* and used in situations where an ordinary oil lamp could not be applied; the rushes last a considerable time. This description was accidentally omitted in the body of the work, and refers to page 449 as well as to 440.

Ehd Of Ise Appendix Io Ihjb Fibsi Volume HOLTZAPFFEL & Co.,

N-64,

Engine, Lathe, & TOOL Manufacturers, TURNING, PLANING, SCREW AND WHEEL CUTTING, FRAMING, 4c IN METAL AND WOOD TO DRAWINGS OR MODELS.

, SUPPLIED WITH THE APPARATUS, TOOLS, AND MATERIALS, THAT ARE REQUIRED

IN TURNING AND THE MECHANICAL ARTS GENERALLY, AND ARE

ALSO PRACTICALLY INSTRUCTED IN THEIR USB.

AN EXTENSIVE ASSORTMENT OF TOOL CUESTS, DRESSING CASES, DRAWING AND MEASURING INSTRUMENTS,

PRINTING PRESSES, GARDEN TOOLS, &C. MANUFACTORY, 127, LONG ACRE. FOREIGN ORDERS, RECEIVED EITHER DIRECT OR THROUGH AGENCY HOUSES, EXECUTED

WITH EXACTNESS AND DISPATCH.

ADVERTISEMENT.

Holtzappfel & Co., in presenting to the public their new and enlarged Catalogue, feel it to be their first duty, to return their most grateful thanka to their numerous and distinguished Customers, for the kind support their house has experienced during a period which, at this present time, extends to half a century, the business having been established in 1794.

A printed Catalogue was first issued by H. & Co. about fifty years back—the second impression of this Catalogue was greatly enlarged, the articles were then arranged alphabetically, and numbered from 1 to 740—and in three successive reprints, many additions were made to their earlier Catalogues, so as to include the more important of the mechanical tools, and turning machinery, known at the respective periods. In order to retain the usefulness of every impression, the introductions were in each case distinguished by letters attached to the original numbers, as 330 A, 330 B, &c.; so that, a reference to the number and letter in any of the copies, accurately distinguished the particular article alluded to. Certain inconveniences were, however, found to be inseparable from this attempt, strictly to maintain an alphabetical and numerical scheme.

The difficulties of distinct and perspicuous arrangement, continually increased with every introduction of new articles; and as many such had now to be inserted, it was considered desirable, with a new epoch to commence a new Catalogue, in which, although the principal arrangement is alphabetical, the subsidiary parts are classified in a manner denoted by the several headings introduced in *italics.* The descriptions are also given more at length, and are interspersed with numerous explanatory notes, which it is hoped will be found useful.

To avoid interfering with the usefulness of the old Catalogues, the present carries a new series of numbers, commencing at 1000, and ending at 2078, so that confusion with the old numbers cannot possibly occur; and to avoid the prospective inconvenience, of the breaking up of this present series of numbers by future additions to the list, it is intended, as occasion may require, to publish supplements; the first of which will bo commenced with the number next following, or 2079.

The prices of the several articles aro partially annexed in two columns, tho lowest and highest of the ordinary prices being in general quoted. When only one price appears, it is meant to express that only one article of that particular description is at present made. It is to be further observed, that the desire to introduce prices, so far as possible, has induced H. & Co. to attach some few of them rather from surmise than experience; and, consequently, a little latitude may occasionally be required.

It is *a* matter of some regret to Holtzapffel & Co. that they could not entirely fill out the columns with prices; but they have found, from long experience, that, in numerous cases, the fixed prices have acted with inconvenienco; becauso tho articles to which they referred were, in many instances, more or less open to changes of construction, which changes necessarily influenced their cost; and, therefore, the obvious tendency of such fixed prices was, to cripple tho emendation of the several articles so particularised. These remarks apply more particularly to the Lathe Apparatus, described on pages 35 to 44.

So far as possible to remedy this inconvenience, H. & Co. will bo happy to supply, on application, the prices of any of the articles specified, agreeably to the several constructions at the time of tho inquiry; and they will be also happy to furnish any additional explanations that the Cataloguo may not bo found to convey.

In selecting thos6 parts of their stock which they do not manufacture, H. & Co. employ the utmost caro to obtain none but those which aro of tho very best workmanship; and in tho cxtensivo portion of tho stock, the work of their own manufactory, they aim at combining the advantages of their former experience, with tho adoption of every improvement, in tho application of machinery to manufactures, likely to insuro or to increase accuracy of result.

APPENDIX (A.)—TO HOLTZAPFFEL AND CO.'S GENERAL CATALOGUE.

HOLTZAPFFEL & CO.'S LIST OF TABLE CUTLERY.

A COMPLETE LIST OF GENERAL CUTLERY, IXCLODfNO Pen, Pocket, Sportsmen's, And Other Knives; Razors, Scissors, And Miscellaneous Articles, Will Be Founo On Packs 13 To 1G or Tub General Catalooub. *The Cases are oj Mahegany, or Oak, and bound with brass thep are lined with baize, and have separate compartments fur each piece. « The Cases are charged a little extra if lined with cloth or with cotton velvet All the Knives supplied in these Cases have balanced handles; but Table Forks, and Dessert Forte, are not inclnded in the annexed Prices.*

Sets of 9 dozen Table Knives, 2 dozen Hossert Knives, 2 Pairs of
Table Carvers, and 1 Knife Steel....

Sets of 4 dozen Tablo Knives, 3 dozen Dessert Knives, 2 Pairs of 1
Table, and 2 Pairs of Poultry Carvers, and 1 Knifo Steel.. J

Sots of.6 dozen Table Knives, 4 dozen Dessert Knives, 3 Pairs of Table, and 3 Pairs of Poultry Carvers, 1 Knifo Steel, and
1 Chceso Scoop.........

Sets of S dozen Tablo Knives, G dozen Dessert Knives, 4 Pairs of Tablo and 4 Pairs of Poultry Carvers, 2 Knifo Steels, and 1 Cheese Scoop .t Knives And Forks Of Steel, Plated With Silver, And Contained Im Cases,

Knives with balanced handles 3t. the dozon extra.

Knives witheut Forks, one third less the dozen than Knives and Fork!

New blades to old handles 12t. to 18t. the dozen.

A general assortment of fowl, ham, and scimitar Carving Knives, Fork,, and Btcels, In Ivory, track.

and stag-hern handles, plain, or with silver mountings. Also sets of Knives for the kitchen, e., namely, asparagus, butchers', bread, butter, cheese, cooks' sad root knives, and 6tools; and cutlet, mincing, and steak cheppers, with ifon, hern, or wood bandies. Crests, Initials, or Names engraved on the handles to order. Cutlery of cvory description repaired, ground, or set, ami mado to any given pattern.

WAni:nousis, 64, CnAiuNO Cross. Tzunvfaotort, *127,* Loso Aot

APPENDIX (R)-TO HOLTZAPFFEL AND CO.'S GENERAL CATALOGUE.

HOLTZAPFFEL AND CO.'S ENGINE-DIVIDED SCALES, APPLICABLE TO

©ngmeermg, architectural, *aritt* General Science.

As the least expensive fabric, each scale is ruled in the Dividing Engine, on a different slip of card paper, 18 inches long, the figures and inscription having been previously printed dry. By this arrangement the confusion of crowded scales is entirely avoided, and any of them may be applied directly to the drawing, or compared with one another, without the employment of the compasses. The material of the scales and of the drawing paper being Identical, they will be found well adapted to the majority of the drawings used in common practice. Numerous experiments on this head are detailed in the pamphlet, published by Holtzaf-ffel and Co., price *2s. 6d.* Mr. Iioltz Pffkl *could not have done a better service for the profession, than turning hit attention to the construction of rcales suitable for their pu»poses.— We have for many years been in the habit of uting scales made of paiier, both for est nutting and drawing, on account of their convenience.— We have very carefully examimd rtveral of the scahs, and have much pleasure in testifying their accuracy and utility."*—Tuo Civil Euiuecr aud Architect's Journal.

ORDINARY DRAWING SCALES.

A series of 24 scales, consisting of the usual reductions of the foot, viz., *fa fa* J, *fa* , iV 5 *i h* i! 1, 1J» 14, 2, 21, 3, 4, 5, 6 inches to the foot, three lines of inches divided into eighths, tenths, and twelfths, and the English foot decimally divided. *The Set of* 24 *Scales in case, price £1 Is.* CHAIN SCALES.

A series of 24 scales, in Chains and Links, viz., 1, 1, 2, % 3, 4, 5, 6, 7, 8, 10, 12, 13.33, 15, 18, 20, 25, 30, 40, 50, 60, 70, 80, 100 chains to the inch.

The Set of 24 *Scales in case, price £1 Is.*

Single Scales of any of the above, and oj many other varieties, kept in stock, price 1. *each.*

Single Scales, graduated to order, in English or French measures, on separate slips of
cardboard, 18 *inches long, price 2s. each.*

Single Scales of other measures, graduated to order, price 8s. *each.*

CIRCULAR CURVES,

For Railway Plans and analogous purposes, made as arcs of circles of any required radii between the two extremes of one inch and one mile.

Holtzapffel and Co. cut these curves in stout cardboard, as this material combines in the most suitable manner, for the purpose, economy with convenience and permanence of form. The convex and concave edges of these curves are both mado to the same radius, and they have received the approval, and permanent patronage, of the great majority of the Civil Engineers, Architects and Surveyors of the United Kingdom.

The usual set consists of 00 curves, of radii increasing in duo proportion from 1 to 120 inches. Those above 12 inches radius measure 3 inches in width and 18 inches in length.

The Set of 60 *Curves in Plain Deal Case, price £2 8s. 6d. Single Curves oj any Jladius kept in stock, price Is. each. Single Curves of any required Jladius in English inches, feet, or chains. French metre, kc, cut to order, price 2s. each.*

THE ODONTAGRAPH,

Invented by Professor R. Willis, A.M. , F.R.S., etc. This is an instrument of easy application, used for describing the teeth of wheels by circular arcs, so that any two wheels of a set may work truly together. The theoretical explanation of this system of setting out the teeth of wheels, which has been most extensively adopted by practical men, will be found in the Trans. Inst. Civil Engineers, vol. ii., and in Willis's Principles of Mechanism.

In Cardboard, tvith tables and instructions for use, rai-nisltcd, price 5s.

In Brass, icith tables atid instructions for use, mounted, price £1 6s. Od.

APPENDIX (C.)-TO HOLTZAPFFEL & CO.

'S GENERAL CATALOGUE. COWPLR'S PARLOUR PRINTING PRESS. MADE ONLY BY HOLTZAPFFEL & CO.,
64, Charing Caoss, And 127, Loxq Aces, London.
Tin!) little Printing Press is made of mahogany, and stands in the small space of 11 by 8 inches. I t is capable of printing a page 7 by 6 inches, and works so easily that a child may use it on the parlour table. A small type case accompanies it, containing a font of about 2500 types, neatly arranged in throe drawers with appropriate divisions; a fourth drawer serves for the furniture, inking tablet, cc; and to these are added tho necessary tools, so as to render tho whole complete. Should it bo required, tho type-case will contain a duplicate supply of type in addition to that usually furnished, and which doubles tho efficiency of the apparatus at a slight additional cost.

The above apparatus is well adapted to tho amusement nnd education of yonth, and also to various applications of tho inestimable typographic art to the common concerns of mankind.

For example.—Companies, institutions, and individuals, have found it convenient for circular letters, invoices, and papers, subservient to the despatch and methodical arrangement of business; naturalists and travellers for short memoirs of scientific researches, or labels for specimens; the friends of educa tion, for disseminating original and other papers; wood-engravers for examining the progress of their blocks: practical printers, for proofs of title-pages, stereotype plates, or cards; and nearly every different pursuit will suggest eome new application of this little Press.

LIST OF PRICES.

SECTION I.—Cowper'3 Parlour Presses And Apparatus.

COWPERB r AHLOUR PRINTING PRESS, with a galley-clmsc, a box of ink, a vulcanized inking roller, and a distributing tray......

SMALL DEAL TYPE CASE, painted, with four drawers; threo of them partitioned to contain, an assortment of about 2500 types, nnd a proportionate supply of loads and brass rulo; the fourth drawer contains roglet, furniture, aide and foot sticks, quoins, flic...........

BET OP EXTRAS—comprising transfer composing stick, bodkin, forceps, mallet, shooting-fitlok, planer, bnish, and turpentine for cleaning tho type, two quires of demy printing paper, cut Into suitable sizes for the press, nnd one pair of damping slates

CrALLEY CHASE seven inches square Inside,.....

Total charge for the Plain Parlour Preti ami Apparatut complete COWPER'S PARLOUR PRINTING PRESS, japanned and finished In the best manner, and fitted with a drawer, In other respects as above.... SMALL MAHOGANY TYPE CASE, with brass lock and handles, in other respects as abovo............ SET OP EXTRAS, comprising Transfer Composing-stick, c., as above (I ALLEY-CHASE *Total charge for the Beit Parlour Prat and Apparatut complete* DUPLICATE SET OF 250O TYPES, nnd which may bo contained In either of the stove APPENDIX (C.)—TO IIOLTZAPFFEL & CO.'S GENERAL CATALOGUE. PRICE LIST OF PRINTING APPARATUS—*(continual)* SECTION II.—Folio Foolscap Presses And Apparatus. I £ . *d.* FOLIO FOOLSCAP PRINTING PRESS, on the principle of Cowper's Parlour Press, suitable to printing the half sheet of Foolscap, or tho quarto sheet of Imperial. Measurement of tho bed 15 by 10 inches. The press varnished and japanned, complete, with two iron chases, register points, &c.. . . 4 14 C LARGE DEAL TYPE CASE, with six drawers, and measuring externally 24 inches by eighteen, and 11 inches high, with iron haudlcs, lock and key. Four of the drawers arc partitioned after the Printer's method for helding 9000 types of the following varieties.—Great Primer Roman, Specimen No. 9— viz. , capitals, figures, points, spaces and quadrats. Pica Roman, No. 13; large and small capitals, lower case (small letters), with accented vowels for printing the foreign languages, figures, points, spaces, quadrats, and space-line leads, complete. Bourokois Roman, No. 17; capitals, figures, points, spaces and quadrats. Bourgeois Antique, No. 23; capitals, figures, points, spaces and quadrats, Two of the drawers contain pace-line leads, furniture, side and foot-sticks, quoins, and rcglct; also a mallet, sheoting-stick, planer, bodkin, printer's composing stick 91 inches long, brush for cleaning the type, a pair of thick damping slates, *&c,* all proportioned to the size of the Foolscap Press

Six Inch vulcanised inking roller in frame and case

Largo box of superfine printing Ink *Total charge/or the Folio Fooltcap Press and Apparatus in the less complete form* FOLIO FOOLSCAP PRINTING PRESS, exactly like the one last described, but with the following additions, namely, on iron bed half an inch thick, planed quite level and true, to increase the permanent accuracy of the Foolscap Press, and an iron counterpoise, to facilitate the working of tbe samo, . I 7 7 0 LARGE DEAL TYPE CASE with eight drawers, similar to tho case with six drawers above described, but three inches higher, and containing a considerably greater supply of each of the kinds of type specified in the foregoing description, together with the addition of Great Primer No, 9, lower case letters, Pica Italio No. 14, capitals, lower case letters, points, and spaces, and Bourgeois Antique No. 23., lower case letters, making the total number of types about 17,000; together with a proportionate increase of space lino leads, furniture &c., and with the addition of 2I pieces of brass rule of three varieties,, 16 in 0

Six inch vulcanized Inking roller, In frame and case 0 16 ti

Large box of superfine printing Ink.. 0 S 0

Composing frame of deal, to receive the drawers of the type-case when In uso. 0 12 0

Inclined galley of mahegany with moveable bottom 0 18 0

Four extra chases, two of them with crosses. I 0 10 0 *Total charge/or the Folio Foolscap Press and Apparatus in the more complete form* 27 7 ; LARGER PRE &SKS ON' THE SAME PRINCIPLE.

FOLIO DEMY PRINTING PRESS, with iron bed and counterpoise; suitablo to printing the half sheet of Demy. Measurement of the bed 2O by 13 inches. The pressvornishedand jnpanned,complcte,with two-chascs,rogister points, Ac. 19 12 0 DUOADSIOE FOOLSCAP PRINTING PRESS, made entirely of iron, and provided with screw adjustments to regulate the pressure as required; suitable to printing tho whele shcct of Foolscap. Measurement of the bed 20 by 15 inches, complete, with four chases, register points, &o...... 18 18 0

Type cases of various sizes, and with any required selection of types. Moulds for casting rollers. Tools, eta, to order.

SECTION III.—Casks For Additional Typm. SMALL TYrE TRAY, 10 by 6 Inches, with a selection of about G00 Roman or Italio types of small sixe, of cither of the numbers 17 to 2O TYPE BOOK 15 by 12 inches, with a selection of about 1500 types, comprising 0 varieties of small types for headings, cards, dec., as described in page 55 of pamphlet LARGE TYPE TRAY 22 by 24 Inches, partitioned after the mode of the printing office, for containing larger quantities of typo of any kind.... 0 7 u MUSIC TYPE CASE of deal, painted, uniform in slzo with the Small Deal Type Case. The Mutic Typo Case contains four drawers, partitioned to receive an assortment of 2'100 music types, of 200 different kinds, as described on page 49 of the pamphlet. Tho case with muiiio types complete.. . j 8 Ifl 0 IIAND CHASE, in a painted case, with cushion, roller, ink, and Inking tray.. 0 18 0 PRINTING APPARATUS FOR THE USE OF AMATEURS. *A pamphlet containing full and practical instructions for the vse* o/Cowpxns Parlour Printing Prkss, *also the description of larger presses on Vic tame principle, and various other apparatus for the Amateur Typographer.—The pamphlet contains likewise numerous specimens tf plain and ornamental types, Wass rules, chechs, borders, ornaments, corners, arms,* $c. Js *Third Editiontfrwtly enlarged* 8vo. *ctotht Prite 9s. 6d.* APPENDIX (D.)—TO IIOLTZArFFEL & CO.'S GENERAL CATALOGUE. TURNING AND MECHANICAL MANIPULATION. BY CHARLES HOLTZAPFFEL,

Asckiati OF TUP. INSTITUTION OF CIVIL KNQlNIr.BS, LONDON J HONORARY MKMBER OF TOE

ROYAL &COTriSII *n l* iV O AKTS J COHRESPONI.INQ M!.Mi:Ml OF THE AMERICAN INSTITUTE OF NEW YORK, ETC.

INTENDED AS A WORK OF GENERAL REFERENCE AND PRACTICAL INSTRUCTION ON THE LATHE, AND THE VARIOUS MECHANICAL PURSUITS FOLLOWED BY AMATEURS.

To be comprised in Six Volumcs, Octavo. Published by Holtzapffel & Co., 64, Charing Cross, and 127, Long Acre, London.

VOL. I. MATERIALS, THEIR DIFFERENCES, CHOICE, AND PREPARATION; VAR10U8 MODES OF WORKING THEM, GENERALLY WITHOUT CUTTING TOOLS. Introduction—Materials from the Vegetable, the Animal, and the Mineral Kingdoms.—Their uses in the Mechanical Arts depend on their structural differences, and physical characters. The modes of severally preparing, working, and Joining the materials, with the practical description of a variety of Processes, which do not, generally, require the use of Tools with cutting edges. VOL. n. TUB PRINCIPLES OF CONSTRUCTION, ACTION, AND APPLICATION. OF CUTTINO TOOLS USED BY HANI); AND ALSO OF MACHINES DERIVED FROM TUE HAND TOOLS. The principles and descriptions of Cutting Tools generally—namely, Chisels and Planes, Turning Tools, Boring Tools, Screw-cutting Tools, Saws, Files, Shears, and Punches. The band tools and their modes of ute are first described; and subsequently various machines in which the hand processes are more or less closely followed.

Vol. m.

ABRASIVE AND MISCELLANEOUS PROCESSES, WHICH CANNOT BE ACCOMPLISHED WITn CUTTING TOOLS. Grinding and Polishing, viewed as extremes of the same process, and as applied both to the production of form, and the embellishment of surface, iu numerous cases to which, from lh« i naturo of the materials operated upon, and other causes, Cutting Tools are altogether inapplicable. Preparation and Application of Varnishes, Lackers, Jtc. VOL. IV. THE PRINCIPLES AND PRACTICE OF HAND OR SIMPLE TURNING. Descriptions of various Lathes;—applications of numerous Chucks, or apparatus for fixing works in the Lathe. Elementary instructions in turning the soft and hard woods, ivory and metals, and also in Screw-cntting. With numerous Practical Examples, some plain and simple, others difficult and complex, to shew hew much may be done with hand tools alone. VOL. V. THE PRINCIPLES AND PRACTICE OF ORNAMENTAL OR COMPLEX TURNING.

Sliding Rest with Fixed Tools—Revolving Cutters, used in the Sliding Rest with the Division

Plate and Overhead Motion. Various kinds of Eccentric, Oval, Spherical, Right-line and otbor

Chucks. Ibbetson's Geometric Chuck. The Rose Engine, and analogous contrivances, &c.

With numerous Practical Examples,

VOL. VI.

THE PRINCIPLES AND PRACTICE OF AMATEUR MECHANICAL ENGINEERING.

Lathes with Sliding Rests for metal turning, Self-acting and Screw-cutting Lathes—Drilling

Machines—Planing Engines—Key-groove, Slotting and Paring Machines—Wheel-cutting and

Shaping Engines, &c.

With numerous Practical Examples.

The Ffrst, Second, and Third Volumes oj this worh, are written as accompanying books, and have one Index in common, so at to constitute a general and preliminary wrk, the addition to which oj any of the other volumes, will render the subject complete for the three classes oj Amateurs referred to in the Introductory Chapter. A few additional copies of Vie Index have been printed for the convenience of these who may desire to bind the Index with Vols. I. and II. HOLTZAPFFEL & CO., 64, CHARING CROSS, And 127, LONG ACRE, LONDON.

J'HE Elliptical Cutting Fbame, invented by Captain Ash, H.E.I.C.S., is employed in the lathe, for ornamenting turned surfaces with elliptical figures, after the same general method that the eccentric cutting frame is employed for producing circular figures on similar surfaces; viz., the object to be ornamented is fixed on the lathe mandrel, and motion is given to the tool by the Elliptical Cutting Frame, which is fitted to the receptacle of the sliding rest, and driven by a band leading from the overhead motion.

The Elliptical Cutting Frame is capable of producing ellipses of all proportions, from a right line to a circle, according as it may be adjusted; and the ellipses may be arranged either in circular order by the employment of the division plate, or in rectilinear order by the motion of the sliding rest; or the two movements may be combined at pleasure. An almost infinite variety of patterns of a highly ornamental APPENDIX (E.)-TO HOLTZAPFFEL AND CO.'S GENERAL CATALOGUE.

ELLIPTICAL CUTTING FRAME. character will, therefore, be produced by the elliptical movement alone. In addition to which, the instrument is adapted to produce epicycloids! patterns of 4 loops, similar to those produced in the geometric chuck; these looped figures may likewise be made in all proportions, and be placed in any positions.

From these comprehensive powers of the Elliptical Cutting Frame, it results that any desired arrangements may be produced of circles, ellipses, right lines, or 4-looped figures: the instrument is, therefore, a most desirablo addition to all lathes for ornamental turning, and, if required, its powers may be still further increased by the addition of other epicycloidal patterns, or by combining its movements with those of the eccentric chuck or other apparatus for ornamental turning.

Qtneral Rcmarh on the Action of the Elliptical Cettino Frame.

The elliptical movement of the tool is produced as in the geometric pen, and in Ibbetson's geometric chuck, by the combination of two circular movements in opposite directions, the one of which travels at double the angular velocity of the other. In the Elliptical Cutting Frame this is effected by the train of wheels seen in the front of the instrument, which are so arranged that the eccentric frame carrying the tool A makes two revolutions to the right, while the radial flange B makes one revolution to the left; and the proportions of the ellipse described by the tool depend upon the relative degrees of eccentricity given to A and B. Thus, when A and B are both placed central, the tool has no eccentricity, and merely produces a dot. When eccentricity is given to A alone, the tool describes a circle, the radius of which will depend upon the movement given to the screw of the eccentric frame, under the guidance of the micrometer head C, which has ten divisions. Supposing the eccentricity to be equal to 4 turns of the screw, or 4 0 divisions of the micrometer head, and that it is desired to convert the circle into a straight line, the flange B is also moved 40 divisions, by means of the adjusting screw D, upon which a winch handle is temporarily fitted.

Any series of ellipses between the straight line and the circle may be described by reducing the eccentricity of the radial flange B. Thus, if it be shifted 5 divisions between each figure, a series of 7 ellipses will be produced, gradually advancing from the right line to the circle. Any other number of divisions may be adopted in the same manner; the instrument being so adjusted that equal numbers of divisions on B and C always produce the straight line. Series of concentric ellipses are produced by adjusting both A and B; thus, in pattern 1, A was shifted 4 divisions, and B two divisions between every cut APPENDIX (E.)—TO HOLTZA.PFFEL AND CO.'S GENERAL CATALOGUE.

ELLIPTICAL CUTTING FRAME.

The radial actiou of Uie flange B, however, lias the effect of placing the ellipses oblique to each other, instead of parallel, and this requires some compensation to be introduced. Captain Asu compensated the obliquity by shifting the division plate of the lathe a proportionate quantity. Subsequently, tho spindle was extended through the stem of the instrument, and a graduated disk fixed on tho end of the spindle was employed for the compensation, but wnich is more conveniently and accurately effected by means of the worm wheel and tangent screw movement, suggested by H. Perioal, Esq., F.R.A.S. The tangent screw E is moved by a winch handle, and has a micrometer so arranged, that tho movement indicated by one division exactly compensates the obliquity produced by moving the flange B ono division; and therefore, to ensure the parallelism of the ellipses, it is only necessary to employ the same number of divisions on B and E. The tangent screw movement may also be employed to give any angular position to the ellipses that may bo required. Thus, the worm wheel having 150 teeth, 37J turns of the tangent screw will place any of the figures at right angles to their former positions.

Tho 4-looped figures are produced by changing the train of wheels. For the ellipses, the train consists of a fixed wheel of 48 teeth, leading into one of 24, to which is attached a 36 wheel, leading into another 36 wheel fixed to tho axis carrying tho eccentrio frame A. For 4-looped figures, the relativo velocity of the eccentric frame is doubled by employing wheels of 48 and 24 teeth, instead of tho pair of 36 wheels. Tho adjustment of tho 4-loopcd figures, for eccentricity and position, is effected in the same general manner as the adjustment of the elliptical figures.

In tho illustrations shown on the previous pages, Pattern 1 consists of ellipses only; the two central series are placed at right angles by the division plate of the lathe. The positions of the outer series are also determined by the division plate, and the ellipses arc made to intersect, by moving the slido rest screw between every cut. Pattern 2 is produced by the right lino and 4-looped movements. The positions of the right lines are determined by the division plate, and the series of ten 4-loopcd figures are interposed either by the tangent screw E, or by the division plate. Pattern 3 consists of ellipses and looped figures. The positions of the ellipses constituting the central portion of the pattern, and the small 4-looped figures of which the border is composed, are given by the division plate and sliding rest. The general position of tho intermediate series of 4-loopcd figures is also given in the same manner; but the figures being eccentric, are duplicated by shifting the tangent screw E 18 turns. Pattern 4 shows a modification of the 4-looped figure in which ilie eccentricities of A and B are in the proportion of 1 to 8. Thus, in producing each of the sixteen squares of the central figure, A was 3, and B 24, divisions eccentric, and the lines were doubled by reducing tho eccentricity of A j, and B 2 divisions. The local positions of the squares were determined by the division plate, but 8 were worked with E central, and for the other 8, E was shifted 18 turns. The border consists of *Ti* squares, for which A was one, and B 8 divisions eccentric, 36 of the squares were worked with E inclined 9 divisions to the right, and for the intermediate 36 squares, E was inclined an equal quantity to the loft.

HE Pen-holder for enfeebled bands was invented for the use of those persons who, from age, rheumatism, gout, stiffness in the joints of the fingers, defects in the nerves of the hands, paralysis, or other infirmity, are deprived of the free use of the fingers, so that they cannot hold a pen in the customary position.

The instrument is represented in three views: in the center as closed for the pocket; on the left as opened for use; and on the right in the act of being used. The shaft of the Pen-holder for enfeebled hands is held quite vertically in the central part of the hand, and grasped by the whole of the fingers; this position the most infirm can usually command. The lower extremity of the shaft is allowed to rest firmly upon the paper, and thereby support the hand, whilst the socket that actually receives the pen or nib is jointed to the vertical shaft at about the angle of 45 degrees, and is pressed on the paper by a feeble spring, so as to assimilate in the closest manner to the action of the ordinary quill pen. The Pen-holder for enfeebled hands will be used with more freedom when neither the hand nor the arm rest upon the paper, but the little finger should almost touch the sloping socket.

The Pen-holder is adapted to receive a gold, steel, or quill pen, at the option of the individual; and the instrument may be carried in the pocket as an ordinary pencil-case. The purpose of the screw at the bottom of the holder is to adapt the length of the vertical shaft to the projection of the pen, as when the latter touches the paper, the length of the central shaft should be such as just to give the shaft the vertical position. Whereas, if the pen should project too much, or too little, it will be needful to incline the shaft to, or from, the individual, which it is desirable to avoid.

Price of the Pen-holder /or Enfeebled Ilandt, in Silver, vitli Extricating Gold Pen, It 16.

APPENDIX (G.)—TO HOLTZAPFFEL AND CO.'S GENERAL CATALOGUE. ROSE CUTTING FRAME FOB rotmctng ftose ngtne patterns IN ORDINARY TURNING LATHES, MADE BY HOLTZAPFFEL & CO., ENGINE, LATHE, AND TOOL MANUFACTURERS, 64, CHARING CROSS, AND 127, LONG ACRE,

This Instrument is employed in the sliding rest in the samo manner as the Eccentric Cutting Frame, and its management is almost as simple. It is capable of producing upon *surfaces,* nearly the whole of the ornamental patterns that have been hitherto confined to the Eose Engine, tho use of which has been much retarded by the expense and size of the inachino, and the numerous apparatus required for its development..

A more compact and less expensive apparatus for Rose Engino Turning has long been a desideratum with Amateurs. Holtzapffel & Co. have thcreforo much pleasure in submitting the present instrument to the notice of their patrons.

The variety of the ornamental patterns produced by the Rose Cutting Frame, is limited solely by the number of Rosettes; these may be increased to any desired extent, but as the instrument admits of numerous variations being produced from ono rosette, only a moderate number will gonerally bo required.

In illustration of tho performance of this instrument, six simplo patterns, from six different rosettes, aro printed on the following pago, and on tho opposito, six variations of ono of these rosettes. Many other patterns may be produced from tho same rosetto, and in thoso offered it will be observed that only the most simplo apparatus have been employed in combination with the Eose Cutting Frame APPENDIX (G.)—TO HOLTZAPFFEL AND CO.'S GENERAL CATALOGUE.

ILLUSTRATION S or HOLTZAPFFEL & CO.'S ROSE CUTTING FRAME.

Simple Patterns produced from Eosettes A, B, C, D, E, F.

Rosette A. Rosette B.

I APPENDIX (O.)-TO HOLTZAPFFEL *k* CO.'S GENERAL CATALOGUE.

ILLUSTRATION S

Oir HOLTZAPFFEL & CO.'S EOSE CUTTING FRAME.

Compound Patterns produced from Eosette B.

Tlio upper figure represents the Knife in a compact form as intended to be carried in the pocket or cartouche box. The four lower figuris show the different pans detached, A—is a turnscrcw, sufficiently powerful to remove the serens from the regulation rifle, n—is a steel worm, adapted either for drawing a bullet or to assist in cleaning the barrel. It screws on to the regulation ramrod, c— drift for oiling the lock. This drift screws iuto the oil-bottle—n. Dod— cramp for securing the main spring of the lock, E—picker. F—twcczers. c— nipplewrench. H—oil-boltle. I—large blade of knife. K— small blade or buttou-hook. *h*—socket in the kni'e to enable it to fit into the main spring cramp—D D D. M— serened ho'e fitting small end of ramrod.

r»ICK, FIIEE BY EOiT—10s. *rosl Office Orders to he made payaUe at the Charing Crost branch.*

This instrument affords a ready means of measuring distances, and is intended to teach the eye readily to estimate them, so necessary an acquirement in Hiflo. booting.

The Stadium has a slide, over a central aperture, the sides of which are divided with scales of distances ranging from SO to 800 yards—the one side for objects 6 ft. and the other for 8 ft. high. The instrument is held in the right hand at 25 inches from the eye, that distance bein: determined hy *a* bead on a line, which is held in the left hand against the face, then looking through the aperture, the slide is rinsed hy the thumb and forefinger of ihe right hand, unlil thc iibject just fits vertically therein, Ihe top edge of the s'ide will then show on the scale the exact distance of the object from the observer using fhe Stadium. ParCE In Electbcm, Free By Post—10s. 4d.

ANTI-CORROSIVE OIL FOB RIFLES, TOOLS, INSTRUMENTS, AND MACHINES.

This oil is eminently suitable for Hides, &c., being made under a process which insures its purity; it is prepared from an animal oil, and is absolutely free from any vegetable or mineral admixture.

Price, *Veti* 3-tz. Bottle-2s.

Lightning Source UK Ltd.
Milton Keynes UK
UKOW021951110213

206132UK00007B/925/P